CAMBRIDGE LIBRARY COLLECTION

Books of enduring scholarly value

Physical Sciences

From ancient times, humans have tried to understand the workings of
the world around them. The roots of modern physical science go back to
the very earliest mechanical devices such as levers and rollers, the mixing
of paints and dyes, and the importance of the heavenly bodies in early
religious observance and navigation. The physical sciences as we know them
today began to emerge as independent academic subjects during the early
modern period, in the work of Newton and other 'natural philosophers',
and numerous sub-disciplines developed during the centuries that followed.
This part of the Cambridge Library Collection is devoted to landmark
publications in this area which will be of interest to historians of science
concerned with individual scientists, particular discoveries, and advances in
scientific method, or with the establishment and development of scientific
institutions around the world.

The Theory of Sound

John William Strutt, third Baron Rayleigh (1842–1919) was an English
physicist best known as the co-discoverer of the element argon, for which
he received the Nobel Prize in Physics in 1904. Rayleigh graduated from
Trinity College, Cambridge, in 1865 and after conducting private research
was appointed Cavendish Professor of Experimental Physics in 1879, a post
which he held until 1884. These highly influential volumes, first published
between 1877 and 1878, contain Rayleigh's classic account of acoustic theory.
Bringing together contemporary research and his own experiments, Rayleigh
clearly describes the origins and transmission of sound waves through
different media. This textbook was considered the standard work on the
subject for many years and provided the foundations of modern acoustic
theory. Volume 1 discusses the origin and transmission of sound waves in
harmonic vibrations, the vibrations of bars, stretched strings, plates and
membranes, through mathematical models and experimental discussions.

Cambridge University Press has long been a pioneer in the reissuing of out-of-print titles from its own backlist, producing digital reprints of books that are still sought after by scholars and students but could not be reprinted economically using traditional technology. The Cambridge Library Collection extends this activity to a wider range of books which are still of importance to researchers and professionals, either for the source material they contain, or as landmarks in the history of their academic discipline.

Drawing from the world-renowned collections in the Cambridge University Library, and guided by the advice of experts in each subject area, Cambridge University Press is using state-of-the-art scanning machines in its own Printing House to capture the content of each book selected for inclusion. The files are processed to give a consistently clear, crisp image, and the books finished to the high quality standard for which the Press is recognised around the world. The latest print-on-demand technology ensures that the books will remain available indefinitely, and that orders for single or multiple copies can quickly be supplied.

The Cambridge Library Collection will bring back to life books of enduring scholarly value (including out-of-copyright works originally issued by other publishers) across a wide range of disciplines in the humanities and social sciences and in science and technology.

The Theory of Sound

VOLUME 1

LORD RAYLEIGH

CAMBRIDGE
UNIVERSITY PRESS

CAMBRIDGE UNIVERSITY PRESS

Cambridge, New York, Melbourne, Madrid, Cape Town,
Singapore, São Paolo, Delhi, Tokyo, Mexico City

Published in the United States of America by Cambridge University Press, New York

www.cambridge.org
Information on this title: www.cambridge.org/9781108032209

© in this compilation Cambridge University Press 2011

This edition first published 1877
This digitally printed version 2011

ISBN 978-1-108-03220-9 Paperback

THE

THEORY OF SOUND.

THE

THEORY OF SOUND.

BY

JOHN WILLIAM STRUTT, BARON RAYLEIGH, M.A., F.R.S.

FORMERLY FELLOW OF TRINITY COLLEGE, CAMBRIDGE.

VOLUME I.

London:

MACMILLAN AND CO.

1877

Cambridge:
PRINTED BY C. J. CLAY, M.A.
AT THE UNIVERSITY PRESS.

PREFACE.

In the work, of which the present volume is an instalment, my endeavour has been to lay before the reader a connected exposition of the theory of sound, which should include the more important of the advances made in modern times by Mathematicians and Physicists. The importance of the object which I have had in view will not, I think, be disputed by those competent to judge. At the present time many of the most valuable contributions to science are to be found only in scattered periodicals and transactions of societies, published in various parts of the world and in several languages, and are often practically inaccessible to those who do not happen to live in the neighbourhood of large public libraries. In such a state of things the mechanical impediments to study entail an amount of unremunerative labour and consequent hindrance to the advancement of science which it would be difficult to overestimate.

Since the well-known Article on Sound in the *Encyclopædia Metropolitana*, by Sir John Herschel (1845), no complete work has been published in which the subject is treated mathematically. By the premature death of Prof. Donkin the scientific world was deprived of one whose mathematical attainments in combination with a practical knowledge of music qualified him in a

R.   *b* special manner to write on Sound. The first part of his *Acoustics* (1870), though little more than a fragment, is sufficient to shew that my labours would have been unnecessary had Prof. Donkin lived to complete his work.

In the choice of topics to be dealt with in a work on Sound, I have for the most part followed the example of my predecessors. To a great extent the theory of Sound, as commonly understood, covers the same ground as the theory of Vibrations in general; but, unless some limitation were admitted, the consideration of such subjects as the Tides, not to speak of Optics, would have to be included. As a general rule we shall confine ourselves to those classes of vibrations for which our ears afford a ready made and wonderfully sensitive instrument of investigation. Without ears we should hardly care much more about vibrations than without eyes we should care about light.

The present volume includes chapters on the vibrations of systems in general, in which, I hope, will be recognised some novelty of treatment and results, followed by a more detailed consideration of special systems, such as stretched strings, bars, membranes, and plates. The second volume, of which a considerable portion is already written, will commence with aërial vibrations.

My best thanks are due to Mr H. M. Taylor of Trinity College, Cambridge, who has been good enough to read the proofs. By his kind assistance several errors and obscurities have been eliminated, and the volume generally has been rendered less imperfect than it would otherwise have been.

Any corrections, or suggestions for improvements, with which my readers may favour me will be highly appreciated.

TERLING PLACE, WITHAM,
April, 1877.

CONTENTS.

CHAPTER I.

CHAPTER II.

CHAPTER IX.

CHAPTER X.

CHAPTER I.

INTRODUCTION.

1. The sensation of sound is a thing *sui generis*, not comparable with any of our other sensations. No one can express the relation between a sound and a colour or a smell. Directly or indirectly, all questions connected with this subject must come for decision to the ear, as the organ of hearing; and from it there can be no appeal. But we are not therefore to infer that all acoustical investigations are conducted with the unassisted ear. When once we have discovered the physical phenomena which constitute the foundation of sound, our explorations are in great measure transferred to another field lying within the dominion of the principles of Mechanics. Important laws are in this way arrived at, to which the sensations of the ear cannot but conform.

2. Very cursory observation often suffices to shew that sounding bodies are in a state of vibration, and that the phenomena of sound and vibration are closely connected. When a vibrating bell or string is touched by the finger, the sound ceases at the same moment that the vibration is damped. But, in order to affect the sense of hearing, it is not enough to have a vibrating instrument; there must also be an uninterrupted communication between the instrument and the ear. A bell rung *in vacuo*, with proper precautions to prevent the communication of motion, remains inaudible. In the air of the atmosphere, however, sounds have a universal vehicle, capable of conveying them without break from the most variously constituted sources to the recesses of the ear.

3. The passage of sound is not instantaneous. When a gun is fired at a distance, a very perceptible interval separates the

report from the flash. This represents the time occupied by sound in travelling from the gun to the observer, the retardation of the flash due to the finite velocity of light being altogether negligible. The first accurate experiments were made by some members of the French Academy, in 1738. Cannons were fired, and the retardation of the reports at different distances observed. The principal precaution necessary is to reverse alternately the direction along which the sound travels, in order to eliminate the influence of the motion of the air in mass. Down the wind, for instance, sound travels relatively to the earth faster than its proper rate, for the velocity of the wind is added to that proper to the propagation of sound in still air. For still dry air at a temperature of 0°C., the French observers found a velocity of 337 metres per second. Observations of the same character were made by Arago and others in 1822; by the Dutch physicists Moll, van Beek and Kuytenbrouwer at Amsterdam; by Bravais and Martins between the top of the Faulhorn and a station below; and by others. The general result has been to give a somewhat lower value for the velocity of sound—about 332 metres per second. The effect of alteration of temperature and pressure on the propagation of sound will be best considered in connection with the mechanical theory.

4. It is a direct consequence of observation, that within wide limits, the velocity of sound is independent, or at least very nearly independent, of its intensity, and also of its pitch. Were this otherwise, a quick piece of music would be heard at a little distance hopelessly confused and discordant. But when the disturbances are very violent and abrupt, so that the alterations of density concerned are comparable with the whole density of the air, the simplicity of this law may be departed from.

5. An elaborate series of experiments on the propagation of sound in long tubes (water-pipes) has been made by Regnault[1]. He adopted an automatic arrangement similar in principle to that used for measuring the speed of projectiles. At the moment when a pistol is fired at one end of the tube a wire conveying an electric current is ruptured by the shock. This causes the withdrawal of a tracing point which was previously marking a line on a revolving drum. At the further end of the pipe is a stretched membrane so arranged that when on the arrival of the sound it yields to the

_____
[1] *Mémoires de l'Académie de France*, t. XXXVII.

impulse, the circuit, which was ruptured during the passage of the sound, is recompleted. At the same moment the tracing point falls back on the drum. The blank space left unmarked corresponds to the time occupied by the sound in making the journey, and, when the motion of the drum is known, gives the means of determining it. The length of the journey between the first wire and the membrane is found by direct measurement. In these experiments the velocity of sound appeared to be not quite independent of the diameter of the pipe, which varied from $0^{m}\cdot108$ to $1^{m}\cdot100$. The discrepancy is perhaps due to friction, whose influence would be greater in smaller pipes.

6.   Although, in practice, air is usually the vehicle of sound, other gases, liquids and solids are equally capable of conveying it. In most cases, however, the means of making a direct measurement of the velocity of sound are wanting, and we are not yet in a position to consider the indirect methods. But in the case of water the same difficulty does not occur. In the year 1826, Colladon and Sturm investigated the propagation of sound in the Lake of Geneva. The striking of a bell at one station was simultaneous with a flash of gunpowder. The observer at a second station measured the interval between the flash and the arrival of the sound, applying his ear to a tube carried beneath the surface. At a temperature of 8°C., the velocity of sound in water was thus found to be 1435 metres per second.

7.   The conveyance of sound by solids may be illustrated by a pretty experiment due to Wheatstone. One end of a metallic wire is connected with the sound-board of a pianoforte, and the other taken through the partitions or floors into another part of the building, where naturally nothing would be audible. If a resonance-board (such as a violin) be now placed in contact with the wire, a tune played on the piano is easily heard, and the sound seems to emanate from the resonance-board.

8.   In an open space the intensity of sound falls off with great rapidity as the distance from the source increases. The same amount of motion has to do duty over surfaces ever increasing as the squares of the distance. Anything that confines the sound will tend to diminish the falling off of intensity. Thus over the flat surface of still water, a sound carries further than over broken ground; the corner between a smooth pavement and a vertical wall is still better; but the most effective of all is a tube-like enclosure,

which prevents spreading altogether. The use of speaking tubes to facilitate communication between the different parts of a building is well known. If it were not for certain effects (frictional and other) due to the sides of the tube, sound might be thus conveyed with little loss to very great distances.

9. Before proceeding further we must consider a distinction, which is of great importance, though not free from difficulty. Sounds may be classed as musical and unmusical; the former for convenience may be called *notes* and the latter *noises*. The extreme cases will raise no dispute; every one recognises the difference between the note of a pianoforte and the creaking of a shoe. But it is not so easy to draw the line of separation. In the first place few notes are free from all unmusical accompaniment. With organ pipes especially, the hissing of the wind as it escapes at the mouth may be heard beside the proper note of the pipe. And, secondly, many noises so far partake of a musical character as to have a definite pitch. This is more easily recognised in a sequence, giving, for example, the common chord, than by continued attention to an individual instance. The experiment may be made by drawing corks from bottles, previously tuned by pouring water into them, or by throwing down on a table sticks of wood of suitable dimensions. But, although noises are sometimes not entirely unmusical, and notes are usually not quite free from noise, there is no difficulty in recognising which of the two is the simpler phenomenon. There is a certain smoothness and continuity about the musical note. Moreover by sounding together a variety of notes—for example, by striking simultaneously a number of consecutive keys on a pianoforte—we obtain an approximation to a noise; while no combination of noises could ever blend into a musical note.

10. We are thus led to give our attention, in the first instance, mainly to musical sounds. These arrange themselves naturally in a certain order according to *pitch*—a quality which all can appreciate to some extent. Trained ears can recognise an enormous number of gradations—more than a thousand, probably, within the compass of the human voice. These gradations of pitch are not, like the degrees of a thermometric scale, without special mutual relations. Taking any given note as a starting point, musicians can single out certain others, which bear a definite relation to the first, and are known as its octave, fifth, &c. The corresponding differences of pitch are called *intervals*, and are

spoken of as always the same for the same relationship. Thus, wherever they may occur in the scale, a note and its octave are separated by *the interval of the octave*. It will be our object later to explain, so far as it can be done, the origin and nature of the consonant intervals, but we must now turn to consider the physical aspect of the question.

Since sounds are produced by vibrations, it is natural to suppose that the simpler sounds, viz. musical notes, correspond to *periodic* vibrations, that is to say, vibrations which after a certain interval of time, called the *period*, repeat themselves with perfect regularity. And this, with a limitation presently to be noticed, is true.

11. Many contrivances may be proposed to illustrate the generation of a musical note. One of the simplest is a revolving wheel whose milled edge is pressed against a card. Each projection as it strikes the card gives a slight tap, whose regular recurrence, as the wheel turns, produces a note of definite pitch, *rising in the scale, as velocity of rotation increases.* But the most appropriate instrument for the fundamental experiments on notes is undoubtedly the Siren, invented by Cagniard de la Tour. It consists essentially of a stiff disc, capable of revolving about its centre, and pierced with one or more sets of holes, arranged at equal intervals round the circumference of circles concentric with the disc. A windpipe in connection with bellows is presented perpendicularly to the disc, its open end being opposite to one of the circles, which contains a set of holes. When the bellows are worked, the stream of air escapes freely, if a hole is opposite to the end of the pipe; but otherwise it is obstructed. As the disc turns, a succession of puffs of air escape through it, until, when the velocity is sufficient, they blend into a note, whose pitch rises continually with the rapidity of the puffs. We shall have occasion later to describe more elaborate forms of the Siren, but for our immediate purpose the present simple arrangement will suffice.

12. One of the most important facts in the whole science is exemplified by the Siren—namely, that the pitch of a note depends upon the period of its vibration. The size and shape of the holes, the force of the wind, and other elements of the problem may be varied; but if the number of puffs in a given time, such as one second, remains unchanged, so also does the pitch. We may even dispense with wind altogether, and produce a note by allowing the corner of a card to tap against the edges of the holes, as they

revolve; the pitch will still be the same. Observation of other sources of sound, such as vibrating solids, leads to the same conclusions, though the difficulties are often such as to render necessary rather refined experimental methods.

But in saying that pitch depends upon period, there lurks an ambiguity, which deserves attentive consideration, as it will lead us to a point of great importance. If a variable quantity is periodic in any time τ, it is also periodic in the times 2τ, 3τ, &c. Conversely, a recurrence within a given period τ, does not exclude a more rapid recurrence within periods which are the aliquot parts of τ. It would appear accordingly that a vibration really recurring in the time $\frac{1}{2}\tau$ (for example) may be regarded as having the period τ, and therefore by the law just laid down as producing a note of the pitch defined by τ. The force of this consideration cannot be entirely evaded by defining as the period the *least* time required to bring about a repetition. In the first place, the necessity of such a restriction is in itself almost sufficient to shew that we have not got to the root of the matter; for although a right to the period τ may be denied to a vibration repeating itself rigorously within a time $\frac{1}{2}\tau$, yet it must be allowed to a vibration that may differ indefinitely little therefrom. In the Siren experiment, suppose that in one of the circles of holes containing an even number, every alternate hole is displaced along the arc of the circle by the same amount. The displacement may be made so small that no change can be detected in the resulting note; but the periodic time on which the pitch depends has been doubled. And secondly it is evident from the nature of periodicity, that the superposition on a vibration of period τ, of others having periods $\frac{1}{2}\tau, \frac{1}{3}\tau$...&c., does not disturb the period τ, while yet it cannot be supposed that the addition of the new elements has left the quality of the sound unchanged. Moreover, since the pitch is not affected by their presence, how do we know that elements of the shorter periods were not there from the beginning?

13.   These considerations lead us to expect remarkable relations between the notes whose periods are as the reciprocals of the natural numbers. Nothing can be easier than to investigate the question by means of the Siren. Imagine two circles of holes, the inner containing any convenient number, and the outer twice as many. Then at whatever speed the disc may turn, the period of the vibration engendered by blowing the first set will necessarily

be the double of that belonging to the second. On making the experiment the two notes are found to stand to each other in the relation of octaves; and we conclude that *in passing from any note to its octave, the frequency of vibration is doubled*. A similar method of experimenting shews, that to the ratio of periods 3 : 1 corresponds the interval known to musicians as the *twelfth*, made up of an octave and a fifth; to the ratio of 4 : 1, the double octave; and to the ratio 5 : 1, the interval made up of two octaves and a *major third*. In order to obtain the intervals of the fifth and third themselves, the ratios must be made 3 : 2 and 5 : 4 respectively.

14. From these experiments it appears that if two notes stand to one another in a fixed relation, then, no matter at what part of the scale they may be situated, their periods are in a certain constant ratio characteristic of the relation. The same may be said of their *frequencies*[1], or the number of vibrations which they execute in a given time. The ratio 2 : 1 is thus characteristic of the octave interval. If we wish to combine two intervals,—for instance, starting from a given note, to take a step of an octave and then another of a fifth in the same direction, the corresponding ratios must be compounded:

$$\frac{2}{1} \times \frac{3}{2} = \frac{3}{1}.$$

The twelfth part of an octave is represented by the ratio $\sqrt[12]{2} : 1$, for this is the step which repeated twelve times leads to an octave above the starting point. If we wish to have a measure of intervals in the proper sense, we must take not the characteristic ratio itself, but the logarithm of that ratio. Then, and then only, will the measure of a compound interval be the *sum* of the measures of the components.

15. From the intervals of the octave, fifth, and third considered above, others known to musicians may be derived. The difference of an octave and a fifth is called a *fourth*, and has the ratio $2 \div \frac{3}{2} = \frac{4}{3}$. This process of subtracting an interval from the octave is called *inverting* it. By inverting the major third

[1] A single word to denote the number of vibrations executed in the unit of time is indispensable: I know no better than ‘frequency,’ which was used in this sense by Young. The same word is employed by Prof. Everett in his excellent edition of Deschanel’s *Natural Philosophy*.

we obtain the minor sixth. Again, by subtraction of a major third from a fifth we obtain the minor third; and from this by inversion the major sixth. The following table exhibits side by side the names of the intervals and the corresponding ratios of frequencies:

Octave	2 : 1
Fifth	3 : 2
Fourth	4 : 3
Major Third	5 : 4
Minor Sixth	8 : 5
Minor Third	6 : 5
Major Sixth	5 : 3

These are all the consonant intervals comprised within the limits of the octave. It will be remarked that the corresponding ratios are all expressed by means of *small* whole numbers, and that this is more particularly the case for the more consonant intervals.

The notes whose frequencies are multiples of that of a given one, are called its *harmonics*, and the whole series constitutes a *harmonic scale*. As is well known to violinists, they may all be obtained from the same string by touching it lightly with the finger at certain points, while the bow is drawn.

The establishment of the connection between musical intervals and definite ratios of frequency—a fundamental point in Acoustics —is due to Mersenne (1636). It was indeed known to the Greeks in what ratios the lengths of strings must be changed in order to obtain the octave and fifth; but Mersenne demonstrated the law connecting the length of a string with the period of its vibration, and made the first determination of the actual rate of vibration of a known musical note.

16. On any note taken as a key-note, or *tonic*, a *diatonic* scale may be founded, whose derivation we now proceed to explain. If the key-note, whatever may be its absolute pitch, be called Do, the fifth above or dominant is Sol, and the fifth below or subdominant is Fa. The common chord on any note is produced by combining it with its major third, and fifth, giving the ratios of frequency $1 : \frac{5}{4} : \frac{3}{2}$ or $4 : 5 : 6$. Now if we take the common chord on the tonic, on the dominant, and on the subdominant, and transpose them when necessary into the octave

lying immediately above the tonic, we obtain notes whose frequencies arranged in order of magnitude are:

$$\begin{array}{cccccccc} \text{Do} & \text{Re} & \text{Mi} & \text{Fa} & \text{Sol} & \text{La} & \text{Si} & \text{Do} \\ 1, & \dfrac{9}{8}, & \dfrac{5}{4}, & \dfrac{4}{3}, & \dfrac{3}{2}, & \dfrac{5}{3}, & \dfrac{15}{8}, & 2. \end{array}$$

Here the common chord on Do is Do—Mi—Sol, with the ratios $1 : \dfrac{5}{4} : \dfrac{3}{2}$; the chord on Sol is Sol—Si—Re, with the ratios $\dfrac{3}{2} : \dfrac{15}{8} : 2 \times \dfrac{9}{8} = 1 : \dfrac{5}{4} : \dfrac{3}{2}$; and the chord on Fa is Fa—La—Do, still with the same ratios. The scale is completed by repeating these notes above and below at intervals of octaves.

If we take as our Do, or key-note, the lower c of a tenor voice, the diatonic scale will be

$$\begin{array}{ccccccc} c & d & e & f & g & a & c' \end{array}$$

Usage differs slightly as to the mode of distinguishing the different octaves; in what follows I adopt the notation of Helmholtz. The octave below the one just referred to is written with capital letters—C, D, &c.; the next below that with a suffix—$C_{,}, D_{,}$, &c.; and the one beyond that with a double suffix—$C_{,,}$, &c. On the other side accents denote elevation by an octave—c', c", &c. The notes of the four strings of a violin are written in this notation, g—d'—a'—e". The middle c of the pianoforte is c'.

17. With respect to an absolute standard of pitch there has been no uniform practice. At the Stuttgard conference in 1834, c' = 264 complete vibrations per second was recommended. This corresponds to a' = 440. The French pitch makes a' = 435. In Handel's time the pitch was much lower. If c' were taken at 256 or 2^8, all the c's would have frequencies represented by powers of 2. This pitch is usually adopted by physicists and acoustical instrument makers, and has the advantage of simplicity.

The determination *ab initio* of the frequency of a given note is an operation requiring some care. The simplest method in principle is by means of the Siren, which is driven at such a rate as to give a note in unison with the given one. The number of turns effected by the disc in one second is given by a counting apparatus, which can be thrown in and out of gear at the beginning and end of a measured interval of time. This multiplied by the number of effective holes gives the required frequency. The consideration of other methods admitting of greater accuracy must be deferred.

18. So long as we keep to the diatonic scale of c, the notes above written are all that are required in a musical composition. But it is frequently desired to change the key-note. Under these circumstances a singer with a good natural ear, accustomed to perform without accompaniment, takes an entirely fresh departure, constructing a new diatonic scale on the new key-note. In this way, after a few changes of key, the original scale will be quite departed from, and an immense variety of notes be used. On an instrument with fixed notes like the piano and organ such a multiplication is impracticable, and some compromise is necessary in order to allow the same note to perform different functions. This is not the place to discuss the question at any length; we will therefore take as an illustration the simplest, as well as the commonest case—modulation into the key of the dominant.

By definition, the diatonic scale of c consists of the common chords founded on c, g and f. In like manner the scale of g consists of the chords founded on g, d and c. The chords of c and g are then common to the two scales; but the third and fifth of d introduce new notes. The third of d written f♯ has a frequency $\frac{9}{8} \times \frac{5}{4} = \frac{45}{32}$, and is far removed from any note in the scale of c. But the fifth of d, with a frequency $\frac{9}{8} \times \frac{3}{2} = \frac{27}{16}$, differs but little from a, whose frequency is $\frac{5}{3}$. In ordinary keyed instruments the interval between the two, represented by $\frac{81}{80}$, and called a *comma*, is neglected, and the two notes by a suitable compromise or *temperament* are identified.

19. Various systems of temperament have been used; the simplest and that now most generally used, or at least aimed at, is the *equal* temperament. On referring to the table of frequencies for the diatonic scale, it will be seen that the intervals from Do to Re, from Re to Mi, from Fa to Sol, from Sol to La, and from La to Si, are nearly the same, being represented by $\frac{9}{8}$ or $\frac{10}{9}$; while the intervals from Mi to Fa and from Si to Do, represented by $\frac{16}{15}$, are about half as much. The equal temperament treats these approximate relations as exact, dividing the octave into twelve equal

parts called mean semitones. From these twelve notes the diatonic scale belonging to any key may be selected according to the following rule. Taking the key-note as the first, fill up the series with the third, fifth, sixth, eighth, tenth, twelfth and thirteenth notes, counting upwards. In this way all difficulties of modulation are avoided, as the twelve notes serve as well for one key as for another. But this advantage is obtained at a sacrifice of true intonation. The equal temperament third, being the third part of an octave, is represented by the ratio $\sqrt[3]{2} : 1$, or approximately 1·2599, while the true third is 1·25. The tempered third is thus higher than the true by the interval 126 : 125. The ratio of the tempered fifth may be obtained from the consideration that seven semitones make a fifth, while twelve go to an octave. The ratio is therefore $2^{\frac{7}{12}} : 1$, which $= 1·4983$. The tempered fifth is thus too low in the ratio 1·4983 : 1·5, or approximately 881 : 882. This error is insignificant; and even the error of the third is not of much consequence in quick music on instruments like the pianoforte. But when the notes are *held*, as in the harmonium and organ, the consonance of chords is materially impaired.

20. The following Table, giving the twelve notes of the chromatic scale according to the system of equal temperament, will be convenient for reference[1]. The standard employed is $a' = 440$; in order to adapt the Table to any other absolute pitch, it is only necessary to multiply throughout by the proper constant.

	$C_{\prime\prime}$	$C_{\prime}$	C	c	c'	c''	c'''	c''''
C	16·35	32·70	65·41	130·8	261·7	523·3	1046·6	2093·2
C♯	17·32	34·65	69·30	138·6	277·2	544·4	1108·8	2217·7
D	18·35	36·71	73·42	146·8	293·7	587·4	1174·8	2349·6
D♯	19·44	38·89	77·79	155·6	311·2	622·3	1244·6	2489·3
E	20·60	41·20	82·41	164·8	329·7	659·3	1318·6	2637·3
F	21·82	43·65	87·31	174·6	349·2	698·5	1397·0	2794·0
F♯	23·12	46·25	92·50	185·0	370·0	740·0	1480·0	2960·1
G	24·50	49·00	98·00	196·0	392·0	784·0	1568·0	3136·0
G♯	25·95	51·91	103·8	207·6	415·3	830·6	1661·2	3322·5
A	27·50	55·00	110·0	220·0	440·0	880·0	1760·0	3520·0
A♯	29·13	58·27	116·5	233·1	466·2	932·3	1864·6	3729·2
B	30·86	61·73	123·5	246·9	493·9	987·7	1975·5	3951·0

[1] Zamminer, *Die Musik und die musikalischen Instrumente.* Giessen, 1855.

The ratios of the intervals of the equal temperament scale are given below (Zamminer) :—

Note.	Frequency.		Note.	Frequency.
c	$= 1{\cdot}00000$		f$\sharp$	$2^{\frac{6}{12}} = 1{\cdot}41421$
c$\sharp$	$2^{\frac{1}{12}} = 1{\cdot}05946$		g	$2^{\frac{7}{12}} = 1{\cdot}49831$
d	$2^{\frac{2}{12}} = 1{\cdot}12246$		g$\sharp$	$2^{\frac{8}{12}} = 1{\cdot}58740$
d$\sharp$	$2^{\frac{3}{12}} = 1{\cdot}18921$		a	$2^{\frac{9}{12}} = 1{\cdot}68179$
e	$2^{\frac{4}{12}} = 1{\cdot}25992$		a$\sharp$	$2^{\frac{10}{12}} = 1{\cdot}78180$
f	$2^{\frac{5}{12}} = 1{\cdot}33484$		b	$2^{\frac{11}{12}} = 1{\cdot}88775$

$$c' = 2{\cdot}000$$

21. Returning now for a moment to the physical aspect of the question, we will assume, what we shall afterwards prove to be true within wide limits,—that, when two or more sources of sound agitate the air simultaneously, the resulting disturbance at any point in the external air, or in the ear-passage, is the simple sum (in the extended geometrical sense) of what would be caused by each source acting separately. Let us consider the disturbance due to a simultaneous sounding of a note and any or all of its harmonics. By definition, the complex whole forms a note having the same period (and therefore pitch) as its gravest element. We have at present no criterion by which the two can be distinguished, or the presence of the higher harmonics recognised. And yet—in the case, at any rate, where the component sounds have an independent origin—it is usually not difficult to detect them by the ear, so as to effect an analysis of the mixture. This is as much as to say that a strictly periodic vibration may give rise to a sensation which is not simple, but susceptible of further analysis. In point of fact, it has long been known to musicians that under certain circumstances the harmonics of a note may be heard along with it, even when the note is due to a single source, such as a vibrating string; but the significance of the fact was not understood. Since attention has been drawn to the subject, it has been proved (mainly by the labours of Ohm and Helmholtz) that almost all musical notes are highly compound, consisting in fact of the notes of a harmonic scale, from which in particular cases one or more members may be missing. The reason of the uncertainty and difficulty of the analysis will be touched upon presently.

22. That kind of note which the ear cannot further resolve is called by Helmholtz in German a '*ton.*' Tyndall and other recent writers on Acoustics have adopted 'tone' as an English equivalent, —a practice which will be followed in the present work. The thing is so important, that a convenient word is almost a matter of necessity. *Notes* then are in general made up of *tones*, the pitch of the note being that of the gravest tone which it contains.

23. In strictness the quality of pitch must be attached in the first instance to simple tones only ; otherwise the difficulty of discontinuity before referred to presents itself. The slightest change in the nature of a note may lower its pitch by a whole octave, as was exemplified in the case of the Siren. We should now rather say that the effect of the slight displacement of the alternate holes in that experiment was to introduce a new feeble tone an octave lower than any previously present. This is sufficient to alter the period of the whole, but the great mass of the sound remains very nearly as before.

In most musical notes, however, the fundamental or gravest tone is present in sufficient intensity to impress its character on the whole. The effect of the harmonic overtones is then to modify the *quality* or *character*[1] of the note, independently of pitch. That such a distinction exists is well known. The notes of a violin, tuning fork, or of the human voice with its different vowel sounds, &c., may all have the same pitch and yet differ independently of loudness; and though a part of this difference is due to accompanying noises, which are extraneous to their nature as notes, still there is a part which is not thus to be accounted for. Musical notes may thus be classified as variable in three ways : First, *pitch*. This we have already sufficiently considered. Secondly, *character*, depending on the proportions in which the harmonic overtones are combined with the fundamental: and thirdly, *loudness*. This has to be taken last, because the ear is not capable of comparing (with any precision) the loudness of two notes which differ much in pitch or character. We shall indeed in a future chapter give a mechanical measure of the intensity of sound, including in one system all gradations of pitch; but this is nothing to the point. We are here concerned with the intensity of the sensation of sound, not with a measure of its physical cause. The difference of loudness is, however, at once recognised as one of more or less ; so that we

[1] German, 'Klangfarbe —French,'timbre.' The word 'character' is used in this sense by Everett.

have hardly any choice but to regard it as dependent *cæteris paribus* on the magnitude of the vibrations concerned.

24. We have seen that a musical note, as such, is due to a vibration which is necessarily periodic ; but the converse, it is evident, cannot be true without limitation. A periodic repetition of a noise at intervals of a second—for instance, the ticking of a clock—would not result in a musical note, be the repetition ever so perfect. In such a case we may say that the fundamental tone lies outside the limits of hearing, and although some of the harmonic overtones would fall within them, these would not give rise to a musical note or even to a chord, but to a noisy mass of sound like that produced by striking simultaneously the twelve notes of the chromatic scale. The experiment may be made with the Siren by distributing the holes quite irregularly round the circumference of a circle, and turning the disc with a moderate velocity. By the construction of the instrument, everything recurs after each complete revolution.

25. The principal remaining difficulty in the theory of notes and tones, is to explain why notes are sometimes analysed by the ear into tones, and sometimes not. If a note is really complex, why is not the fact immediately and certainly perceived, and the components disentangled ? The feebleness of the harmonic overtones is not the reason, for, as we shall see at a later stage of our inquiry, they are often of surprising loudness, and play a prominent part in music. On the other hand, if a note is sometimes perceived as a whole, why does not this happen always ? These questions have been carefully considered by Helmholtz[1], with a tolerably satisfactory result. The difficulty, such as it is, is not peculiar to Acoustics, but may be paralleled in the cognate science of Physiological Optics.

The knowledge of external things which we derive from the indications of our senses, is for the most part the result of inference. When an object is before us, certain nerves in our retinæ are excited, and certain sensations are produced, which we are accustomed to associate with the object, and we forthwith infer its presence. In the case of an unknown object the process is much the same. We interpret the sensations to which we are subject so as to form a pretty good idea of their exciting cause. From the slightly different perspective views received by the two eyes we infer, often by a highly elaborate process, the actual relief and

---

[1] *Tonempfindungen*, 3rd edition, p. 98.

distance of the object, to which we might otherwise have had no
clue. These inferences are made with extreme rapidity and quite
unconsciously. The whole life of each one of us is a continued
lesson in interpreting the signs presented to us, and in drawing
conclusions as to the actualities outside. Only so far as we succeed
in doing this, are our sensations of any use to us in the ordinary
affairs of life. This being so, it is no wonder that the study of our
sensations themselves falls into the background, and that subjective
phenomena, as they are called, become exceedingly difficult of
observation. As an instance of this, it is sufficient to mention the
'blind spot' on the retina, which might *a priori* have been
expected to manifest itself as a conspicuous phenomenon, though
as a fact probably not one person in a hundred million would find
it out for themselves. The application of these remarks to the
question in hand is tolerably obvious. In the daily use of our ears
our object is to disentangle from the whole mass of sound that
may reach us, the parts coming from sources which may interest
us at the moment. When we listen to the conversation of a friend,
we fix our attention on the sound proceeding from him and
endeavour to grasp that as a whole, while we ignore, as far as
possible, any other sounds, regarding them as an interruption.
There are usually sufficient indications to assist us in making this
partial analysis. When a man speaks, the whole sound of his
voice rises and falls together, and we have no difficulty in recognising its unity. It would be no advantage, but on the contrary
a great source of confusion, if we were to carry the analysis further,
and resolve the whole mass of sound present into its component
tones. Although, as regards sensation, a resolution into tones
might be expected, the necessities of our position and the practice
of our lives lead us to stop the analysis at the point, beyond
which it would cease to be of service in deciphering our sensations, considered as signs of external objects[1].

But it may sometimes happen that however much we may
wish to form a judgment, the materials for doing so are absolutely
wanting. When a note and its octave are sounding close together
and with perfect uniformity, there is nothing in our sensations to
enable us to distinguish, whether the notes have a double or a
single origin. In the mixture stop of the organ, the pressing down
of each key admits the wind to a group of pipes, giving a note and

---

[1] Most probably the power of attending to the important and ignoring the
unimportant part of our sensations is to a great extent inherited—to how great an
extent we shall perhaps never know.

its first three or four harmonics. The pipes of each group always sound together, and the result is usually perceived as a single note, although it does not proceed from a single source.

26. The resolution of a note into its component tones is a matter of very different difficulty with different individuals. A considerable effort of attention is required, particularly at first; and, until a habit has been formed, some external aid in the shape of a suggestion of what is to be listened for, is very desirable.

The difficulty is altogether very similar to that of learning to draw. From the machinery of vision it might have been expected that nothing would be easier than to make, on a plane surface, a representation of surrounding solid objects; but experience shews that much practice is generally required.

We shall return to the question of the analysis of notes at a later stage, after we have treated of the vibrations of strings, with the aid of which it is best elucidated; but a very instructive experiment, due originally to Ohm and improved by Helmholtz, may be given here. Helmholtz[1] took two bottles of the shape represented in the figure, one about twice as large as the other. These were blown by streams of air directed across the mouth and issuing from gutta-percha tubes, whose ends had been softened and pressed flat, so as to reduce the bore to the form of a narrow slit, the tubes being in connection with the same bellows. By pouring in water when the note is too low and by partially obstructing the mouth when the note is too high, the bottles may be made to give notes with the exact interval of an octave, 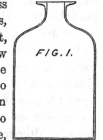 such as b and b. The larger bottle, blown alone, gives a somewhat muffled sound similar in character to the vowel U; but, when both bottles are blown, the character of the resulting sound is sharper, resembling rather the vowel O. For a short time after the notes had been heard separately Helmholtz was able to distinguish them in the mixture; but as the memory of their separate impressions faded, the higher note seemed by degrees to amalgamate with the lower, which at the same time became louder and acquired a sharper character. This blending of the two notes may take place even when the high note is the louder.

27. Seeing now that notes are usually compound, and that only a particular sort called tones are incapable of further analysis,

[1] *Tonempfindungen*, p. 109.

we are led to inquire what is the physical characteristic of tones, to which they owe their peculiarity? What sort of periodic vibration is it, which produces a simple tone? According to what mathematical function of the time does the pressure vary in the passage of the ear? No question in Acoustics can be more important.

The simplest periodic functions with which mathematicians are acquainted are the circular functions, expressed by a sine or cosine; indeed there are no others at all approaching them in simplicity. They may be of any period, and admitting of no other variation (except magnitude), seem well adapted to produce simple tones. Moreover it has been proved by Fourier, that the most general single-valued periodic function can be resolved into a series of circular functions, having periods which are submultiples of that of the given function. Again, it is a consequence of the general theory of vibration that the particular type, now suggested as corresponding to a simple tone, is the only one capable of preserving its integrity among the vicissitudes which it may have to undergo. Any other kind is liable to a sort of physical analysis, one part being differently affected from another. If the analysis within the ear proceeded on a different principle from that effected according to the laws of dead matter outside the ear, the consequence would be that a sound originally simple might become compound on its way to the observer. There is no reason to suppose that anything of this sort actually happens. When it is added that according to all the ideas we can form on the subject, the analysis within the ear must take place by means of a physical machinery, subject to the same laws as prevail outside, it will be seen that a strong case has been made out for regarding tones as due to vibrations expressed by circular functions. We are not however left entirely to the guidance of general considerations like these. In the chapter on the vibration of strings, we shall see that in many cases theory informs us beforehand of the nature of the vibration executed by a string, and in particular whether any specified simple vibration is a component or not. Here we have a decisive test. It is found by experiment that, whenever according to theory any simple vibration is present, the corresponding tone can be heard, but, whenever the simple vibration is absent, then the tone cannot be heard. We are therefore justified in asserting that simple tones and vibrations of a circular type are indissolubly connected. This law was discovered by Ohm.

CHAPTER II.

HARMONIC MOTIONS.

28. THE vibrations expressed by a circular function of the time and variously designated as *simple, pendulous,* or *harmonic,* are so important in Acoustics that we cannot do better than devote a chapter to their consideration, before entering on the dynamical part of our subject. The quantity, whose variation constitutes the 'vibration,' may be the displacement of a particle measured in a given direction, the pressure at a fixed point in a fluid medium, and so on. In any case denoting it by u, we have

$$u = a \cos\left(\frac{2\pi t}{\tau} - \epsilon\right) \dotfill (1),$$

in which a denotes the *amplitude*, or extreme value of u; τ is the *periodic time*, or *period*, after the lapse of which the values of u recur; and ϵ determines the *phase* of the vibration at the moment from which t is measured.

Any number of harmonic vibrations *of the same period* affecting a variable quantity, compound into another of the same type, whose elements are determined as follows:

$$u = \Sigma a \cos\left(\frac{2\pi t}{\tau} - \epsilon\right)$$

$$= \cos\frac{2\pi t}{\tau} \Sigma a \cos\epsilon + \sin\frac{2\pi t}{\tau} \Sigma a \sin\epsilon$$

$$= r \cos\left(\frac{2\pi t}{\tau} - \theta\right) \dotfill (2),$$

if
$$r = \{(\Sigma a \cos\epsilon)^2 + (\Sigma a \sin\epsilon)^2\}^{\frac{1}{2}} \dotfill (3),$$

and
$$\tan\theta = \Sigma a \sin\epsilon \div \Sigma a \cos\epsilon \dotfill (4).$$

For example, let there be two components,

$$u = a \cos\left(\frac{2\pi t}{\tau} - \epsilon\right) + a' \cos\left(\frac{2\pi t}{\tau} - \epsilon'\right);$$

then

$$r = \{a^2 + a'^2 + 2aa' \cos(\epsilon - \epsilon')\}^{\frac{1}{2}}\dots\dots\dots\dots(5),$$

$$\tan\theta = \frac{a \sin\epsilon + a' \sin\epsilon'}{a \cos\epsilon + a' \cos\epsilon'},\dots\dots\dots\dots\dots\dots(6).$$

Particular cases may be noted. If the phases of the two components agree,

$$u = (a + a') \cos\left(\frac{2\pi t}{\tau} - \epsilon\right)$$

If the phases differ by half a period,

$$u = (a - a') \cos\left(\frac{2\pi t}{\tau} - \epsilon\right),$$

so that if $a' = a$, u vanishes. In this case the vibrations are often said to *interfere*, but the expression is rather misleading. Two sounds may very properly be said to interfere, when they together cause silence; but the mere superposition of two vibrations (whether rest is the consequence, or not) cannot properly be so called. At least if this be interference, it is difficult to say what non-interference can be. It will appear in the course of this work that when vibrations exceed a certain intensity they no longer compound by mere addition; *this* mutual action might more properly be called interference, but it is a phenomenon of a totally different nature from that with which we are now dealing.

Again, if the phases differ by a quarter or by three-quarters of a period, $\cos(\epsilon - \epsilon') = 0$, and

$$r = \{a^2 + a'^2\}^{\frac{1}{2}}$$

Harmonic vibrations of given period may be represented by lines drawn from a pole, the lengths of the lines being proportional to the amplitudes, and the inclinations to the phases of the vibrations. The resultant of any number of harmonic vibrations is then represented by the geometrical resultant of the corresponding lines. For example, if they are disposed symmetrically round the pole, the resultant of the lines, or vibrations, is zero.

29. If we measure off along an axis of x distances proportional to the time, and take u for an ordinate, we obtain the harmonic curve, or curve of sines,

$$u = a \cos\left(\frac{2\pi x}{\lambda} - \epsilon\right),$$

where λ, called the wave-length, is written in place of τ, both quantities denoting the range of the independent variable corresponding to a complete recurrence of the function. The harmonic curve is thus the locus of a point subject at once to a uniform motion, and to a harmonic vibration in a perpendicular direction. In the next chapter we shall see that the vibration of a tuning fork is simple harmonic; so that if an excited tuning fork is moved with uniform velocity parallel to the line of its handle, a tracing point attached to the end of one of its prongs describes a harmonic curve, which may be obtained in a permanent form by allowing the tracing point to bear gently on a piece of smoked paper. In Fig. 2 the continuous lines are two harmonic curves of the same wave-length and amplitude, but of different

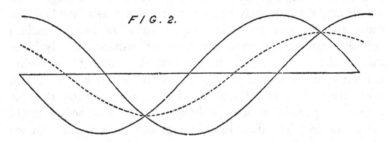

FIG. 2.

phases; the dotted curve represents half their resultant, being the locus of points midway between those in which the two curves are met by any ordinate.

30.   If two harmonic vibrations of different periods coexist,

$$u = a \cos\left(\frac{2\pi t}{\tau} - \epsilon\right) + a' \cos\left(\frac{2\pi t}{\tau'} - \epsilon'\right).$$

The resultant cannot here be represented as a simple harmonic motion with other elements. If τ and τ' be incommensurable, the value of u never recurs; but, if τ and τ' be in the ratio of two whole numbers, u recurs after the lapse of a time equal to the least common multiple of τ and τ'; but the vibration is not simple harmonic. For example, when a note and its fifth are sounding together, the vibration recurs after a time equal to twice the period of the graver.

One case of the composition of harmonic vibrations of different periods is worth special discussion, namely, when the difference of the periods is small. If we fix our attention on the course of things during an interval of time including merely a few periods, we see that the two vibrations are nearly the same as if their periods were absolutely equal, in which case they would, as we know, be equivalent to another simple harmonic vibration of the same period. For a few periods then the resultant motion is approximately simple harmonic, but the same harmonic will not continue to represent it for long. The vibration having the shorter period continually gains on its fellow, thereby altering the difference of phase on which the elements of the resultant depend. For simplicity of statement let us suppose that the two components have equal amplitudes, frequencies represented by m and n, where $m - n$ is small, and that when first observed their phases agree. At this moment their effects conspire, and the resultant has an amplitude double of that of the components. But after a time $1 \div 2 (m - n)$ the vibration m will have gained half a period relatively to the other; and the two, being now in complete disagreement, neutralize each other. After a further interval of time equal to that above named, m will have gained altogether a whole vibration, and complete accordance is once more re-established. The resultant motion is therefore approximately simple harmonic, with an amplitude not constant, but varying from zero to twice that of the components, the frequency of these alterations being $m - n$. If two tuning forks with frequencies 500 and 501 be equally excited, there is every second a rise and fall of sound corresponding to the coincidence or opposition of their vibrations. This phenomenon is called beats. We do not here fully discuss the question how the ear behaves in the presence of vibrations having nearly equal frequencies, but it is obvious that if the motion in the neighbourhood of the ear almost ceases for a considerable fraction of a second, the sound must appear to fall. For reasons that will afterwards appear, beats are best heard when the interfering sounds are simple tones. Consecutive notes of the stopped diapason of the organ shew the phenomenon very well, at least in the lower parts of the scale. A permanent interference of two notes may be obtained by mounting two stopped organ pipes of similar construction and identical pitch side by side on the same wind chest. The vibrations of the two pipes adjust themselves to complete opposition, so that at a little distance nothing can be heard, except the hissing of the wind. If by a rigid wall between the two pipes one sound could be cut off, the other would be instantly restored. Or the balance, on which silence depends, may be upset by connecting the ear with a tube, whose other end lies close to the mouth of one of the pipes.

By means of beats two notes may be tuned to unison with great exactness. The object is to make the beats as slow as possible, since the number of beats in a second is equal to the difference of the frequencies of the notes. Under favourable circumstances beats so slow as one in 30 seconds may be recognised, and would indicate that the higher note gains only two vibrations a *minute* on the lower. Or it might be desired merely to ascertain the difference of the frequencies of two notes nearly in unison, in which case nothing more is necessary than to count the number of beats. It will be remembered that the difference of frequencies does not determine the *interval* between the two notes; that depends on the *ratio* of frequencies. Thus the rapidity of the beats given by two notes nearly in unison is doubled, when both are taken an exact octave higher.

Analytically
$$u = a \cos (2\pi mt - \epsilon) + a' \cos (2\pi nt - \epsilon'),$$
where $m - n$ is small.

Now $\cos (2\pi nt - \epsilon')$ may be written
$$\cos \{2\pi mt - 2\pi (m - n) t - \epsilon'\},$$
and we have
$$u = r \cos (2\pi mt - \theta) \quad\dots\dots\dots\dots\dots\dots(1),$$
where
$$r^2 = a^2 + a'^2 + 2aa' \cos \{2\pi (m - n) t + \epsilon' - \epsilon\} \dots\dots (2),$$
$$\tan \theta = \frac{a \sin \epsilon + a' \sin \{2\pi (m - n) t + \epsilon'\}}{a \cos \epsilon + a' \cos \{2\pi (m - n) t + \epsilon'\}} \dots\dots\dots\dots(3).$$

The resultant vibration may thus be considered as harmonic with elements r and θ, which are not constant but slowly varying functions of the time, having the frequency $m - n$. The amplitude r is at its maximum when
$$\cos \{2\pi (m - n) t + \epsilon' - \epsilon\} = + 1,$$
and at its minimum when
$$\cos \{2\pi (m - n) t + \epsilon' - \epsilon\} = - 1,$$
the corresponding values being $a + a'$ and $a - a'$ respectively.

31. Another case of great importance is the composition of vibrations corresponding to a tone and its harmonics. It is known that the most general single-valued finite periodic function can be expressed by a series of simple harmonics—

$$u = a_0 + \Sigma_{n=1}^{n=\infty} a_n \cos\left(\frac{2\pi nt}{\tau} - \epsilon_n\right) \dots\dots\dots\dots (1),$$

a theorem usually quoted as Fourier's. Analytical proofs will be found in Todhunter's *Integral Calculus* and Thomson and Tait's *Natural Philosophy;* and a line of argument almost if not quite amounting to a demonstration will be given later in this work. A few remarks are all that will be required here.

Fourier's theorem is not obvious. A vague notion is not uncommon that the infinitude of arbitrary constants in the series of necessity endows it with the capacity of representing an arbitrary periodic function. That this is an error will be apparent, when it is observed that the same argument would apply equally, if one term of the series were omitted; in which case the expansion would not in general be possible.

Another point worth notice is that simple harmonics are not the only functions, in a series of which it is possible to expand one arbitrarily given. Instead of the simple elementary term

$$\cos\left(\frac{2\pi nt}{\tau} - \epsilon_n\right),$$

we might take

$$\cos\left(\frac{2\pi nt}{\tau} - \epsilon_n\right) + \frac{1}{2}\cos\left(\frac{4\pi nt}{\tau} - \epsilon_n\right),$$

formed by adding a similar one in the same phase of half the amplitude and period. It is evident that these terms would serve as well as the others; for

$$\cos\left(\frac{2\pi nt}{\tau} - \epsilon_n\right) = \left\{\cos\left(\frac{2\pi nt}{\tau} - \epsilon_n\right) + \frac{1}{2}\cos\left(\frac{4\pi nt}{\tau} - \epsilon_n\right)\right\}$$

$$- \frac{1}{2}\left\{\cos\left(\frac{4\pi nt}{\tau} - \epsilon_n\right) + \frac{1}{2}\cos\left(\frac{8\pi nt}{\tau} - \epsilon_n\right)\right\}$$

$$+ \frac{1}{4}\left\{\cos\left(\frac{8\pi nt}{\tau} - \epsilon_n\right) + \frac{1}{2}\cos\left(\frac{16\pi nt}{\tau} - \epsilon_n\right)\right\}$$

$$- \dots\dots ad\ infin.,$$

so that each term in Fourier's series, and therefore the sum of the series, can be expressed by means of the double elementary terms now suggested. This is mentioned here, because students, not being acquainted with other expansions, may imagine that simple harmonic functions are by nature the only ones qualified to be the elements in the development of a periodic function. The reason of the preeminent importance of Fourier's series in Acoustics is the mechanical one referred to in the preceding chapter, and to be explained more fully hereafter, namely, that, in general, simple harmonic vibrations are the only kind that are propagated through a vibrating system without suffering decomposition.

32. As in other cases of a similar character, e.g. Taylor's theorem, if the possibility of the expansion be known, the coefficients may be determined by a comparatively simple process. We may write (1) of § 31

$$u = A_0 + \Sigma_{n=1}^{n=\infty} A_n \cos \frac{2n\pi t}{\tau} + \Sigma_{n=1}^{n=\infty} B_n \sin \frac{2n\pi t}{\tau} \dots \dots (1).$$

Multiplying by $\cos \dfrac{2n\pi t}{\tau}$ or $\sin \dfrac{2n\pi t}{\tau}$, and integrating over a complete period from $t = 0$ to $t = \tau$, we find

$$\left. \begin{aligned} A_n &= \frac{2}{\tau} \int_0^\tau u \cos \frac{2n\pi t}{\tau} \, dt \\ B_n &= \frac{2}{\tau} \int_0^\tau u \sin \frac{2n\pi t}{\tau} \, dt \end{aligned} \right\} \dots \dots \dots \dots \dots (2).$$

An immediate integration gives

$$A_0 = \frac{1}{\tau} \int_0^\tau u \, dt \dots \dots \dots \dots \dots \dots \dots \dots (3),$$

indicating that A_0 is the *mean* value of u throughout the period.

The degree of convergency in the expansion of u depends in general on the continuity of the function and its derivatives. The series formed by successive differentiations of (1) converge less and less rapidly, but still remain convergent, and arithmetical representatives of the differential coefficients of u, so long as these latter are everywhere finite. Thus (Thomson and Tait, § 77), if all the derivatives up to the mth inclusive are free from infinite values, the series for u is more convergent than one with

$$1, \quad \frac{1}{2^m}, \quad \frac{1}{3^m}, \quad \frac{1}{4^m}, \dots \dots \&c.,$$

for coefficients.

33. Another class of compounded vibrations, interesting from
the facility with which they lend themselves to optical observation, occur when two harmonic vibrations affecting the same particle are executed *in perpendicular directions,* more especially
when the periods are not only commensurable, but in the ratio
of two *small* whole numbers. The motion is then completely
periodic, with a period not many times greater than those of the
components, and the curve described is re-entrant. If u and v
be the co-ordinates, we may take

$$u = a \cos (2\pi n t - \epsilon), \quad v = b \cos 2\pi n't \ldots\ldots\ldots\ldots(1).$$

First let us suppose that the periods are equal, so that $n' = n$;
the elimination of t gives for the equation of the curve described,

$$\frac{u^2}{a^2} + \frac{v^2}{b^2} - \frac{2uv}{ab} \cos\epsilon - \sin^2\epsilon = 0 \ \ldots\ldots\ldots\ldots(2),$$

representing in general an ellipse, whose position and dimensions
depend upon the amplitudes of the original vibrations and upon
the difference of their phases. If the phases differ by a quarter
period, $\cos\epsilon = 0$, and the equation becomes

$$\frac{u^2}{a^2} + \frac{v^2}{b^2} = 1.$$

In this case the axes of the ellipse coincide with those of
co-ordinates. If further the two components have equal amplitudes, the locus degenerates into the circle

$$u^2 + v^2 = a^2,$$

which is described with uniform velocity. This shews how a
uniform circular motion may be analysed into two rectilinear
harmonic motions, whose directions are perpendicular.

If the phases of the components agree, $\epsilon = 0$, and the ellipse
degenerates into the coincident straight lines

$$\left(\frac{u}{a} - \frac{v}{b}\right)^2 = 0;$$

or if the difference of phase amount to half a period, into

$$\left(\frac{u}{a} + \frac{v}{b}\right)^2 = 0.$$

When the unison of the two vibrations is exact, the elliptic
path remains perfectly steady, but in practice it will almost
always happen that there is a slight difference between the
periods. The consequence is that though a fixed ellipse represents

the curve described with sufficient accuracy for a few periods, the ellipse itself gradually changes in correspondence with the alteration in the magnitude of ϵ. It becomes therefore a matter of interest to consider the system of ellipses represented by (2), supposing a and b constants, but ϵ variable.

Since the extreme values of u and v are $\pm a$, $\pm b$ respectively, the ellipse is in all cases inscribed in the rectangle whose sides are $2a$, $2b$. Starting with the phases in agreement, or $\epsilon = 0$, we have the ellipse coincident with the diagonal $\dfrac{u}{a} - \dfrac{v}{b} = 0$. As ϵ increases from 0 to $\tfrac{1}{2}\pi$, the ellipse opens out until its equation becomes

$$\frac{u^2}{a^2} + \frac{v^2}{b^2} = 1.$$

From this point it closes up again, ultimately coinciding with the other diagonal $\dfrac{u}{a} + \dfrac{v}{b} = 0$, corresponding to the increase of ϵ from $\tfrac{1}{2}\pi$ to π. After this, as ϵ ranges from π to 2π, the ellipse retraces its course until it again coincides with the first diagonal. The sequence of changes is exhibited in Fig. 3.

FIG.3.

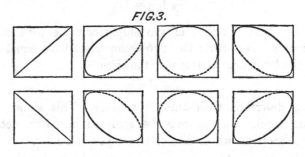

The ellipse, having already four given tangents, is completely determined by its point of contact P (Fig. 4) with the line $v = b$.

FIG. 4.

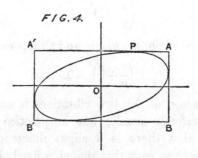

In order to connect this with ϵ, it is sufficient to observe that when $v = b$, $\cos 2\pi nt = 1$; and therefore $u = a \cos \epsilon$. Now if the elliptic paths be the result of the superposition of two harmonic vibrations of nearly coincident pitch, ϵ varies uniformly with the time, so that P itself executes a harmonic vibration along AA' with a frequency equal to the difference of the two given frequencies.

34. Lissajous[1] has shewn that this system of ellipses may be regarded as the different aspects of one and the same ellipse described on the surface of a transparent cylinder. In Fig. 5

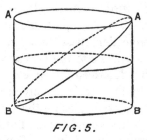

FIG. 5.

$AA'B'B$ represents the cylinder, of which AB' is a plane section. Seen from an infinite distance in the direction of the common tangent at A to the plane sections, the cylinder is projected into a rectangle, and the ellipse into its diagonal. Suppose now that the cylinder turns upon its axis, carrying the plane section with it. Its own projection remains a constant rectangle in which the pro-

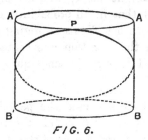

FIG. 6.

jection of the ellipse is inscribed. Fig. 6 represents the position of the cylinder after a rotation through a right angle. It appears therefore that by turning the cylinder round we obtain in succession all the ellipses corresponding to the paths described by a point subject to two harmonic vibrations of equal period and fixed amplitudes. Moreover if the cylinder be turned continuously

[1] *Annales de Chimie* (3) LI. 147.

with uniform velocity, which insures a harmonic motion for P, we obtain a complete representation of the varying orbit described by the point when the periods of the two components differ slightly, each complete revolution answering to a gain or loss of a single vibration[1]. The revolutions of the cylinder are thus synchronous with the beats which would result from the composition of the two vibrations, if they were to act in the same direction.

35. Vibrations of the kind here considered are very easily realized experimentally. A heavy pendulum-bob, hung from a fixed point by a long wire or string, describes ellipses under the action of gravity, which may in particular cases, according to the circumstances of projection, pass into straight lines or circles. But in order to see the orbits to the best advantage, it is necessary that they should be described so quickly that the impression on the retina made by the moving point at any part of its course has not time to fade materially, before the point comes round again to renew its action. This condition is fulfilled by the vibration of a silvered bead (giving by reflection a luminous point), which is attached to a straight metallic wire (such as a knitting-needle), firmly clamped in a vice at the lower end. When the system is set into vibration, the luminous point describes ellipses, which appear as fine lines of light. These ellipses would gradually contract in dimensions under the influence of friction until they subsided into a stationary bright point, without undergoing any other change, were it not that in all probability, owing to some want of symmetry, the wire has slightly differing periods according to the plane in which the vibration is executed. Under these circumstances the orbit is seen to undergo the cycle of changes already explained.

36. So far we have supposed the periods of the component vibrations to be equal, or nearly equal; the next case in order of simplicity is when one is the double of the other. We have

$$u = a \cos (4n\pi t - \epsilon), \quad v = b \cos 2n\pi t.$$

The locus resulting from the elimination of t may be written

$$\frac{u}{a} = \cos \epsilon \left(2 \frac{v^2}{b^2} - 1 \right) + 2 \sin \epsilon \frac{v}{b} \sqrt{1 - \frac{v^2}{b^2}} \quad \ldots \ldots \ldots \ldots (1),$$

[1] By a vibration will always be meant in this work a *complete* cycle of changes.

which for all values of ϵ represents a curve inscribed in the rectangle $2a$, $2b$. If $\epsilon = 0$, or π, we have

$$v^2 = \frac{b^2}{2}\left(1 \pm \frac{u}{a}\right),$$

representing parabolas. Fig. 7 shews the various curves for the intervals of the octave, twelfth, and fifth.

FIG. 7.

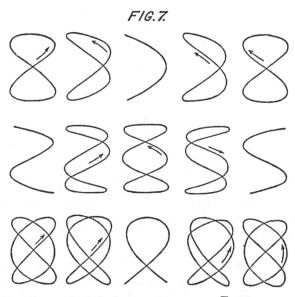

To all these systems Lissajous' method of representation by the transparent cylinder is applicable, and when the relative phase is altered, whether from the different circumstances of projection in different cases, or continuously owing to a slight deviation from exactness in the ratio of the periods, the cylinder will appear to turn, so as to present to the eye different aspects of the same line traced on its surface.

37. There is no difficulty in arranging a vibrating system so that the motion of a point shall consist of two harmonic vibrations in perpendicular planes, with their periods in any assigned ratio. The simplest is that known as Blackburn's pendulum. A wire ACB is fastened at A and B, two fixed points at the same level. The bob P is attached to its middle point by another wire CP. For vibrations in the plane of the diagram, the point of suspension is practically C, provided that the wires are sufficiently stretched;

but for a motion perpendicular to this plane, the bob turns about
D, carrying the wire ACB with it. The periods of vibration in

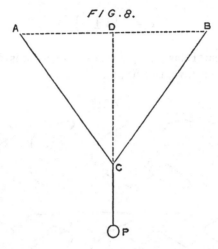

FIG. 8.

the principal planes are in the ratio of the square roots of CP and
DP. Thus if $DC = 3CP$, the bob describes the figures of the
octave. To obtain the sequence of curves corresponding to
approximate unison, ACB must be so nearly tight, that CD is
relatively small.

38.   Another contrivance called the kaleidophone was originally invented by Wheatstone. A straight thin bar of steel carrying
a bead at its upper end is fastened in a vice, as explained in a
previous paragraph. If the section of the bar is square, or circular,
the period of vibration is independent of the plane in which it is
performed. But let us suppose that the section is a rectangle
with unequal sides. The stiffness of the bar—the force with
which it resists bending—is then greater in the plane of greater
thickness, and the vibrations in this plane have the shorter period.
By a suitable adjustment of the thicknesses, the two periods of
vibration may be brought into any required ratio, and the corresponding curve exhibited.

The defect in this arrangement is that the same bar will give
only one set of figures. In order to overcome this objection
the following modification has been devised. A slip of steel is
taken whose rectangular section is very elongated, so that as
regards bending in one plane the stiffness is so great as to amount
practically to rigidity. The bar is divided into two parts, and the

broken ends reunited, the two pieces being turned on one another through a right angle, so that the plane, which contains the small thickness of one, contains the great thickness of the other. When the compound rod is clamped in a vice at a point below the junction, the period of the vibration in one direction, depending almost entirely on the length of the upper piece, is nearly constant; but that in the second direction may be controlled by varying the point at which the lower piece is clamped.

39. In this arrangement the luminous point itself executes the vibrations which are to be observed; but in Lissajous' form of the experiment, the point of light remains really fixed, while its *image* is thrown into apparent motion by means of successive reflection from two vibrating mirrors. A small hole in an opaque screen placed close to the flame of a lamp gives a point of light, which is observed after reflection in the mirrors by means of a small telescope. The mirrors, usually of polished steel, are attached to the prongs of stout tuning forks, and the whole is so disposed that when the forks are thrown into vibration the luminous point appears to describe harmonic motions in perpendicular directions, owing to the angular motions of the reflecting surfaces. The amplitudes and periods of these harmonic motions depend upon those of the corresponding forks, and may be made such as to give with enhanced brilliancy any of the figures possible with the kaleidophone. By a similar arrangement it is possible to project the figures on a screen. In either case they gradually contract as the vibrations of the forks die away.

40. The principles of this chapter have received an important application in the investigation of rectilinear periodic motions. When a point, for instance a particle of a sounding string, is vibrating with such a period as to give a note within the limits of hearing, its motion is much too rapid to be followed by the eye; so that, if it be required to know the character of the vibration, some indirect method must be adopted. The simplest, theoretically, is to compound the vibration under examination with a uniform motion of translation in a perpendicular direction, as when a tuning fork draws a harmonic curve on smoked paper. Instead of moving the vibrating body itself, we may make use of a revolving mirror, which provides us with an *image* in motion. In this way we obtain a representation of the function characteristic of the vibration, with the abscissa proportional to time.

But it often happens that the application of this method would be difficult or inconvenient. In such cases we may substitute for the uniform motion a harmonic vibration of suitable period in the same direction. To fix our ideas, let us suppose that the point, whose motion we wish to investigate, vibrates vertically with a period τ, and let us examine the result of combining with this a horizontal harmonic motion, whose period is some multiple of τ, say, $n\tau$. Take a rectangular piece of paper, and with axes parallel to its edges draw the curve representing the vertical motion (by setting off abscissæ proportional to the time) on such a scale that the paper just contains n repetitions or waves, and then bend the paper round so as to form a cylinder, with a re-entrant curve running round it.. A point describing this curve in such a manner that it revolves uniformly about the axis of the cylinder will appear from a distance to combine the given vertical motion of period τ, with a horizontal harmonic motion of period $n\tau$. Conversely therefore, in order to obtain the representative curve of the vertical vibrations, the cylinder containing the apparent path must be imagined to be divided along a generating line, and developed into a plane. There is less difficulty in conceiving the cylinder and the situation of the curve upon it, when the adjustment of the periods is not quite exact, for then the cylinder appears to turn, and the contrary motions serve to distinguish those parts of the curve which lie on its nearer and further face.

41. The auxiliary harmonic motion is generally obtained optically, by means of an instrument called a vibration-microscope invented by Lissajous. One prong of a large tuning fork carries a lens, whose axis is perpendicular to the direction of vibration; and which may be used either by itself, or as the object-glass of a compound microscope formed by the addition of an eye-piece independently supported. In either case a stationary point is thrown into apparent harmonic motion along a line parallel to that of the fork's vibration.

The vibration-microscope may be applied to test the rigour and universality of the law connecting *pitch* and *period*. Thus it will be found that any point of a vibrating body which gives a pure musical note will appear to describe a re-entrant curve, when examined with a vibration-microscope whose note is in strict unison with its own. By the same means the ratios of frequencies characteristic of the consonant intervals may be verified; though for this latter purpose a more thoroughly acoustical method, to be described in a future chapter, may be preferred.

42. Another method of examining the motion of a vibrating body depends upon the use of intermittent illumination. Suppose, for example, that by means of suitable apparatus a series of electric sparks are obtained at regular intervals τ. A vibrating body, whose period is also τ, examined by the light of the sparks must appear at rest, because it can be seen only in one position. If, however, the period of the vibration differ from τ ever so little, the illuminated position varies, and the body will appear to vibrate slowly with a frequency which is the difference of that of the spark and that of the body. The type of vibration can then be observed with facility.

The series of sparks can be obtained from an induction-coil, whose primary circuit is periodically broken by a vibrating fork, or by some other interrupter of sufficient regularity. But a better result is afforded by sunlight rendered intermittent with the aid of a fork, whose prongs carry two small plates of metal, parallel to the plane of vibration and close together. In each plate is a slit parallel to the prongs of the fork, and so placed as to afford a free passage through the plates when the fork is at rest, or passing through the middle point of its vibrations. On the opening so formed, a beam of sunlight is concentrated by means of a burning-glass, and the object under examination is placed in the cone of rays diverging on the further side[1]. When the fork is made to vibrate by an electro-magnetic arrangement, the illumination is cut off except when the fork is passing through its position of equilibrium, or nearly so. The flashes of light obtained by this method are not so instantaneous as electric sparks (especially when a jar is connected with the secondary wire of the coil), but in my experience the regularity is more perfect. Care should be taken to cut off extraneous light as far as possible, and the effect is then very striking.

A similar result may be arrived at by looking at the vibrating body through a series of holes arranged in a circle on a revolving disc. Several series of holes may be provided on the same disc, but the observation is not satisfactory without some provision for securing uniform rotation.

[1] Töpler, *Phil. Mag.* Jan. 1867.

Except with respect to the sharpness of definition, the result is the same when the period of the light is any multiple of that of the vibrating body. This point must be attended to when the revolving wheel is used to determine an unknown frequency.

When the frequency of intermittence is an exact multiple of that of the vibration, the object is seen without apparent motion, but generally in more than one position. This condition of things is sometimes advantageous.

Similar effects arise when the frequencies of the vibrations and of the flashes are in the ratio of two small whole numbers. If, for example, the number of vibrations in a given time be half as great again as the number of flashes, the body will appear stationary, and in general double.

CHAPTER III.

43. THE material systems, with whose vibrations Acoustics is concerned, are usually of considerable complication, and are susceptible of very various modes of vibration, any or all of which may coexist at any particular moment. Indeed in some of the most important musical instruments, as strings and organ-pipes, the number of independent modes is theoretically infinite, and the consideration of several of them is essential to the most practical questions relating to the nature of the consonant chords. Cases, however, often present themselves, in which one mode is of paramount importance; and even if this were not so, it would still be proper to commence the consideration of the general problem with the simplest case—that of one degree of freedom. It need not be supposed that the mode treated of is the only one possible, because so long as vibrations of other modes do not occur their possibility under other circumstances is of no moment.

44. The condition of a system possessing one degree of freedom is defined by the value of a single co-ordinate u, whose origin may be taken to correspond to the position of equilibrium. The kinetic and potential energies of the system for any given position are proportional respectively to $\dot{u}^2$ and u^2:—

$$T = \tfrac{1}{2}\, m\dot{u}^2, \quad V = \tfrac{1}{2}\, \mu u^2 \dotfill (1),$$

where m and μ are in general functions of u. But if we limit ourselves to the consideration of positions *in the immediate neighbourhood of that corresponding to equilibrium*, u is a small quantity, and m and μ are sensibly constant. On this understanding we

3—2

now proceed. If there be no forces, either resulting from internal friction or viscosity, or impressed on the system from without, the whole energy remains constant. Thus

$$T + V = \text{constant.}$$

Substituting for T and V their values, and differentiating with respect to the time, we obtain the equation of motion

$$m\ddot{u} + \mu u = 0 \dots\dots\dots\dots\dots\dots (2) ;$$

of which the complete integral is

$$u = a \cos (nt - \alpha) \dots\dots\dots\dots\dots (3),$$

where $n^2 = \mu \div m$, representing a *harmonic* vibration. It will be seen that the period alone is determined by the nature of the system itself; the amplitude and phase depend on collateral circumstances. If the differential equation were exact, that is to say, if T were strictly proportional to $\dot{u}^2$, and V to u^2, then, without any restriction, the vibrations of the system about its configuration of equilibrium would be accurately harmonic. But in the majority of cases the proportionality is only approximate, depending on an assumption that the displacement u is always small—how small depends on the nature of the particular system and the degree of approximation required; and then of course we must be careful not to push the application of the integral beyond its proper limits.

But, although not to be stated without a limitation, the principle that the vibrations of a system about a configuration of equilibrium have a period depending on the structure of the system and not on the particular circumstances of the vibration, is of supreme importance, whether regarded from the theoretical or the practical side. If the pitch and the loudness of the note given by a musical instrument were not within wide limits independent, the art of the performer on many instruments, such as the violin and pianoforte, would be revolutionized.

The periodic time

$$\tau = \frac{2\pi}{n} = 2\pi \sqrt{\frac{m}{\mu}} \dots\dots\dots\dots\dots (4),$$

so that an increase in m, or a decrease in μ, protracts the duration of a vibration. By a generalization of the language employed in the case of a material particle urged towards a position of equilibrium by a spring, m may be called the inertia of the system, and

μ the force of the equivalent spring. Thus an augmentation of mass, or a relaxation of spring, increases the periodic time. By means of this principle we may sometimes obtain limits for the value of a period, which cannot, or cannot easily, be calculated exactly.

45. The absence of all forces of a frictional character is an ideal case, never realized but only approximated to in practice. The original energy of a vibration is always dissipated sooner or later by conversion into heat. But there is another source of loss, which though not, properly speaking, dissipative, yet produces results of much the same nature. Consider the case of a tuning-fork vibrating *in vacuo*. The internal friction will in time stop the motion, and the original energy will be transformed into heat. But now suppose that the fork is transferred to an open space. In strictness the fork and the air surrounding it constitute a single system, whose parts cannot be treated separately. In attempting, however, the exact solution of so complicated a problem, we should generally be stopped by mathematical difficulties, and in any case an approximate solution would be desirable. The effect of the air during a few periods is quite insignificant, and becomes important only by accumulation. We are thus led to consider its effect as a *disturbance* of the motion which would take place *in vacuo*. The disturbing force is periodic (to the same approximation that the vibrations are so), and may be divided into two parts, one proportional to the acceleration, and the other to the velocity. The former produces the same effect as an alteration in the mass of the fork, and we have nothing more to do with it at present. The latter is a force arithmetically proportional to the velocity, and always acts in opposition to the motion, and therefore produces effects of the same character as those due to friction. In many similar cases the loss of motion by communication may be treated under the same head as that due to dissipation proper, and is represented in the differential equation with a degree of approximation sufficient for acoustical purposes by a term proportional to the velocity. Thus

$$\ddot{u} + \kappa \dot{u} + n^2 u = 0 \dots\dots\dots\dots\dots\dots (1)$$

is the equation of vibration for a system with one degree of freedom subject to frictional forces. The solution is

$$u = A e^{-\frac{1}{2}\kappa t} \cos \left\{ \sqrt{n^2 - \tfrac{1}{4}\kappa^2} \cdot t - \alpha \right\} \dots\dots\dots\dots (2).$$

If the friction be so great that $\frac{1}{4}\kappa^2 > n^2$, the solution changes its
form, and no longer corresponds to an oscillatory motion; but in
all acoustical applications κ is a small quantity. Under these
circumstances (2) may be regarded as expressing a harmonic vibration, whose amplitude is not constant, but diminishes in geometrical progression, when considered after equal intervals of
time. The difference of the logarithms of successive extreme
excursions is nearly constant, and is called the Logarithmic Decre-
ment. It is expressed by $\frac{1}{4}\kappa\tau$, if τ be the periodic time.

The frequency, depending on $n^2 - \frac{1}{4}\kappa^2$, involves only the second
power of κ; so that to the first order of approximation *the friction
has no effect on the period*,—a principle of very general application.

The vibration here considered is called the *free* vibration. It
is that executed by the system, when disturbed from equilibrium,
and then *left to itself.*

46. We must now turn our attention to another problem, not
less important,—the behaviour of the system, when subjected to a
force varying as a harmonic function of the time. In order to save
repetition, we may take at once the more general case including
friction. If there be no friction, we have only to put in our results
$\kappa = 0$. The differential equation is

$$\ddot{u} + \kappa\dot{u} + n^2 u = E \cos pt \dots\dots\dots\dots (1).$$

Assume

$$u = a \cos (pt - \epsilon)\dots\dots\dots\dots\dots(2),$$

and substitute:

$$a (n^2 - p^2) \cos (pt - \epsilon) - \kappa pa \sin (pt - \epsilon)$$
$$= E \cos \epsilon \cos (pt - \epsilon) - E \sin \epsilon \sin (pt - \epsilon);$$

whence, on equating coefficients of $\cos (pt - \epsilon)$, $\sin (pt - \epsilon)$,

$$\left.\begin{array}{c} a (n^2 - p^2) = E \cos \epsilon \\ a . p\kappa = E \sin \epsilon \end{array}\right\}\dots\dots\dots\dots(3),$$

so that the solution may be written

$$u = \frac{E \sin \epsilon}{p\kappa} \cos (pt - \epsilon)\dots\dots\dots\dots(4),$$

where

$$\tan \epsilon = \frac{p\kappa}{n^2 - p^2}\dots\dots\dots\dots(5).$$

This is called a *forced* vibration; it is the response of the system
to a force imposed upon it from without, and is maintained by the
continued operation of that force. The amplitude is proportional

to E—the magnitude of the force, and the period is the same as that of the force.

Let us now suppose E given, and trace the effect on a given system of a variation in the period of the force. The effects produced in different cases are not strictly similar; because the frequency of the vibrations produced is always the same as that of the force, and therefore variable in the comparison which we are about to institute. We may, however, compare the energy of the system in different cases at the moment of passing through the position of equilibrium. It is necessary thus to specify the moment at which the energy is to be computed in each case, because the total energy is not invariable throughout the vibration. During one part of the period the system receives energy from the impressed force, and during the remainder of the period yields it back again.

From (4), if $u = 0$,

$$\text{energy} \propto \dot{u}^2 \propto \sin^2 \epsilon,$$

and is therefore a maximum, when $\sin \epsilon = 1$, or, from (5), $p = n$. If the maximum kinetic energy be denoted by T_0, we have

$$T = T_0 \sin^2 \epsilon \dots\dots\dots\dots\dots\dots (6).$$

The kinetic energy of the motion is therefore the greatest possible, when the period of the force is that in which the system would vibrate freely under the influence of its own elasticity (or other internal forces), *without friction*. The vibration is then by (4) and (5),

$$u = \frac{E}{n\kappa} \sin nt,$$

and, if κ be small, its amplitude is very great. Its phase is a quarter of a period behind that of the force.

The case, where $p = n$, may also be treated independently. Since the period of the actual vibration is the same as that natural to the system,

$$\ddot{u} + n^2 u = 0,$$

so that the differential equation (1) reduces to

$$\kappa \dot{u} = E \cos pt,$$

whence by integration

$$u = \frac{E}{\kappa} \int \cos pt \, dt = \frac{E}{p\kappa} \sin pt,$$

as before.

If p be less than n, the retardation of phase relatively to the force lies between zero and a quarter period, and when p is greater than n, between a quarter period and a half period.

In the case of a system devoid of friction, the solution is

$$u = \frac{E}{n^2 - p^2} \cos pt \dots\dots\dots\dots\dots\dots(7).$$

When p is smaller than n, the phase of the vibration agrees with that of the force, but when p is the greater, the sign of the vibration is changed. The change of phase from complete agreement to complete disagreement, which is gradual when friction acts, here takes place abruptly as p passes through the value n. At the same time the expression for the amplitude becomes infinite. Of course this only means that, in the case of equal periods, friction *must* be taken into account, however small it may be, and however insignificant its result when p and n are not approximately equal. The limitation as to the magnitude of the vibration, to which we are all along subject, must also be borne in mind.

That the excursion should be at its maximum in one direction while the generating force is at its maximum in the opposite direction, as happens, for example, in the canal theory of the tides, is sometimes considered a paradox. Any difficulty that may be felt will be removed by considering the extreme case, in which the "spring" vanishes, so that the natural period is infinitely long. In fact we need only consider the force acting on the bob of a common pendulum swinging freely, in which case the excursion on one side is greatest when the action of gravity is at its maximum in the opposite direction. When on the other hand the inertia of the system is very small, we have the other extreme case in which the so-called equilibrium theory becomes applicable, the force and excursion being in the same phase.

When the period of the force is longer than the natural period, the effect of an increasing friction is to introduce a retardation in the phase of the displacement varying from zero up to a quarter period. If, however, the period of the natural vibration be the longer, the original retardation of half a period is diminished by something short of a quarter period; or the effect of friction is to *accelerate* the phase of the displacement estimated from that corresponding to the absence of friction. In either case the influence of friction is to cause an approximation to the state of things that would prevail if friction were paramount.

If a force of nearly equal period with the free vibrations vary slowly to a maximum and then slowly decrease, the displacement does not reach its maximum until after the force has begun to diminish. Under the operation of the force at its maximum, the vibration continues to increase until a certain limit is approached, and this increase continues for a time even although the force, having passed its maximum, begins to diminish. On this principle the retardation of spring tides behind the days of new and full moon has been explained[1].

47. From the linearity of the equations it follows that the motion resulting from the simultaneous action of any number of forces is the simple sum of the motions due to the forces taken separately. Each force causes the vibration proper to itself, without regard to the presence or absence of any others. The peculiarities of a force are thus in a manner transmitted into the motion of the system. For example, if the force be periodic in time τ, so will be the resulting vibration. Each harmonic element of the force will call forth a corresponding harmonic vibration in the system. But since the retardation of phase ϵ, and the ratio of amplitudes $a : E$, is not the same for the different components, the resulting vibration, though periodic in the same time, is different in *character* from the force. It may happen, for instance, that one of the components is isochronous, or nearly so, with the free vibration, in which case it will manifest itself in the motion out of all proportion to its original importance. As another example we may consider the case of a system acted on by two forces of nearly equal period. The resulting vibration, being compounded of two nearly in unison, is intermittent, according to the principles explained in the last chapter.

To the motions, which are the immediate effects of the impressed forces, must always be added the term expressing free vibrations, if it be desired to obtain the most general solution. Thus in the case of one impressed force,

$$u = \frac{E \sin \epsilon}{p\kappa} \cos (pt - \epsilon) + A e^{-\frac{1}{2}\kappa t} \cos \{\sqrt{n^2 - \tfrac{1}{4}\kappa^2} \cdot t - \alpha\} \ldots\ldots\ldots(1),$$

where A and α are arbitrary.

48. The distinction between *forced* and *free* vibrations is very

---

[1] Airy's *Tides and Waves*, Art. 328.

important, and must be clearly understood.   The period of the former is determined solely by the force which is supposed to act on the system from without; while that of the latter depends only on the constitution of the system itself.   Another point of difference is that so long as the external influence continues to operate, a forced vibration is permanent, being represented strictly by a harmonic function; but a free vibration gradually dies away, becoming negligible after a time.   Suppose, for example, that the system is at rest when the force $E \cos pt$ begins to operate.   Such finite values must be given to the constants A and α in (1) of § 47, that both u and $\dot{u}$ are initially zero.   At first then there is a free vibration not less important than its rival, but after a time friction reduces it to insignificance, and the forced vibration is left in complete possession of the field.   This condition of things will continue so long as the force operates.   When the force is removed, there is, of course, no discontinuity in the values of u or $\dot{u}$, but the forced vibration is at once converted into a free vibration, and the period of the force is exchanged for that natural to the system.

During the coexistence of the two vibrations in the earlier part of the motion, the curious phenomenon of beats may occur, in case the two periods differ but slightly.   For, n and p being nearly equal, and κ small, the initial conditions are approximately satisfied by

$$u = a \cos (pt - \epsilon) - a e^{-\frac{1}{2}\kappa t} \cos \{\sqrt{n^2 - \tfrac{1}{4}\kappa^2} \, . \, t - \epsilon\}.$$

There is thus a rise and fall in the motion, so long as $e^{-\frac{1}{2}\kappa t}$ remains sensible.   This intermittence is very conspicuous in the earlier stages of the motion of forks driven by electro-magnetism (§ 63).

49.   Vibrating systems of one degree of freedom may vary in two ways according to the values of the constants n and κ.   The distinction of pitch is sufficiently intelligible; but it is worth while to examine more closely the consequences of a greater or less degree of damping.   The most obvious is the more or less rapid extinction of a free vibration.   The effect in this direction may be measured by the number of vibrations which must elapse before the amplitude is reduced in a given ratio.   Initially the amplitude may be taken as unity; after a time t, let it be θ.   Then $\theta = e^{-\frac{1}{2}\kappa t}$.

If $t = x\tau$, we have $x = -\dfrac{2}{\kappa\tau} \log \theta$. In a system subject to only a moderate degree of damping, we may take approximately,

$$\tau = 2\pi \div n\,;$$

so that
$$x = -\frac{n}{\kappa\pi} \log \theta \quad\dotfill(1).$$

This gives the number of vibrations which are performed, before the amplitude falls to θ.

The influence of damping is also powerfully felt in a forced vibration, when there is a near approach to isochronism. In the case of an exact equality between p and n, it is the damping alone which prevents the motion becoming infinite. We might easily anticipate that when the damping is small, a comparatively slight deviation from perfect isochronism would cause a large falling off in the magnitude of the vibration, but that with a larger damping, the same precision of adjustment would not be required. From the equations

$$T = T_0 \sin^2 \epsilon, \quad \tan \epsilon = \frac{\kappa p}{n^2 - p^2}\,,$$

we get
$$\frac{n^2 - p^2}{\kappa p} = \sqrt{\frac{T_0 - T}{T}} \quad\dotfill(2)\,;$$

so that if κ be small, p must be very nearly equal to n, in order to produce a motion not greatly less than the maximum.

The two principal effects of damping may be compared by eliminating κ between (1) and (2). The result is

$$\frac{\log \theta}{x} = \pi \left(\frac{p}{n} - \frac{n}{p}\right)\sqrt{\frac{T}{T_0 - T}} \quad\dotfill(3),$$

where the sign of the square root must be so chosen as to make the right-hand side negative.

If, when a system vibrates freely, the amplitude be reduced in the ratio θ after x vibrations; then, when it is acted on by a force (p), the energy of the resulting motion will be less than in the case of perfect isochronism in the ratio $T : T_0$. It is a matter of indifference whether the forced or the free vibration be the higher; all depends on the *interval*.

In most cases of interest the interval is small; and then, putting $p = n + \delta n$, the formula may be written,

$$\frac{\log \theta}{x} = \frac{2\pi\delta n}{n} \sqrt{\frac{T}{T_0 - T}} \quad\dotfill(4).$$

The following table calculated from these formulæ has been given by Helmholtz[1]:

Interval corresponding to a reduction of the resonance to one-tenth. $T : T_0 = 1 : 10.$	Number of vibrations after which the intensity of a free vibration is reduced to one-tenth. $\theta^2 = \frac{1}{10}.$
$\frac{1}{8}$ tone.	38·00
$\frac{1}{4}$ tone.	19·00
$\frac{1}{2}$ tone.	9·50
$\frac{3}{4}$ tone.	6·33
Whole tone.	4·75
$\frac{5}{4}$ tone.	3·80
$\frac{6}{4}$ tone = minor third.	3·17
$\frac{7}{4}$ tone.	2·71
Two whole tones = major third.	2·37

Formula (4) shews that, when δn is small, it varies *cæteris paribus* as $\frac{1}{x}$.

50. From observations of forced vibrations due to known forces, the natural period and damping of a system may be determined. The formulæ are

$$u = \frac{E \sin \epsilon}{p\kappa} \cos (pt - \epsilon),$$

where

$$\tan \epsilon = \frac{p\kappa}{n^2 - p^2}.$$

On the equilibrium theory we should have

$$u = \frac{E}{n^2} \cos pt.$$

The ratio of the actual amplitude to this is

$$\frac{E \sin \epsilon}{p\kappa} : \frac{E}{n^2} = \frac{n^2 \sin \epsilon}{p\kappa}$$

If the equilibrium theory be known, the comparison of amplitudes tells us the value of $\frac{n^2 \sin \epsilon}{p\kappa}$, say

$$\frac{n^2 \sin \epsilon}{p\kappa} = a,$$

[1] *Tonempfindungen*, p. 221.

and ϵ is also known, whence

$$n^2 = p^2 \div \left(1 - \frac{\cos \epsilon}{a}\right), \text{ and } \kappa = \frac{p \sin \epsilon}{a - \cos \epsilon} \ \ldots\ldots\ldots (1).$$

51.   As has been already stated, the distinction of forced and free vibrations is important; but it may be remarked that most of the forced vibrations which we shall have to consider as affecting a system, take their origin ultimately in the motion of a second system, which influences the first, and is influenced by it. A vibration may thus have to be reckoned as forced in its relation to a system whose limits are fixed arbitrarily, even when that system has a share in determining the period of the force which acts upon it.   On a wider view of the matter embracing both the systems, the vibration in question will be recognized as free.   An example may make this clearer.   A tuning-fork vibrating in air is part of a compound system including the air and itself, and in respect of this compound system the vibration is free.   But although the fork is influenced by the reaction of the air, yet the amount of such influence is small.   For practical purposes it is convenient to consider the motion of the fork as given, and that of the air as forced.   No error will be committed if the *actual* motion of the fork (as influenced by its surroundings) be taken as the basis of calculation.   But the peculiar advantage of this mode of conception is manifested in the case of an approximate solution being required.   It may then suffice to substitute for the actual motion, what would be the motion of the fork in the absence of air, and afterwards introduce a correction, if necessary.

52.   Illustrations of the principles of this chapter may be drawn from all parts of Acoustics.   We will give here a few applications which deserve an early place on account of their simplicity or importance.

A string or wire ACB is stretched between two fixed points A and B, and at its centre carries a mass M, which is supposed to be so considerable as to render the mass of the string itself negligible.   When M is pulled aside from its position of equilibrium, and then let go, it executes along the line CM vibrations, which are the subject of inquiry. $AC = CB = a$. $CM = x$.   The tension of the string in the position of equilibrium depends on the amount of the stretching to which it has been subjected.   In any other position the tension is greater; but we limit ourselves to the case of vibrations so small that the additional stretching is a negligible fraction of the whole. On this condition the tension may be treated as constant. We denote it by T.

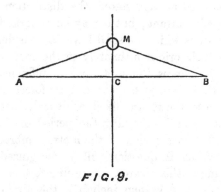

FIG. 9.

Thus, kinetic energy $= \frac{1}{2} M \dot{x}^2$,

and potential energy $= 2T \{ \sqrt{a^2 + x^2} - a \} = T \dfrac{x^2}{a}$ approximately.

The equation of motion (which may be derived also independently) is therefore

$$M\ddot{x} + 2T \frac{x}{a} = 0 \quad\dotfill\quad (1),$$

from which we infer that the mass M executes harmonic vibrations, whose period

$$\tau = 2\pi \div \sqrt{\frac{2T}{aM}} \dotfill (2).$$

The amplitude and phase depend of course on the initial circumstances, being arbitrary so far as the differential equation is concerned.

Equation (2) expresses the manner in which τ varies with each of the independent quantities T, M, a: results which may all be obtained by consideration of the *dimensions* (in the technical sense) of the quantities involved. The argument from dimensions is so often of importance in Acoustics that it may be well to consider this first instance at length.

In the first place we must assure ourselves that of all the quantities on which τ may depend, the only ones involving a reference to the three fundamental units—of length, time, and mass—are a, M, and T. Let the solution of the problem be written

$$\tau = f(a, M, T) \dots\dots\dots\dots\dots\dots (3).$$

This equation must retain its form unchanged, whatever may be the fundamental units by means of which the four quantities are numerically expressed, as is evident, when it is considered that in deriving it no assumptions would be made as to the magnitudes of those units. Now of all the quantities on which f depends, T is the only one involving time; and since its dimensions are (Mass) (Length) (Time)$^{-2}$, it follows that when a and M are constant, $\tau \propto T^{-\frac{1}{2}}$; otherwise a change in the unit of time would necessarily disturb the equation (3). This being admitted, it is easy to see that in order that (3) may be independent of the unit of length, we must have $\tau \propto T^{-\frac{1}{2}} . a^{\frac{1}{2}}$, when M is constant; and finally, in order to secure independence of the unit of mass,

$$\tau \propto T^{-\frac{1}{2}} . M^{\frac{1}{2}} . a^{\frac{1}{2}}.$$

To determine these indices we might proceed thus:—assume

$$\tau \propto T^x . M^y . a^z;$$

then by considering the dimensions in time, space, and mass, we obtain respectively

$$1 = -2x, \quad 0 = x + z, \quad 0 = x + y,$$

whence as above

$$x = -\frac{1}{2}, \quad y = \frac{1}{2}, \quad z = \frac{1}{2}.$$

There must be no mistake as to what this argument does and does not prove. We have *assumed* that there is a definite periodic time depending on no other quantities, having dimensions in space, time, and mass, than those above mentioned. For example, we have not proved that τ is independent of the amplitude of vibration. That, so far as it is true at all, is a consequence of the linearity of the approximate differential equation.

From the necessity of a complete enumeration of all the quantities on which the required result may depend, the method of dimensions is somewhat dangerous; but when used with proper care it is unquestionably of great power and value.

53. The solution of the present problem might be made the foundation of a method for the absolute measurement of pitch. The principal impediment to accuracy would probably be the difficulty of making M sufficiently large in relation to the mass of the wire, without at the same time lowering the note too much in the musical scale.

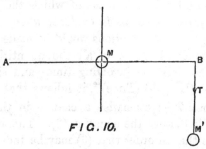

FIG. 10.

The wire may be stretched by a weight M' attached to its further end beyond a bridge or pulley at B. The periodic time would be calculated from

$$\tau = 2\pi . \sqrt{\frac{aM}{2gM'}} \dots\dots\dots(1).$$

The ratio of $M' : M$ is given by the balance. If a be measured in feet, and $g = 32\cdot2$, the periodic time is expressed in seconds.

54. In an ordinary musical string the weight, instead of being concentrated in the centre, is uniformly distributed over its length. Nevertheless the present problem gives some idea of the nature of the gravest vibration of such a string. Let us compare the two cases more closely, supposing the amplitudes of vibration the same at the middle point.

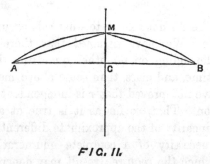

FIG. 11.

When the uniform string is straight, at the moment of passing through the position of equilibrium, its different parts are animated with a variable velocity, increasing from either end towards the centre. If we attribute to the whole mass the velocity of the centre, it is evident that the kinetic energy will be considerably over-estimated. Again, at the moment of maximum excursion, the uniform string is more stretched than its substitute, which follows the straight courses AM, MB, and accordingly the potential energy is diminished by the substitution. The concentration of the mass at the middle point at once increases the kinetic energy when $x = 0$, and decreases the potential energy when $\dot{x} = 0$, and therefore, according to the principle explained in § 44, prolongs the periodic time. For a string then the period is less than that calculated from the formula of the last section, on the supposition that M denotes the mass of the string. It will afterwards appear that in order to obtain a correct result we should have to take instead of M only $\dfrac{4}{\pi^2} M$. Of the factor $\dfrac{4}{\pi^2}$ by far the more important part, viz. $\dfrac{1}{2}$, is due to the difference of the kinetic energies.

55. As another example of a system possessing practically but one degree of freedom, let us consider the vibration of a spring, one end of which is clamped in a vice or otherwise held fast, while the other carries a heavy mass.

In strictness, this system like the last has an infinite number of independent modes of vibration; but, when the mass of the spring is relatively small, that vibration which is nearly independent of its inertia becomes so much the 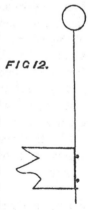 most important that the others may be ignored. Pushing this idea to its limit, we may regard the spring merely as the origin of a force urging the attached mass towards the position of equilibrium, and, if a certain point be not exceeded, in simple proportion to the displacement. The result is a harmonic vibration, with a period dependent on the stiffness of the spring and the mass of the load.

FIG 12.

56. In consequence of the oscillation of the centre of inertia, there is a constant tendency towards the communication of motion to the supports, to resist which adequately the latter must be very firm and massive. In order to obviate this inconvenience,

two precisely similar springs and loads may be mounted on
the same frame-work in a symmetrical manner.
If the two loads perform vibrations of equal amplitude in such a manner that the motions are always
opposite, or, as it may otherwise be expressed, with
a phase-difference of half a period, the centre of
inertia of the whole system remains at rest, and
there is no tendency to set the frame-work into
vibration. We shall see in a future chapter that
this peculiar relation of phases will quickly establish itself, whatever may be the original disturbance. In fact, any part of the motion which does
not conform to the condition of leaving the centre
of inertia unmoved is soon extinguished by damping, unless indeed the supports of the system are
more than usually firm.

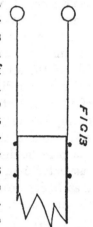

57. As in our first example we found a rough illustration of
the fundamental vibration of a musical string, so here with the
spring and attached load we may compare a uniform slip, or bar,
of elastic material, one end of which is securely fastened, such for
instance as the *tongue* of a *reed* instrument. It is true of course
that the mass is not concentrated at one end, but distributed
over the whole length; yet on account of the smallness of
the motion near the point of support, the inertia of that part
of the bar is of but little account. We infer that the fundamental
vibration of a uniform rod cannot be very different in character
from that which we have been considering. Of course for purposes requiring precise calculation, the two systems are sufficiently
distinct; but where the object is to form clear ideas, precision may
often be advantageously exchanged for simplicity.

In the same spirit we may regard the combination of two
springs and loads shewn in Fig. 13 as a representation of a
tuning fork. This instrument, which has been much improved
of late years, is indispensable to the acoustical investigator. On
a large scale and for rough purposes it may be made by welding
a cross piece on the middle of a bar of steel, so as to form a T, and
then bending the bar into the shape of a horse shoe. On the
handle a screw should be cut. But for the better class of tuning
forks it is preferable to shape the whole out of one piece of steel.
A division running from one end down the middle of a bar is first

made, the two parts opened out to form the prongs of the fork, and the whole worked by the hammer and file into the required shape. The two prongs must be exactly symmetrical with respect to a plane passing through the axis of the handle, in order that during the vibration the centre of inertia may remain unmoved, —unmoved, that is, in the direction in which the prongs vibrate.

The tuning is effected thus. To make the note higher, the equivalent inertia of the system must be reduced. This is done by filing away the ends of the prongs, either diminishing their thickness, or actually shortening them. On the other hand, to lower the pitch, the substance of the prongs near the bend may be reduced, the effect of which is to diminish the force of the spring, leaving the inertia practically unchanged; or the inertia may be increased (a method which would be preferable for temporary purposes) by loading the ends of the prongs with wax, or other material. Large forks are sometimes provided with moveable weights, which slide along the prongs, and can be fixed in any position by screws. As these approach the ends (where the velocity is greatest) the equivalent inertia of the system increases. In this way a considerable range of pitch may be obtained from one fork. The number of vibrations per second for any position of the weights may be marked on the prongs.

The relation between the pitch and the size of tuning forks is remarkably simple. In a future chapter it will be proved that, provided the material remains the same and the shape constant, the period of vibration varies directly as the linear dimension. Thus, if the linear dimensions of a tuning fork be doubled, its note falls an octave.

58. The note of a tuning fork is a nearly pure tone. Immediately after a fork is struck, high tones may indeed be heard, corresponding to modes of vibration, whose nature will be subsequently considered; but these rapidly die away, and even while they exist, they do not blend with the proper tone of the fork, partly on account of their very high pitch, and partly because they do not belong to its harmonic scale. In the forks examined by Helmholtz the first of these overtones had a frequency from 5·8 to 6·6 times that of the proper tone.

Tuning forks are now generally supplied with resonance cases, whose effect is greatly to augment the volume and purity of the sound, according to principles to be hereafter developed. In order to excite them, a violin or cello bow, well supplied with rosin, is drawn across the prongs in the direction of vibration. The sound so produced will last a minute or more.

59. As standards of pitch tuning forks are invaluable. The pitch of organ-pipes varies with the temperature and with the pressure of the wind; that of strings with the tension, which can never be retained constant for long; but a tuning fork kept clean and not subjected to violent changes of temperature or magnetization, preserves its pitch with great fidelity.

By means of beats a standard tuning fork may be copied with very great precision. The number of beats heard in a second is the difference of the frequencies of the two tones which produce them; so that if the beats can be made so slow as to occupy half a minute each, the frequencies differ by only 1-30th of a vibration. Still greater precision might be obtained by Lissajous' optical method.

Very slow beats being difficult of observation, in consequence of the uncertainty whether a falling off in the sound is due to interference or to the gradual dying away of the vibrations, Scheibler adopted a somewhat modified plan. He took a fork slightly different in pitch from the standard—whether higher or lower is not material, but we will say, lower,—and counted the number of beats, when they were sounded together. About four beats a second is the most suitable, and these may be counted for perhaps a minute. The fork to be adjusted is then made slightly higher than the auxiliary fork, and tuned to give with it precisely the same number of beats, as did the standard. In this way a copy as exact as possible is secured. To facilitate the counting of the beats Scheibler employed pendulums, whose periods of vibration could be adjusted.

60. The method of beats was also employed by Scheibler to determine the absolute pitch of his standards. Two forks were tuned to an octave, and a number of others prepared to bridge over the interval by steps so small that each fork gave with its immediate neighbours in the series a number of beats that could be easily counted. The difference of frequency corresponding to each step was observed with all possible accuracy. Their sum, being the difference of frequencies for the interval of the octave, was equal to the frequency of that fork which formed the starting

point at the bottom of the series. The pitch of the other forks could be deduced.

If consecutive forks give four beats per second, 65 in all will be required to bridge over the interval from c' (256) to c'' (512). On this account the method is laborious; but it is probably the most accurate for the original determination of pitch, as it is liable to no errors but such as care and repetition will eliminate. It may be observed that the essential thing is the measurement of the *difference* of frequencies for two notes, whose *ratio* of frequencies is independently known. If we could be sure of its accuracy, the interval of the fifth, fourth, or even major third, might be substituted for the octave, with the advantage of reducing the number of the necessary interpolations. It is probable that with the aid of optical methods this course might be successfully adopted, as the corresponding Lissajous' figures are easily recognised, and their steadiness is a very severe test of the accuracy with which the ratio is attained.

The frequency of large tuning forks may be determined by allowing them to trace a harmonic curve on smoked paper, which may conveniently be mounted on the circumference of a revolving drum. The number of waves executed in a second of time gives the frequency.

In many cases the use of intermittent illumination described in § 42 gives a convenient method of determining an unknown frequency.

61. A series of forks ranging at small intervals over an octave is very useful for the determination of the frequency of any musical note, and is called Scheibler's Tonometer. It may also be used for tuning a note to any desired pitch. In either case the frequency of the note is determined by the number of beats which it gives with the forks, which lie nearest to it (on each side) in pitch.

For tuning pianofortes or organs, a set of twelve forks may be used giving the notes of the chromatic scale on the equal temperament, or any desired system. The corresponding notes are adjusted to unison, and the others tuned by octaves. It is better, however, to prepare the forks so as to give four vibrations per second less than is above proposed. Each note is then tuned a little higher than the corresponding fork, until they give when sounded together exactly four beats in the second. It will be observed that the addition (or subtraction) of a constant number to the frequencies is not the same thing as a mere displacement of the scale in absolute pitch.

In the ordinary practice of tuners a' is taken from a fork, and the other notes determined by estimation of fifths. It will be remembered that twelve true fifths are slightly in excess of seven octaves, so that on the equal temperament system each fifth is a little flat. The tuner proceeds upwards from a' by successive fifths, coming down an octave after about every alternate step, in order to remain in nearly the same part of the scale. Twelve fifths should bring him back to a. If this be not the case, the work must be readjusted, until all the twelve fifths are too flat by, as nearly as can be judged, the same small amount. The inevitable error is then impartially distributed, and rendered as little sensible as possible. The octaves, of course, are all tuned true. The following numbers indicate the order in which the notes may be taken:

$$a\sharp \quad b \quad c' \quad c'\sharp \quad d' \quad d'\sharp \quad e' \quad f' \quad f'\sharp \quad g' \quad g'\sharp \quad a' \quad a'\sharp \quad b' \quad c'' \quad c''\sharp \quad d'' \quad d''\sharp \quad e''$$
$$13 \quad 5 \quad 16 \quad 8 \quad 19 \quad 11 \quad 3 \quad 14 \quad 6 \quad 17 \quad 9 \quad 1 \quad 12 \quad 4 \quad 15 \quad 7 \quad 18 \quad 10 \quad 2$$

In practice the equal temperament is only approximately attained; but this is perhaps not of much consequence, considering that the system aimed at is itself by no means perfection.

Violins and other instruments of that class are tuned by true fifths from a'.

62. In illustration of *forced* vibration let us consider the case of a pendulum whose point of support is subject to a small horizontal harmonic motion. Q is the bob attached by a fine wire to a moveable point P. $OP = x_0$, $PQ = l$, and x is the horizontal co-ordinate of Q. Since the vibrations are supposed small, the vertical motion may be neglected, and the tension of the wire equated to the weight of Q. Hence for the horizontal

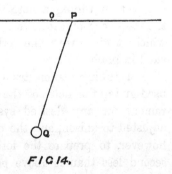

FIG 14.

motion $\ddot{x} + \kappa\dot{x} + \dfrac{g}{l}(x - x_0) = 0$.

Now $x_0 \propto \cos pt$; so that putting $g \div l = n^2$, our equation takes the form already treated of, viz.

$$\ddot{x} + \kappa\dot{x} + n^2 x = E \cos pt.$$

If p be equal to n, the motion is limited only by the friction. The assumed horizontal harmonic motion for P may be realized by means of a second pendulum of massive construction, which carries P with it in its motion. An efficient arrangement is shewn in the figure. A, B are iron rings screwed into a beam, or other firm

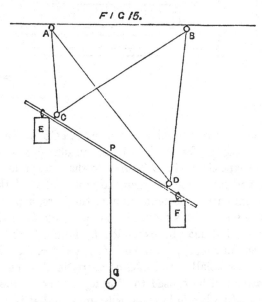

FIG 15.

support; C, D similar rings attached to a stout bar, which carries equal heavy weights E, F, attached near its ends, and is supported in a horizontal position at right angles to the beam by a wire passing through the four rings in the manner shewn. When the pendulum is made to vibrate, a point in the rod midway between C and D executes a harmonic motion in a direction parallel to CD, and may be made the point of attachment of another pendulum PQ. If the weights E and F be very great in relation to Q, the upper pendulum swings very nearly in its own proper period, and induces in Q a forced vibration of the same period. When the length PQ is so adjusted that the natural periods of the two pendulums are nearly the same, Q will be thrown into violent motion, even though the vibration of P be of but inconsiderable amplitude. In this case the difference of phase is about a quarter of a period, by which amount the upper pendulum is in advance. If the two periods be very different, the vibrations either agree or are completely opposed in phase, according to equations (4) and (5) of § 46.

63.   A very good example of a forced vibration is afforded by a fork under the influence of an intermittent electric current,

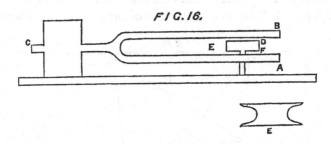

FIG. 16.

whose period is nearly equal to its own.   *ACB* is the fork; *E* a small electro-magnet, formed by winding insulated wire on an iron core of the shape shewn in *E* (similar to that known as 'Siemens' armature'), and supported between the prongs of the fork.   When an intermittent current is sent through the wire, a periodic force acts upon the fork.   This force is not expressible by a simple circular function; but may be expanded by Fourier's theorem in a series of such functions, having periods τ, $\frac{1}{2}\tau$, $\frac{1}{3}\tau$, &c.   If any of these, of not too small amplitude, be nearly isochronous with the fork, the latter will be caused to vibrate; otherwise the effect is insignificant.   In what follows we will suppose that it is the complete period τ which nearly agrees with that of the fork, and consequently regard the series expressing the periodic force as reduced to its first term.

In order to obtain the maximum vibration, the fork must be carefully tuned by a small sliding piece or by wax[1], until its natural period (without friction) is equal to that of the force.   This is best done by actual trial.   When the desired equality is approached, and the fork is allowed to start from rest, the forced and complementary free vibration are of nearly equal amplitudes and frequencies, and therefore (§ 48) in the beginning of the motion produce *beats*, whose slowness is a measure of the accuracy of

---

[1] For this purpose wax may conveniently be softened by melting it with a little turpentine.

the adjustment. It is not until after the free vibration has had time to subside, that the motion assumes its permanent character. The vibrations of a tuning fork properly constructed and mounted are subject to very little damping; consequently a very slight deviation from perfect isochronism occasions a marked falling off in the intensity of the resonance.

The amplitude of the forced vibration can be observed with sufficient accuracy by the ear or eye; but the experimental verification of the relations pointed out by theory between its phase and that of the force which causes it, requires a modified arrangement.

Two similar electro-magnets acting on similar forks, and included in the same circuit are excited by the same intermittent current. Under these circumstances it is clear that the systems will be thrown into similar vibrations, because they are acted on by equal forces. This similarity of vibrations refers both to phase and amplitude. Let us suppose now that the vibrations are effected in perpendicular directions, and by means of one of Lissajous' methods are optically compounded. The resulting figure is necessarily a straight line. Starting from the case in which the amplitudes are a maximum, viz. when the natural periods of both forks are the same as that of the force, let one of them be put a little out of tune. It must be remembered that whatever their natural periods may be, the two forks vibrate in perfect unison with the force, and therefore with one another. The principal effect of the difference of the natural periods is to destroy the synchronism of phase. The straight line, which previously represented the compound vibration, becomes an ellipse, and this remains perfectly steady, so long as the forks are not touched. Originally the forks are both a quarter period behind the force. When the pitch of one is slightly lowered, it falls still more behind the force, and at the same time its amplitude diminishes. Let the difference of phase between the two forks be e', and the ratio of amplitudes of vibration $a : a_0$. Then by (6) of § 46

$$a = a_0 \cos e'.$$

The following table shews the simultaneous values of $a : a_0$ and e'.

$a : a_0$	e'
1·0	0
·9	25° 50′
·8	36° 52′
·7	45° 34′
·6	53° 7′
·5	60°
·4	66° 25′
·3	72° 32′
·2	78° 27′
·1	84° 15′[1]

It appears that a considerable alteration of phase in either direction may be obtained without very materially reducing the amplitude. When one fork is vibrating at its maximum, the other may be made to differ from it on either side by as much as 60° in phase, without losing more than half its amplitude, or by as much as 45°, without losing more than half its *energy*. By allowing one fork to vibrate 45° in advance, and the other 45° in arrear of the phase corresponding to the case of maximum resonance, we obtain a phase difference of 90° in conjunction with an equality of amplitudes. Lissajous' figure then becomes a circle.

64. The intermittent current is best obtained by a fork-interrupter invented by Helmholtz. This may consist of a fork and electro-magnet mounted as before. The wires of the magnet are connected, one with one pole of the battery, and the other with a mercury cup. The other pole of the battery is connected with a second mercury cup. A U-shaped rider of insulated wire is carried by the lower prong just over the cups, at such a height that during the vibration the circuit is alternately made and broken by the passage of one end into and out of the mercury. The other end may be kept permanently immersed. By means of the periodic force thus obtained, the effect of friction is compensated, and the vibrations of the fork permanently maintained. In order to set another fork into forced vibration, its associated electro-magnet may be included, either in the same driving-circuit,

---

[1] *Tonempfindungen*, p. 190.

or in a second, whose periodic interruption is effected by another rider dipping into mercury cups[1].

The *modus operandi* of this kind of self-acting instrument is often imperfectly apprehended. If the force acting on the fork depended only on its position—on whether the circuit were open or closed—the work done in passing through any position would be undone on the return, so that after a complete period there would be nothing outstanding by which the effect of the frictional forces could be compensated. Any explanation which does not take account of the retardation of the current is wholly beside the mark. The causes of retardation are two : irregular contact, and self-induction. When the point of the rider first touches the mercury, the electric contact is imperfect, probably on account of adhering air. On the other hand, in leaving the mercury the contact is prolonged by the adhesion of the liquid in the cup to the amalgamated wire. On both accounts the current is retarded behind what would correspond to the mere position of the fork. But, even if the resistance of the circuit depended only on the position of the fork, the current would still be retarded by its self-induction. However perfect the contact may be, a finite current cannot be generated until after the lapse of a finite time, any more than in ordinary mechanics a finite velocity can be suddenly impressed on an inert body. From whatever causes arising[2], the effect of the retardation is that more work is gained by the fork during the retreat of the rider from the mercury, than is lost during its entrance, and thus a balance remains to be set off against friction.

If the magnetic force depended only on the position of the fork, the phase of its first harmonic component might be considered to be $180°$ in advance of that of the fork's own vibration. The re-

[1] I have arranged several interrupters on the above plan, all the component parts being of home manufacture. The forks were made by the village blacksmith. The cups consisted of iron thimbles, soldered on one end of copper slips, the further end being screwed down on the base board of the instrument. Some means of adjusting the level of the mercury surface is necessary. In Helmholtz' interrupter a horse-shoe electro-magnet embracing the fork is adopted, but I am inclined to prefer the present arrangement, at any rate if the pitch be low. In some cases a greater motive power is obtained by a horse-shoe magnet acting on a soft iron armature carried horizontally by the upper prong and perpendicular to it. I have usually found a single Smee cell sufficient battery power.

[2] Any desired retardation might be obtained, in default of other means, by attaching the rider, not to the prong itself, but to the further end of a light straight spring carried by the prong and set into forced vibration by the motion of its point of attachment.

tardation spoken of reduces this advance. If the phase-difference
be reduced to 90°, the force acts in the most favourable manner,
and the greatest possible vibration is produced.

It is important to notice that (except in the case just referred
to) the actual pitch of the interrupter differs to some extent from
that natural to the fork according to the law expressed in (5) of
§ 46, ϵ being in the present case a prescribed phase-difference
depending on the nature of the contacts and the magnitude of the
self-induction. If the intermittent current be employed to drive
a second fork, the maximum vibration is obtained, when the frequency of the fork coincides, not with the natural, but with the
modified frequency of the interrupter.

The deviation of a tuning-fork interrupter from its natural
pitch is practically very small; but the fact that such a deviation
is possible, is at first sight rather surprising. The explanation (in
the case of a small retardation of current) is, that during that half
of the motion in which the prongs are the most separated, the
electro-magnet acts in aid of the proper recovering power due to
rigidity, and so naturally raises the pitch. Whatever the relation
of phases may be, the force of the magnet may be divided into
two parts respectively proportional to the velocity and displacement (or acceleration). To the first exclusively is due the sustaining power of the force, and to the second the alteration of pitch.

65. The general phenomenon of resonance, though it cannot
be exhaustively considered under the head of one degree of freedom, is in the main referable to the same general principles.
When a forced vibration is excited in one part of a system, all
the other parts are also influenced, a vibration of the same period
being excited, whose amplitude depends on the constitution of the
system considered as a whole. But it not unfrequently happens
that interest centres on the vibration of an outlying part whose
connection with the rest of the system is but loose. In such a case
the part in question, provided a certain limit of amplitude be
not exceeded, is very much in the position of a system possessing
one degree of freedom and acted on by a force, which may be
regarded as *given*, independently of the natural period. The
vibration is accordingly governed by the laws we have already
investigated. In the case of approximate equality of periods to
which the name of resonance is generally restricted, the amplitude may be very considerable, even though in other cases it
might be so small as to be of little account; and the precision

required in the adjustment of the periods in order to bring out the effect, depends on the degree of damping to which the system is subjected.

Among bodies which resound without an extreme precision of tuning, may be mentioned stretched membranes, and strings associated with sounding-boards, as in the pianoforte and the violin. When the proper note is sounded in their neighbourhood, these bodies are caused to vibrate in a very perceptible manner. The experiment may be made by singing into a pianoforte the note given by any of its strings, having first raised the corresponding damper. Or if one of the strings belonging to any note be plucked (like a harp string) with the finger, its fellows will be set into vibration, as may immediately be proved by stopping the first.

The phenomenon of resonance is, however, most striking in cases where a very accurate equality of periods is necessary in order to elicit the full effect. Of this class tuning forks, mounted on resonance boxes, are a conspicuous example. When the unison is perfect the vibration of one fork will be taken up by another across the width of a room, but the slightest deviation of pitch is sufficient to render the phenomenon almost insensible. Forks of 256 vibrations per second are commonly used for the purpose, and it is found that a deviation from unison giving only one beat in a second makes all the difference. When the forks are well tuned and close together, the vibration may be transferred backwards and forwards between them several times, by damping them alternately, with a touch of the finger.

Illustrations of the powerful effects of isochronism must be within the experience of every one. They are often of importance in very different fields from any with which acoustics is concerned. For example, few things are more dangerous to a ship than to lie in the trough of the sea under the influence of waves whose period is nearly that of its own natural rolling.

66. The solution of the equation for free vibration, viz.

$$\ddot{u} + \kappa\dot{u} + n^2 u = 0 \dots\dots\dots (1)$$

may be put into another form by expressing the arbitrary constants of integration A and α in terms of the initial values of u and $\dot{u}$, which we may denote by u_0 and u_0. We obtain at once

$$u = e^{-\frac{1}{2}\kappa t}\left\{\dot{u}_0 \frac{\sin n't}{n'} + u_0\left(\cos n't + \frac{\kappa}{2n'}\sin n't\right)\right\} \dots (2),$$

where $n' = \sqrt{n^2 - \frac{1}{4}\kappa^2}.$

If there be no friction, $\kappa = 0$, and then

$$u = \dot{u}_0 \frac{\sin nt}{n} + u_0 \cos nt \dots\dots\dots\dots(3).$$

These results may be employed to obtain the solution of the complete equation

$$\ddot{u} + \kappa\dot{u} + n^2 u = U \dots\dots\dots\dots\dots(4),$$

where U is an explicit function of the time; for from (2) we see that the effect at time t of a velocity δu communicated at time t' is

$$u = \delta\dot{u}\, e^{-\frac{1}{2}\kappa(t-t')} \frac{\sin n' (t-t')}{n'}.$$

The effect of U is to generate in time dt' a velocity $U dt'$, whose result at time t will therefore be

$$u = \frac{1}{n'} U dt'\, e^{-\frac{1}{2}\kappa(t-t')} \sin n' (t-t'),$$

and thus the solution of (4) will be

$$u = \frac{1}{n'} \int^t e^{-\frac{1}{2}\kappa(t-t')} \sin n' (t-t')\, U\, dt' \dots\dots\dots(5).$$

If there be no friction, we have simply

$$u = \frac{1}{n} \int^t \sin n\, (t-t')\, U\, dt' \dots\dots\dots\dots(6),$$

U being the force at time t'.

The lower limit of the integrals is so far arbitrary, but it will generally be convenient to make it zero.

On this supposition u and $\dot{u}$ as given by (6) vanish, when $t = 0$, and the complete solution is

$$u = e^{-\frac{1}{2}\kappa t} \left\{ \dot{u}_0 \frac{\sin n't}{n'} + u_0 \left(\cos n't + \frac{\kappa}{2n'} \sin n't\right) \right\}$$

$$+ \frac{1}{n'} \int_0^t e^{-\frac{1}{2}\kappa(t-t')} \sin n' (t-t')\, U\, dt' \dots\dots(7),$$

or if there be no friction

$$u = \dot{u}_0 \frac{\sin nt}{n} + u_0 \cos nt + \frac{1}{n} \int_0^t \sin n\, (t-t')\, U\, dt' \dots\dots\dots(8).$$

When t is sufficiently great, the complementary terms tend to vanish on account of the factor $e^{-\frac{1}{2}\kappa t}$, and may then be omitted.

67. For most acoustical purposes it is sufficient to consider the vibrations of the systems, with which we may have to deal, as infinitely small, or rather as similar to infinitely small vibrations. This restriction is the foundation of the important laws of isochronism for free vibrations, and of persistence of period for forced vibrations. There are, however, phenomena of a subordinate but not insignificant character, which depend essentially on the square and higher powers of the motion. We will therefore devote the remainder of this chapter to the discussion of the motion of a sytem of one degree of freedom, the motion not being so small that the squares and higher powers can be altogether neglected.

The approximate expressions for the potential and kinetic energies will be of the form

$$T = \tfrac{1}{2}\,(m_0 + m_1 u)\,\dot{u}^2, \quad V = \tfrac{1}{2}\,(\mu_0 + \mu_1 u)\,u^2.$$

If the sum of T and V be differentiated with respect to the time, we find as the equation of motion

$$m_0 \ddot{u} + \mu_0 u + m_1 u\,\ddot{u} + \frac{1}{2}\,m_1 \dot{u}^2 + \frac{3}{2}\,\mu_1 u^2 = \text{Impressed Force,}$$

which may be treated by the method of successive approximation. For the sake of simplicity we will take the case where $m_1 = 0$, a supposition in no way affecting the essence of the question. The *inertia* of the system is thus constant, while the force of restitution is a composite function of the displacement, partly proportional to the displacement itself and partly proportional to its square—accordingly unsymmetrical with respect to the position of equilibrium. Thus for free vibrations our equation is of the form

$$\ddot{u} + n^2 u + \alpha u^2 = 0 \dotfill (1),$$

with the approximate solution

$$u = A \cos nt \dotfill (2),$$

where A—the amplitude—is to be treated as a small quantity.

Substituting the value of u expressed by (2) in the last term, we find

$$\ddot{u} + n^2 u = -\alpha \frac{A^2}{2}\,(1 + \cos 2nt),$$

whence for a second approximation to the value of u

$$u = A \cos nt - \frac{\alpha A^2}{2n^2} + \frac{\alpha A^2}{6n^2} \cos 2nt \dotfill (3)\ ;$$

shewing that the proper tone (n) of the system is accompanied by its *octave* $(2n)$, whose *relative* importance increases with the amplitude of vibration. A trained ear can generally perceive the octave in the sound of a tuning fork caused to vibrate strongly by means of a bow, and with the aid of appliances, to be explained later, the existence of the octave may be made manifest to any one. By following the same method the approximation can be carried further; but we pass on now to the case of a system in which the recovering power is symmetrical with respect to the position of equilibrium. The equation of motion is then approximately

$$\ddot{u} + n^2u + \beta u^3 = 0 \dots\dots\dots\dots\dots(4),$$

which may be understood to refer to the vibrations of a heavy pendulum, or of a load carried at the end of a straight spring.

If we take as a first approximation $u = A\cos nt$, corresponding to $\beta = 0$, and substitute in the term multiplied by β, we get

$$\ddot{u} + n^2u = -\frac{\beta A^3}{4}\cos 3nt - \frac{3\beta A^3}{4}\cos nt.$$

Corresponding to the last term of this equation, we should obtain in the solution a term of the form $t\sin nt$, becoming greater without limit with t. This, as in a parallel case in the Lunar Theory, indicates that our assumed first approximation is not really an approximation at all, or at least does not *continue* to be such. If, however, we take as our starting point $u = A\cos mt$, with a suitable value for m, we shall find that the solution may be completed with the aid of periodic terms only. In fact it is evident beforehand that all we are entitled to assume is that the motion is approximately simple harmonic, with a period *approximately* the same, as if $\beta = 0$. A very slight examination is sufficient to shew that the term varying as u^3, not only may, but *must* affect the period. At the same time it is evident that a solution, in which the period is assumed wrongly, no matter by how little, must at length cease to represent the motion with any approach to accuracy.

We take then for the approximate equation

$$\ddot{u} + n^2u = -\frac{3\beta A^3}{4}\cos mt - \frac{\beta A^3}{4}\cos 3mt \dots\dots (5),$$

of which the solution will be

$$u = A\cos mt + \frac{\beta A^3}{4}\frac{\cos 3mt}{9m^2 - n^2}\dots\dots\dots\dots(6),$$

provided that m be taken so as to satisfy

$$A\left(-m^2+n^2\right)=-\frac{3\beta A^3}{4},$$

or $$m^2=n^2+\frac{3\beta A^2}{4} \quad\dotsfill (7).$$

The term in β thus produces two effects. It alters the pitch of the fundamental vibration, and it introduces the *twelfth* as a necessary accompaniment. The alteration of pitch is in most cases exceedingly small—depending on the square of the amplitude, but it is not altogether insensible. Tuning forks generally rise a little, though very little, in pitch as the vibration dies away. It may be remarked that the same slight dependence of pitch on amplitude occurs when the force of restitution is of the form $n^2u+\alpha u^2$, as may be seen by continuing the approximation to the solution of (1) one step further than (3). The result in that case is

$$m^2=n^2+\frac{\alpha^2 A^2}{6n^2} \quad\dotsfill (8).$$

The difference m^2-n^2 is of the same order in A in both cases; but in one respect there is a distinction worth noting, namely, that in (8) m^2 is always greater than n^2, while in (7) it depends on the sign of β whether its effect is to raise or lower the pitch. However, in most cases of the unsymmetrical class the change of pitch would depend partly on a term of the form αu^2 and partly on another of the form βu^3, and then

$$m^2=n^2+\frac{\alpha^2 A^2}{6n^2}+\frac{3\beta A^2}{4} \quad\dotsfill (9).$$

68. We now pass to the consideration of the vibrations forced on an unsymmetrical system by two harmonic forces

$$E\cos pt, \quad F\cos(qt-e).$$

The equation of motion is

$$\ddot{u}+n^2u=-\alpha u^2+E\cos pt+F\cos(qt-e) \quad\dots\dots(1).$$

To find a first approximation we neglect the term containing α. Thus

$$u=e\cos pt+f\cos(qt-e)\dotsfill(2),$$

where $$e=\frac{E}{n^2-p^2}, \quad f=\frac{F}{n^2-q^2}\dotsfill(3).$$

Substituting this in the term multiplied by α, we get

$$\ddot{u} + n^2 u = E \cos pt + F \cos (qt - \epsilon)$$

$$- \alpha \left[\frac{e^2 + f^2}{2} + \frac{e^2}{2} \cos 2pt + \frac{f^2}{2} \cos 2(qt - \epsilon) + ef \cos \{(p - q) t + e\} \right.$$

$$\left. + ef \cos \{(p + q) t - e\} \right],$$

whence as a second approximation for u

$$u = e \cos pt + f \cos (qt - \epsilon) - \frac{\alpha (e^2 + f^2)}{2n^2} - \frac{\alpha e^2}{2 (n^2 - 4p^2)} \cos 2pt$$

$$- \frac{\alpha f^2}{2 (n^2 - 4q^2)} \cos 2 (qt - \epsilon) - \frac{\alpha ef}{n^2 - (p - q)^2} \cos \{(p - q) t + e\}$$

$$- \frac{\alpha ef}{n^2 - (p + q)^2} \cos \{(p + q) t - e\} \dots \dots \dots \dots (4).$$

The additional terms represent vibrations having frequencies which are severally the doubles and the sum and difference of those of the primaries. Of the two latter the amplitudes are proportional to the product of the original amplitudes, shewing that the derived tones increase in relative importance with the intensity of their parent tones.

In a future chapter we shall have to consider the important consequences which Helmholtz has deduced from this theory.

CHAPTER IV.

VIBRATING SYSTEMS IN GENERAL.

69. WE have now examined in some detail the oscillations of a system possessed of one degree of freedom, and the results, at which we have arrived, have a very wide application. But material systems enjoy in general more than one degree of freedom. In order to define their configuration at any moment several independent variable quantities must be specified, which, by a generalization of language originally employed for a point, are called the *co-ordinates* of the system, the number of independent co-ordinates being the *index of freedom*. Strictly speaking, the displacements possible to a natural system are infinitely various, and cannot be represented as made up of a finite number of displacements of specified type. To the elementary parts of a solid body any arbitrary displacements may be given, subject to conditions of continuity. It is only by a process of abstraction of the kind so constantly practised in Natural Philosophy, that solids are treated as rigid, fluids as incompressible, and other simplifications introduced so that the position of a system comes to depend on a finite number of co-ordinates. It is not, however, our intention to exclude the consideration of systems possessing infinitely various freedom; on the contrary, some of the most interesting applications of the results of this chapter will lie in that direction. But such systems are most conveniently conceived as limits of others, whose freedom is of a more restricted kind. We shall accordingly commence with a system, whose position is specified by a finite number of independent co-ordinates ψ_1, ψ_2, ψ_3, &c.

70. The main problem of Acoustics consists in the investigation of the vibrations of a system about a position of stable equilibrium, but it will be convenient to commence with the statical part of the subject. By the Principle of Virtual Ve-

locities, if we reckon the co-ordinates ψ_1, ψ_2, &c. from the configuration of equilibrium, the potential energy of any other configuration will be a homogeneous quadratic function of the co-ordinates, provided that the displacement be sufficiently small. This quantity is called V, and represents the work that may be gained in passing from the actual to the equilibrium configuration. We may write

$$V = \tfrac{1}{2}c_{11}\psi_1^2 + \tfrac{1}{2}c_{22}\psi_2^2 + \dots + c_{12}\psi_1\psi_2 + c_{23}\psi_2\psi_3 + \dots\dots(1).$$

Since by supposition the equilibrium is thoroughly stable, the quantities c_{11}, c_{22}, c_{12}, &c. must be such that V is positive for all real values of the co-ordinates.

71. If the system be displaced from the zero configuration by the action of given forces, the new configuration may be found from the Principle of Virtual Velocities. If the work done by the given forces on the hypothetical displacement $\delta\psi_1$, $\delta\psi_2$, &c. be

$$\Psi_1\delta\psi_1 + \Psi_2\delta\psi_2 + \dots\dots\dots\dots\dots\dots(1),$$

this expression must be equivalent to δV, so that since $\delta\psi_1$, $\delta\psi_2$, &c. are independent, the new position of equilibrium is determined by

$$\frac{dV}{d\psi_1} = \Psi_1, \quad \frac{dV}{d\psi_2} = \Psi_2, \text{ &c.}\dots\dots\dots\dots\dots(2),$$

or by (1) of § 70,

$$\left.\begin{array}{l} c_{11}\psi_1 + c_{12}\psi_2 + c_{13}\psi_3 + \dots\dots = \Psi_1 \\ c_{21}\psi_1 + c_{22}\psi_2 + c_{23}\psi_3 + \dots\dots = \Psi_2 \end{array}\right\}\dots\dots\dots(3),$$

where there is no distinction in value between c_{rs} and c_{sr}.

From these equations the co-ordinates may be determined in terms of the forces. If ∇ be the determinant

$$\nabla = \left|\begin{array}{llll} c_{11}, & c_{12}, & c_{13}, & \dots \\ c_{21}, & c_{22}, & c_{23}, & \dots \\ c_{31}, & c_{32}, & c_{33}, & \dots \\ \multicolumn{4}{c}{\dots\dots\dots\dots} \end{array}\right| \dots\dots\dots\dots(4),$$

the solution of (3) may be written

$$\left.\begin{array}{l} \nabla \cdot \psi_1 = \dfrac{d\nabla}{dc_{11}}\Psi_1 + \dfrac{d\nabla}{dc_{12}}\Psi_2 + \dots \\[2mm] \nabla \cdot \psi_2 = \dfrac{d\nabla}{dc_{21}}\Psi_1 + \dfrac{d\nabla}{dc_{22}}\Psi_2 + \dots \end{array}\right\}\dots\dots\dots(5).$$

These equations determine ψ_1, ψ_2, &c. uniquely, since ∇ does not vanish, as appears from the consideration that the equations $\dfrac{dV}{d\psi_1} = 0$, &c. could otherwise be satisfied by finite values of the co-ordinates, provided only that the *ratios* were suitable, which is contrary to the hypothesis that the system is thoroughly stable in the zero configuration.

72. If ψ_1, ... Ψ_1, ... and ψ_1', ... Ψ_1', ... be two sets of displacements and corresponding forces, we have the following reciprocal relation,

$$\Psi_1\psi_1' + \Psi_2\psi_2' + ... = \Psi_1'\psi_1 + \Psi_2'\psi_2 + ... \quad .... (1),$$

as may be seen by substituting the values of the forces, when each side of (1) takes the form,

$$c_{11}\psi_1\psi_1' + c_{22}\psi_2\psi_2' + ...$$
$$+ c_{12}(\psi_2\psi_1' + \psi_2'\psi_1) + c_{23}(\psi_3\psi_2' + \psi_3'\psi_2) + ...$$

Suppose in (1) that all the forces vanish except Ψ_2 and Ψ_1'; then

$$\Psi_2\psi_2' = \Psi_1'\psi_1 \quad .... (2).$$

If the forces Ψ_2 and Ψ_1' be of the same kind, we may suppose them equal, and we then recognise that a force of any type acting alone produces a displacement of a second type equal to the displacement of the first type due to the action of an equal force of the second type. For example, if A and B be two points of a rod supported horizontally in any manner, the vertical deflection at A, when a weight W is attached at B, is the same as the deflection at B, when W is applied at A[1].

73. Since V is a homogeneous quadratic function of the co-ordinates,

$$2V = \frac{dV}{d\psi_1}\psi_1 + \frac{dV}{d\psi_2}\psi_2 + ... (1),$$

or, if Ψ_1, Ψ_2, &c. be the forces necessary to maintain the displacement represented by ψ_1, ψ_2, &c.,

$$2V = \Psi_1\psi_1 + \Psi_2\psi_2 + ... (2).$$

If $\psi_1 + \Delta\psi_1$, $\psi_2 + \Delta\psi_2$, &c. represent another displacement for which the necessary forces are $\Psi_1 + \Delta\Psi_1$, $\Psi_2 + \Delta\Psi_2$, &c., the cor-

---

[1] On this subject, see *Phil. Mag.*, Dec., 1874, and March, 1875.

responding potential energy is given by

$$2\,(V + \Delta V) = (\Psi_1 + \Delta\Psi_1)\,(\psi_1 + \Delta\psi_1) + \ldots$$
$$= 2V + \Psi_1\Delta\psi_1 + \Psi_2\Delta\psi_2 + \ldots$$
$$+ \Delta\Psi_1 \cdot \psi_1 + \Delta\Psi_2 \cdot \psi_2 + \ldots$$
$$+ \Delta\Psi_1 \cdot \Delta\psi_1 + \Delta\Psi_2 \cdot \Delta\psi_2 + \ldots,$$

so that we may write

$$2\,\Delta V = \Sigma\,\Psi \cdot \Delta\psi + \Sigma\,\Delta\Psi \cdot \psi + \Sigma\,\Delta\Psi \cdot \Delta\psi \ldots\ldots\ldots (3),$$

where ΔV is the difference of the potential energies in the two cases, and we must particularly notice that by the reciprocal relation, § 72 (1),

$$\Sigma\,\Psi \cdot \Delta\psi = \Sigma\,\Delta\Psi \cdot \psi \ldots\ldots\ldots\ldots\ldots (4).$$

From (3) and (4) we may deduce two important theorems, relating to the value of V for a system subjected to given displacements, and to given forces respectively.

74. The first theorem is to the effect that, if given displacements (not sufficient by themselves to determine the configuration) be produced in a system by forces of corresponding types, the resulting value of V for the system so displaced, and in equilibrium, is as small as it can be under the given displacement conditions; and that the value of V for any other configuration exceeds this by the potential energy of the configuration which is the difference of the two. The only difficulty in the above statement consists in understanding what is meant by 'forces of corresponding types.' Suppose, for example, that the system is a stretched string, of which a given point P is to be subject to an obligatory displacement; the force of corresponding type is here a force applied at the point P itself. And generally, the forces, by which the proposed displacement is to be made, must be such as would do no work on the system, provided only that that displacement were *not* made.

By a suitable choice of co-ordinates, the given displacement conditions may be expressed by ascribing given values to the first r co-ordinates ψ_1, ψ_2, ... ψ_r, and the conditions as to the forces will then be represented by making the forces of the remaining types Ψ_{r+1}, Ψ_{r+2}, &c. vanish. If $\psi + \Delta\psi$ refer to any other configuration of the system, and $\Psi + \Delta\Psi$ be the corresponding forces, we are to suppose that $\Delta\psi_1$, $\Delta\psi_2$, &c. as far as $\Delta\psi_r$ all vanish. Thus for the first r suffixes $\Delta\psi$ vanishes, and for the remaining suffixes Ψ vanishes. Accordingly $\Sigma \Psi.\Delta\psi$ is zero, and therefore $\Sigma \Delta\Psi.\psi$ is also zero. Hence

$$2 \Delta V = \Sigma \Delta\Psi.\Delta\psi \quad\dots\dots\dots\dots\dots(1),$$

which proves that if the given displacements be made in any other than the prescribed way, the potential energy is increased by the energy of the difference of the configurations.

By means of this theorem we may trace the effect on V of any relaxation in the stiffness of a system, subject to given displacement conditions. For, if after the alteration in stiffness the original equilibrium configuration be considered, the value of V corresponding thereto is by supposition less than before; and, as we have just seen, there will be a still further diminution in the value of V when the system passes to equilibrium under the altered conditions. Hence we conclude that a diminution in V as a function of the co-ordinates entails also a diminution in the actual value of V when a system is subject to given displacements. It will be understood that in particular cases the diminution spoken of may vanish[1].

For example, if a point P of a bar clamped at both ends be displaced laterally to a given small amount by a force there applied, the potential energy of the deformation will be diminished by any relaxation (however local) in the stiffness of the bar.

75. The second theorem relates to a system displaced *by given forces*, and asserts that in this case the value of V in equilibrium is greater than it would be in any other configuration in which the system could be maintained at rest under the given forces, by the operation of mere constraints. We will shew that the *removal* of constraints increases the value of V.

The co-ordinates may be so chosen that the conditions of constraint are expressed by

$$\psi_1 = 0, \quad \psi_2 = 0, \dots\dots \psi_r = 0 \quad\dots\dots\dots\dots(1).$$

We have then to prove that when Ψ_{r+1}, Ψ_{r+2}, &c. are given, the value of V is least when the conditions (1) hold. The second configuration being denoted as before by $\psi_1 + \Delta\psi_1$ &c., we see that for suffixes up to r inclusive ψ vanishes, and for higher suffixes $\Delta\Psi$ vanishes. Hence

$$\Sigma \psi\Delta\Psi = \Sigma \Delta\psi.\Psi = 0,$$

[1] See a paper on General Theorems relating to Equilibrium and Initial and Steady Motions. *Phil. Mag.*, March, 1875.

and therefore

$$2 \Delta V = \Sigma \, \Delta \Psi . \Delta \psi \, \dots\dots\dots\dots\dots (2),$$

shewing that the increase in V due to the removal of the constraints is equal to the potential energy of the difference of the two configurations.

76.   We now pass to the investigation of the initial motion of a system which starts from rest under the operation of given impulses.   The motion thus acquired is independent of any potential energy which the system may possess when actually displaced, since by the nature of impulses we have to do only with the initial configuration itself.   The initial motion is also independent of any forces of a finite kind, whether impressed on the system from without, or of the nature of viscosity.

If P, Q, R be the component impulses, parallel to the axes, on a particle m whose rectangular co-ordinates are x, y, z, we have by D'Alembert's Principle

$$\Sigma m \, (\ddot{x}\delta x + \ddot{y}\delta y + \ddot{z}\delta z) = \Sigma \, (P\delta x + Q\delta y + R\delta z) \dots\dots (1),$$

where $\dot{x}, \dot{y}, \dot{z}$ denote the velocities acquired by the particle in virtue of the impulses, and $\delta x, \delta y, \delta z$ correspond to any arbitrary displacement of the system which does not violate the connection of its parts.   It is required to transform (1) into an equation expressed by the independent generalized co-ordinates.

For the first side,

$$\Sigma m \, (\ddot{x}\delta x + \ddot{y}\delta y + \ddot{z}\delta z) = \delta\psi_1 \, \Sigma m \left(\dot{x}\frac{dx}{d\psi_1} + \dot{y}\frac{dy}{d\psi_1} + \dot{z}\frac{dz}{d\psi_1} \right)$$

$$+ \, \delta\psi_2 \, \Sigma m \left(\dot{x}\frac{dx}{d\psi_2} + \dot{y}\frac{dy}{d\psi_2} + \dot{z}\frac{dz}{d\psi_2} \right) + \dots\dots$$

$$= \delta\psi_1 \, \Sigma m \left(\dot{x}\frac{d\dot{x}}{d\dot{\psi}_1} + \dot{y}\frac{d\dot{y}}{d\dot{\psi}_1} + \dot{z}\frac{d\dot{z}}{d\dot{\psi}_1} \right) + \dots\dots$$

$$= \delta\psi_1 \, . \, \tfrac{1}{2}\Sigma m \frac{d}{d\dot{\psi}_1} \, (\dot{x}^2 + \dot{y}^2 + \dot{z}^2) + \dots\dots$$

$$= \delta\psi_1 \frac{dT}{d\dot{\psi}} + \delta\psi_2 \frac{dT}{d\dot{\psi}_2} + \dots\dots\dots\dots\dots\dots\dots (2),$$

where T, the kinetic energy of the system, is supposed to be expressed as a function of $\dot{\psi}_1, \dot{\psi}_2$, &c.

On the second side,

$$\Sigma\left(P\delta x + Q\delta y + R\delta z\right) = \delta\psi_1\, \Sigma m\left(P\frac{dx}{d\dot\psi_1} + Q\frac{dy}{d\dot\psi_1} + R\frac{dz}{d\dot\psi_1}\right) + \cdots\cdots$$

$$= \xi_1\,\delta\psi_1 + \xi_2\,\delta\psi_2 + \cdots\cdots\cdots\cdots\cdots\cdots (3),$$

if $\qquad\qquad \Sigma m\left(P\frac{dx}{d\dot\psi_1} + Q\frac{dy}{d\dot\psi_1} + R\frac{dz}{d\dot\psi_1}\right) = \xi_1,$ &c.

The transformed equation is therefore

$$\left(\frac{dT}{d\dot\psi_1} - \xi_1\right)\delta\psi_1 + \left(\frac{dT}{d\dot\psi_2} - \xi_2\right)\delta\psi_2 + \ldots = 0 \cdots\cdots\cdots (4),$$

where $\delta\psi_1$, $\delta\psi_2$, &c. are now completely independent. Hence to determine the motion we have

$$\frac{dT}{d\dot\psi_1} = \xi_1, \qquad \frac{dT}{d\dot\psi_2} = \xi_2, \text{ &c.} \cdots\cdots\cdots\cdots\cdots (5),$$

where ξ_1, ξ_2, &c. may be considered as the generalized components of impulse.

77. Since T is a homogeneous quadratic function of the generalized co-ordinates, we may take

$$T = \tfrac{1}{2}a_{11}\dot\psi_1^{\,2} + \tfrac{1}{2}a_{22}\dot\psi_2^{\,2} + \cdots\cdots + a_{12}\dot\psi_1\dot\psi_2 + a_{23}\dot\psi_2\dot\psi_3 + \cdots\cdots(1),$$

whence

$$\left.\begin{aligned}
\xi_1 &= \frac{dT}{d\dot\psi_1} = a_{11}\dot\psi_1 + a_{12}\dot\psi_2 + a_{13}\dot\psi_3 + \cdots\cdots \\
\xi_2 &= \frac{dT}{d\dot\psi_2} = a_{21}\dot\psi_1 + a_{22}\dot\psi_2 + a_{23}\dot\psi_3 + \cdots\cdots \\
&\qquad\cdots\cdots\cdots\cdots\cdots\cdots\cdots\cdots\cdots\cdots\cdots
\end{aligned}\right\}\cdots\cdots\cdots(2),$$

where there is no distinction in value between a_{rs} and a_{sr}.

Again, by the nature of T,

$$2T = \dot\psi_1\frac{dT}{d\dot\psi_1} + \dot\psi_2\frac{dT}{d\dot\psi_2} + \cdots\cdots$$

$$= \xi_1\dot\psi_1 + \xi_2\dot\psi_2 + \cdots\cdots\cdots\cdots\cdots\cdots\cdots\cdots\cdots(3).$$

The theory of initial motion is closely analogous to that of the displacement of a system from a configuration of stable equilibrium by steadily applied forces. In the present theory the initial kinetic energy T bears to the velocities and impulses the same relations as in the former V bears to the displacements and forces respectively. In one respect the theory of initial motions is the more complete, inasmuch as T is exactly, while V is in general only approximately, a homogeneous quadratic function of the variables.

If $\dot{\psi}_1, \dot{\psi}_2, ..., \xi_1, \xi_2, ...$ denote one set of velocities and impulses for a system started from rest, and $\dot{\psi}_1', \dot{\psi}_2', ..., \xi_1', \xi_2', ...$ a second set, we may prove, as in § 72, the following reciprocal relation:

$$\xi_1'\dot{\psi}_1 + \xi_2'\dot{\psi}_2 + ... = \xi_1\dot{\psi}_1' + \xi_2\dot{\psi}_2' + ...(4)^1.$$

This theorem admits of interesting application to fluid motion. It is known, and will be proved later in the course of this work, that the motion of a frictionless incompressible liquid, which starts from rest, is of such a kind that its component velocities at any point are the corresponding differential coefficients of a certain function, called the velocity-potential. Let the fluid be set in motion by a prescribed arbitrary deformation of the surface S of a closed space described within it. The resulting motion is determined by the normal velocities of the elements of S, which, being shared by the fluid in contact with them, are denoted by $\frac{du}{dn}$, if u be the velocity-potential, which interpreted physically denotes the impulsive pressure. Hence by the theorem, if v be the velocity-potential of a second motion, corresponding to another set of arbitrary surface velocities $\frac{dv}{dn}$,

$$\iint u \frac{dv}{dn} dS = \iint v \frac{du}{dn} dS .... (5),$$

—an equation immediately following from Green's theorem, if besides S there be only fixed solids immersed in the fluid. The present method enables us to attribute to it a much higher generality. For example, the immersed solids, instead of being fixed, may be free, altogether or in part, to take the motion imposed upon them by the fluid pressures.

78. A particular case of the general theorem is worthy of special notice. In the first motion let

$$\dot{\psi}_1 = A, \quad \dot{\psi}_2 = 0, \quad \xi_3 = \xi_4 = \xi_5 ...... = 0;$$

and in the second,

$$\dot{\psi}_1' = 0, \quad \dot{\psi}_2' = A, \quad \xi_3' = \xi_4' = \xi_5' ...... = 0.$$

Then $\qquad\qquad\qquad \xi_1' = \xi_2 ....(1).$

---

[1] Thomson and Tait, § 313 (*f*).

In words, if, by means of a suitable impulse of the corresponding type, a given arbitrary velocity of one co-ordinate be impressed on a system, the impulse corresponding to a second co-ordinate necessary in order to prevent it from changing, is the same as would be required for the first co-ordinate, if the given velocity were impressed on the second.

As a simple example, take the case of two spheres A and B immersed in a liquid, whose centres are free to move along certain lines. If A be set in motion with a given velocity, B will naturally begin to move also. The theorem asserts that the impulse required to prevent the motion of B, is the same as if the functions of A and B were exchanged : and this even though there be other rigid bodies, C, D, &c., in the fluid, either fixed, or free in whole or in part.

The case of electric currents mutually influencing each other by induction is precisely similar. Let there be two circuits A and B, in the neighbourhood of which there may be any number of other wire circuits or solid conductors. If a unit current be suddenly developed in the circuit A, the electromotive impulse induced in B is the same as there would have been in A, had the current been forcibly developed in B.

79. The motion of a system, on which given arbitrary velocities are impressed by means of the necessary impulses of the corresponding types, possesses a remarkable property discovered by Thomson. The conditions are that $\dot{\psi}_1,\ \dot{\psi}_2,\ \dot{\psi}_3 ... \dot{\psi}_r$ are given, while $\xi_{r+1},\ \xi_{r+2}, ...$ vanish. Let $\dot{\psi}_1,\ \dot{\psi}_2 ... \xi_1,\ \xi_2$, &c. correspond to the actual motion; and

$$\dot{\psi}_1 + \Delta\dot{\psi}_1,\ \dot{\psi}_2 + \Delta\dot{\psi}_2 ... \ \xi_1 + \Delta\xi_1,\ \xi_2 + \Delta\xi_2, ...$$

to another motion satisfying the same *velocity* conditions. For each suffix either $\Delta\dot{\psi}$ or ξ vanishes. Now for the kinetic energy of the supposed motion,

$$2\,(T + \Delta T) = (\xi_1 + \Delta\xi_1)\,(\dot{\psi}_1 + \Delta\dot{\psi}_1) + ...$$
$$= 2T + \xi_1\Delta\dot{\psi}_1 + \xi_2\Delta\dot{\psi}_2 + ...$$
$$+ \Delta\xi_1 . \dot{\psi}_1 + \Delta\xi_2 . \dot{\psi}_2 + ... + \Delta\xi_1\Delta\dot{\psi}_1 + \Delta\xi_2\Delta\dot{\psi}_2 + ...$$

But by the reciprocal relation (4) of § 77

$$\xi_1\Delta\dot{\psi}_1 + ... \ = \Delta\xi_1 . \dot{\psi}_1 + ...,$$

of which the former by hypothesis is zero; so that

$$2\Delta T = \Delta\xi_1\Delta\dot{\psi}_1 + \Delta\xi_2\Delta\dot{\psi}_2 + ... (1),$$

shewing that the energy of the supposed motion exceeds that of the actual motion by the energy of that motion which would have to be compounded with the latter to produce the former. The motion actually induced in the system has thus less energy than any other satisfying the same velocity conditions. In a subsequent chapter we shall make use of this property to find a superior limit to the energy of a system set in motion with prescribed velocities.

If any diminution be made in the inertia of any of the parts of a system, the motion corresponding to prescribed velocity conditions will in general undergo a change. The value of T will necessarily be less than before; for there would be a decrease even if the motion remained unchanged, and therefore *a fortiori* when the motion is such as to make T an absolute minimum. Conversely any increase in the inertia increases the initial value of T.

This theorem is analogous to that of § 74. The analogue for initial motions of the theorem of § 75, relating to the potential energy of a system displaced by given forces, is that of Bertrand, and may be thus stated :—If a system start from rest under the operation of given impulses, the kinetic energy of the actual motion exceeds that of any other motion which the system might have been guided to take with the assistance of mere constraints, by the kinetic energy of the difference of the motions[1].

80. We will not dwell at any greater length on the mechanics of a system subject to impulses, but pass on to investigate Lagrange's equations for continuous motion. We shall suppose that the connections binding together the parts of the system are not explicit functions of the time ; such cases of forced motion as we shall have to consider will be specially shewn to be within the scope of the investigation.

By D'Alembert's Principle in combination with that of Virtual Velocities,

$$\Sigma m\,(\ddot{x}\delta x + \ddot{y}\delta y + \ddot{z}\delta z) = \Sigma\,(X\delta x + Y\delta y + Z\delta z)\ldots\ldots(1),$$

where δx, δy, δz denote a displacement of the system of the most general kind possible without violating the connections of its parts. Since the displacements of the individual particles of the system are mutually related, $\delta x, \ldots$ are not independent. The object now is to transform to other variables $\psi_1, \psi_2, \ldots$, which shall be independent. We have

$$\ddot{x}\delta x = \frac{d}{dt}\,(\dot{x}\delta x) - \tfrac{1}{2}\,\delta \dot{x}^2,$$

[1] Thomson and Tait, § 311. *Phil. Mag.* March, 1875.

so that

$$\Sigma m\,(\ddot{x}\delta x + \ddot{y}\delta y + \ddot{z}\delta z) = \frac{d}{dt}.\,\Sigma m\,(\dot{x}\delta x + \dot{y}\delta y + \dot{z}\delta z) - \delta T.$$

But (§ 76) we have already found that

$$\Sigma m\,(\dot{x}\delta x + \dot{y}\delta y + \dot{z}\delta z) = \frac{dT}{d\dot{\psi}_1}\,\delta\psi_1 + \frac{dT}{d\dot{\psi}_2}\,\delta\psi_2 + \ldots,$$

while

$$\delta T = \frac{dT}{d\psi_1}\,\delta\psi_1 + \frac{dT}{d\dot{\psi}_1}\,\delta\dot{\psi}_1 + \ldots,$$

if T be expressed as a quadratic function of $\dot{\psi}_1,\,\dot{\psi}_2,\,\ldots$, whose coefficients are in general functions of $\psi_1,\,\psi_2,\ldots$. Also

$$\frac{d}{dt}\left(\frac{dT}{d\dot{\psi}_1}\,\delta\psi_1\right) = \frac{d}{dt}\left(\frac{dT}{d\dot{\psi}_1}\right).\,\delta\psi_1 + \frac{dT}{d\dot{\psi}_1}\,\delta\dot{\psi}_1,$$

inasmuch as

$$\frac{d}{dt}\,\delta\psi_1 = \delta\frac{d}{dt}\,\psi_1.$$

Accordingly

$$\Sigma m\,(\ddot{x}\delta x + \ddot{y}\delta y + \ddot{z}\delta z) = \left\{\frac{d}{dt}\left(\frac{dT}{d\dot{\psi}_1}\right) - \frac{dT}{d\psi_1}\right\}\delta\psi_1$$

$$+ \left\{\frac{d}{dt}\left(\frac{dT}{d\dot{\psi}_2}\right) - \frac{dT}{d\psi_2}\right\}\delta\psi_2 + \ldots\ldots (2).$$

Thus, if the transformation of the second side of (1) be

$$\Sigma\,(X\delta x + Y\delta y + Z\delta z) = \Psi_1\delta\psi_1 + \Psi_2\delta\psi_2 + \ldots\ldots\ldots(3),$$

we have equations of motion of the form

$$\frac{d}{dt}\left(\frac{dT}{d\dot{\psi}}\right) - \frac{dT}{d\psi} = \Psi \ldots\ldots\ldots\ldots\ldots(4).$$

Since $\Psi\delta\psi$ denotes the work done on the system during a displacement $\delta\psi$, Ψ may be regarded as the generalized component of force.

In the case of a conservative system it is convenient to separate from Ψ those parts which depend only on the configuration of the system. Thus, if V denote the potential energy, we may write

$$\frac{d}{dt}\left(\frac{dT}{d\dot{\psi}}\right) - \frac{dT}{d\psi} + \frac{dV}{d\psi} = \Psi \ldots\ldots\ldots\ldots(5),$$

where Ψ is now limited to the forces acting on the system which are not already taken account of in the term $\dfrac{dV}{d\psi}$.

81.   There is also another group of forces whose existence it is often advantageous to recognize specially, namely those arising from friction or viscosity.   If we suppose that each particle of the system is retarded by forces proportional to its component velocities, the effect will be shewn in the fundamental equation (1) § 80 by the addition to the left-hand member of the terms

$$\Sigma \left(\kappa_x \dot{x}\, \delta x + \kappa_y \dot{y}\, \delta y + \kappa_z \dot{z}\, \delta z \right),$$

where κ_x, κ_y, κ_z are coefficients independent of the velocities, but possibly dependent on the configuration of the system.   The transformation to the independent co-ordinates ψ_1, ψ_2, &c. is effected in a similar manner to that of

$$\Sigma m \left(\dot{x}\delta x + \dot{y}\delta y + \dot{z}\delta z \right)$$

considered above (§ 80), and gives

$$\frac{dF}{d\dot{\psi}_1}\, \delta\psi_1 + \frac{dF}{d\dot{\psi}_2}\, \delta\psi_2 + \ldots\ldots\ldots\ldots\ldots\ldots(1),$$

where
$$F = \tfrac{1}{2}\Sigma \left(\kappa_x \dot{x}^2 + \kappa_y \dot{y}^2 + \kappa_z \dot{z}^2 \right)$$
$$= \tfrac{1}{2}b_{11}\dot{\psi}_1{}^2 + \tfrac{1}{2}b_{22}\dot{\psi}_2{}^2 + \ldots + b_{12}\dot{\psi}_1\dot{\psi}_2 + b_{23}\dot{\psi}_2\dot{\psi}_3 + \ldots\ldots (2).$$

F, it will be observed, is like T a homogeneous quadratic function of the velocities, positive for all real values of the variables.   It represents half the rate at which energy is dissipated.

The above investigation refers to retarding forces proportional to the absolute velocities; but it is equally important to consider such as depend on the *relative* velocities of the parts of the system, and fortunately this can be done without any increase of complication.   For example, if a force act on the particle x_1 proportional to $\dot{x}_1 - \dot{x}_2$, there will be at the same moment an equal and opposite force acting on the particle x_2.   The additional terms in the fundamental equation will be of the form

$$\kappa_x \left(\dot{x}_1 - \dot{x}_2 \right) \delta x_1 + \kappa_x \left(\dot{x}_2 - \dot{x}_1 \right) \delta x_2,$$

which may be written

$$\kappa_x \left(\dot{x}_1 - \dot{x}_2 \right) \delta \left(x_1 - x_2 \right) = \delta\psi_1 \frac{d}{d\dot{\psi}_1} \left\{ \tfrac{1}{2} \kappa_x \left(\dot{x}_1 - \dot{x}_2 \right)^2 \right\} + \ldots,$$

and so on for any number of pairs of mutually influencing particles.   The only effect is the addition of new terms to F, which still appears in the form (2)[1].   We shall see presently that

[1] The differences referred to in the text may of course pass into differential coefficients in the case of a body continuously deformed.

the existence of the function F, which may be called the Dissipation Function, implies certain relations among the coefficients of the generalized equations of vibration, which carry with them important consequences[1].

But although in an important class of cases the effects of viscosity are represented by the function F, the question remains open whether such a method of representation is applicable in all cases. I think it probable that it is so; but it is evident that we cannot expect to prove any general property of viscous forces in the absence of a strict definition which will enable us to determine with certainty what forces are viscous and what are not. In some cases considerations of symmetry are sufficient to shew that the retarding forces may be represented as derived from a dissipation function. At any rate whenever the retarding forces are proportional to the absolute or relative velocities of the parts of the system, we shall have equations of motion of the form

$$\frac{d}{dt}\left(\frac{dT}{d\dot{\psi}}\right) - \frac{dT}{d\psi} + \frac{dF}{d\dot{\psi}} + \frac{dV}{d\psi} = \Psi \dots\dots\dots\dots(3).$$

82. We may now introduce the condition that the motion takes place in the immediate neighbourhood of a configuration of thoroughly stable equilibrium ; T and F are then homogeneous quadratic functions of the velocities with coefficients which are to be treated as constant, and V is a similar function of the co-ordinates themselves, provided that (as we suppose to be the case) the origin of each co-ordinate is taken to correspond with the configuration of equilibrium. Moreover all three functions are essentially positive. Since terms of the form $\dfrac{dT}{d\psi}$ are of the second order of small quantities, the equations of motion become linear, assuming the form

$$\frac{d}{dt}\left(\frac{dT}{d\dot{\psi}}\right) + \frac{dF}{d\dot{\psi}} + \frac{dV}{d\psi} = \Psi \dots\dots\dots\dots (1),$$

where under Ψ are to be included all forces acting on the system not already provided for by the differential coefficients of F and V.

---

[1] The Dissipation Function appears for the first time, so far as I am aware, in a paper on General Theorems relating to Vibrations, published in the *Proceedings of the Mathematical Society* for June, 1873.

The three quadratic functions will be expressed as follows:—

$$\left.\begin{array}{l} T = \tfrac{1}{2}a_{11}\dot{\psi}_1{}^2 + \tfrac{1}{2}a_{22}\dot{\psi}_2{}^2 + \ldots + a_{12}\dot{\psi}_1\dot{\psi}_2 + \ldots \\ F = \tfrac{1}{2}b_{11}\dot{\psi}_1{}^2 + \tfrac{1}{2}b_{22}\dot{\psi}_2{}^2 + \ldots + b_{12}\dot{\psi}_1\dot{\psi}_2 + \ldots \\ V = \tfrac{1}{2}c_{11}\psi_1{}^2 + \tfrac{1}{2}c_{22}\psi_2{}^2 + \ldots + c_{12}\psi_1\psi_2 + \ldots \end{array}\right\} \ldots\ldots (2),$$

where the coefficients a, b, c are constants.

From equation (1) we may of course fall back on previous results by supposing F and V, or F and T, to vanish.

A third set of theorems of interest in the application to Electricity may be obtained by omitting T and V, while F is retained, but it is unnecessary to pursue the subject here.

If we substitute the values of T, F and V, and write D for $\dfrac{d}{dt}$, we obtain a system of equations which may be put into the form

$$\left.\begin{array}{l} e_{11}\psi_1 + e_{12}\psi_2 + e_{13}\psi_3 + \ldots = \Psi_1 \\ e_{21}\psi_1 + e_{22}\psi_2 + e_{23}\psi_3 + \ldots = \Psi_2 \\ e_{31}\psi_1 + e_{32}\psi_2 + e_{33}\psi_3 + \ldots = \Psi_3 \\ \ldots\ldots\ldots\ldots\ldots\ldots\ldots\ldots\ldots\ldots\ldots \end{array}\right\} \ldots\ldots\ldots (3),$$

where e_{rs} denotes the quadratic operator

$$e_{rs} = a_{rs}D^2 + b_{rs}D + c_{rs} \ldots\ldots\ldots\ldots\ldots (4).$$

It must be particularly remarked that since

$$a_{rs} = a_{sr}, \quad b_{rs} = b_{sr}, \quad c_{rs} = c_{sr},$$

it follows that

$$e_{rs} = e_{sr} \ldots\ldots\ldots\ldots\ldots\ldots\ldots\ldots\ldots (5).$$

83. Before proceeding further, we may draw an important inference from the *linearity* of our equations. If corresponding respectively to the two sets of forces $\Psi_1, \Psi_2, \ldots, \Psi_1', \Psi_2', \ldots$ two motions denoted by $\psi_1, \psi_2, \ldots, \psi_1', \psi_2', \ldots$ be possible, then must also be possible the motion $\psi_1 + \psi_1', \psi_2 + \psi_2', \ldots$ in conjunction with the forces $\Psi_1 + \Psi_1', \Psi_2 + \Psi_2', \ldots$ Or, as a particular case, when there are no impressed forces, the superposition of any two natural vibrations constitutes also a natural vibration. This is the celebrated principle of the Coexistence of Small Motions, first clearly enunciated by Daniel Bernoulli. It will be understood that its truth depends in general on the justice of the assumption that the motion is so small that its square may be neglected.

84.   To investigate the free vibrations, we must put $\Psi_1, \Psi_2, \ldots$ equal to zero; and we will commence with a system on which no frictional forces act, for which therefore the coefficients e_{rs}, &c. are *even* functions of the symbol D. We have

$$\left.\begin{array}{l} e_{11}\psi_1 + e_{12}\psi_2 + \ldots = 0 \\ e_{21}\psi_1 + e_{22}\psi_2 + \ldots = 0 \\ \cdots\cdots\cdots\cdots\cdots \end{array}\right\} \ldots\ldots\ldots\ldots\ldots (1).$$

From these equations, of which there are as many (m) as the system possesses degrees of liberty, let all but one of the variables be eliminated. The result, which is of the same form whichever be the co-ordinate retained, may be written

$$\nabla\psi = 0 \ldots\ldots\ldots\ldots\ldots\ldots (2),$$

where ∇ denotes the determinant

$$\left|\begin{array}{l} e_{11}, \ e_{12}, \ e_{13}, \ \cdots \\ e_{21}, \ e_{22}, \ e_{23}, \ \cdots \\ e_{31}, \ e_{32}, \ e_{33}, \ \cdots \\ \cdots\cdots\cdots\cdots\cdots \end{array}\right| \ldots\ldots\ldots\ldots\ldots (3),$$

and is (if there be no friction) an even function of D of degree $2m$. Let $\pm\lambda_1, \pm\lambda_2, \ldots, \pm\lambda_m$ be the roots of $\nabla = 0$ considered as an equation in D. Then by the theory of differential equations the most general value of ψ is

$$\psi = Ae^{\lambda_1 t} + A'e^{-\lambda_1 t} + Be^{\lambda_2 t} + B'e^{-\lambda_2 t} + \ldots\ldots (4),$$

where the $2m$ quantities $A, A', B, B',$ &c. are arbitrary constants. This form holds good for each of the co-ordinates, but the constants in the different expressions are not independent. In fact if a particular solution be

$$\psi_1 = A_1 e^{\lambda_1 t}, \quad \psi_2 = A_2 e^{\lambda_1 t}, \quad \&c.,$$

the *ratios* $A_1 : A_2 : A_3 \ldots$ are completely determined by the equations

$$\left.\begin{array}{l} e_{11}A_1 + e_{12}A_2 + e_{13}A_3 + \ldots\ldots = 0 \\ e_{21}A_1 + e_{22}A_2 + e_{23}A_3 + \ldots\ldots = 0 \\ \cdots\cdots\cdots\cdots\cdots\cdots\cdots \end{array}\right\} \ldots\ldots\ldots (5),$$

where in each of the coefficients such as e_{rs}, λ_1 is substituted for D. Equations (5) are necessarily compatible, by the condition that λ_1 is a root of $\nabla = 0$. The ratios $A_1' : A_2' : A_3' \ldots$ corresponding to the root $-\lambda_1$ are the same as the ratios $A_1 : A_2 : A_3 : \ldots$, but for the other pairs of roots $\lambda_2, -\lambda_2$, &c. there are distinct systems of ratios.

R.   6

85.   The nature of the system with which we are dealing imposes an important restriction on the possible values of λ. If λ_1 were real, either λ_1 or $-\lambda_1$ would be real and positive, and we should obtain a particular solution for which the co-ordinates, and with them the kinetic energy denoted by

$$\lambda_1^2 \left\{ \tfrac{1}{2} a_{11} A_1^2 + \dots a_{12} A_1 A_2 + \dots \right\} e^{\pm 2\lambda_1 t},$$

increase without limit.   Such a motion is obviously impossible for a conservative system, whose whole energy can never differ from the sum of the potential and kinetic energies with which it was animated at starting.   This conclusion is not evaded by taking λ_1 negative; because we are as much at liberty to trace the motion backwards as forwards.   It is as certain that the motion never *was* infinite, as that it never *will be*.   The same argument excludes the possibility of a complex value of λ.

We infer that all the values of λ are purely imaginary, corresponding to *real negative* values of λ^2.   Analytically, the fact that the roots of $\nabla = 0$, considered as an equation in D^2, are all real and negative, must be a consequence of the relations subsisting between the coefficients a_{11}, a_{12}, ..., c_{11}, c_{12}, ... in virtue of the fact that for all real values of the variables T and V are positive.   The case of two degrees of liberty will be afterwards worked out in full.

86.   The form of the solution may now be advantageously changed by writing in_1 for λ_1, &c. (where $i = \sqrt{-1}$), and taking new arbitrary constants.   Thus

$$\left. \begin{aligned} \psi_1 &= A_1 \cos (n_1 t - \alpha) + B_1 \cos (n_2 t - \beta) + C_1 \cos (n_3 t - \gamma) + \dots \\ \psi_2 &= A_2 \cos (n_1 t - \alpha) + B_2 \cos (n_2 t - \beta) + C_2 \cos (n_3 t - \gamma) + \dots \\ \psi_3 &= A_3 \cos (n_1 t - \alpha) + B_3 \cos (n_2 t - \beta) + C_3 \cos (n_3 t - \gamma) + \dots \end{aligned} \right\} \dots (1),$$

where n_1^2, n_2^2, &c. are the m roots of the equation of mth degree in n^2 found by writing $-n^2$ for D^2 in $\nabla = 0$.   For each value of n the ratios $A_1 : A_2 : A_{..} ...$ are determinate and real.

This is the complete solution of the problem of the free vibrations of a conservative system.   We see that the whole motion may be resolved into m normal harmonic vibrations of (in general) different periods, each of which is entirely independent of the others.   If the motion, depending on the original disturbance, be such as to reduce itself to one of these (n_1), we have

$$\psi_1 = A_1 \cos (n_1 t - \alpha), \quad \psi_2 = A_2 \cos (n_1 t - \alpha), \quad \&c. \dots\dots (2),$$

where the ratios $A_1 : A_2 : A_3 \ldots$ depend on the constitution of the system, and only the absolute amplitude and phase are arbitrary. The several co-ordinates are always in similar (or opposite) phases of vibration, and the whole system is to be found in the configuration of equilibrium at the same moment.

We perceive here the mechanical foundation of the supremacy of harmonic vibrations. If the motion be sufficiently small, the differential equations become linear with constant coefficients; while circular (and exponential) functions are the only ones which retain their type on differentiation.

87. The m periods of vibration, determined by the equation $\nabla = 0$, are quantities intrinsic to the system, and must come out the same whatever co-ordinates may be chosen to define the configuration. But there is one system of co-ordinates, which is especially suitable, that namely in which the normal types of vibration are defined by the vanishing of all the co-ordinates but one. In the first type the original co-ordinates ψ_1, ψ_2, &c. have given ratios; let the quantity fixing the absolute values be ϕ_1, so that in this type each co-ordinate is a known multiple of ϕ_1. So in the second type each co-ordinate may be regarded as a known multiple of a second quantity ϕ_2, and so on. By a suitable determination of the m quantities ϕ_1, ϕ_2, &c., *any* configuration of the system may be represented as compounded of the m configurations of these types, and thus the quantities ϕ themselves may be looked upon as co-ordinates defining the configuration of the system. They are called the *normal* co-ordinates.

When expressed in terms of the normal co-ordinates, T and V are reduced to sums of squares; for it is easily seen that if the products also appeared, the resulting equations of vibration would not be satisfied by putting any $m-1$ of the co-ordinates equal to zero, while the remaining one was finite.

We might have commenced with this transformation, assuming from Algebra that any two homogeneous quadratic functions can be reduced by linear transformations to sums of squares. Thus

$$T = \tfrac{1}{2} a_1 \dot\phi_1{}^2 + \tfrac{1}{2} a_2 \dot\phi_2{}^2 + \ldots \left. \right\}$$
$$V = \tfrac{1}{2} c_1 \phi_1{}^2 + \tfrac{1}{2} c_2 \phi_2{}^2 + \ldots \left. \right\} \quad \ldots\ldots\ldots\ldots\ldots\ldots (1),$$

where the coefficients (in which the double suffixes are no longer required) are necessarily positive.

Lagrange's equations now become

$$a_1\ddot{\phi}_1 + c_1\phi_1 = 0, \quad a_2\ddot{\phi}_2 + c_2\phi_2 = 0, \quad \&c. \ldots\ldots\ldots\ldots(2),$$

of which the solution is

$$\phi_1 = A \cos(n_1 t - \alpha), \quad \phi_2 = B\cos(n_2 t - \beta), \quad \&c. \ldots\ldots(3),$$

where A, B..., α, β... are arbitrary constants, and

$$n_1^2 = c_1 \div a_1, \quad n_2^2 = c_2 \div a_2, \quad \&c. \ldots\ldots\ldots\ldots(4).$$

88. The interpretation of the equations of motion leads to a theorem of considerable importance, which may be thus stated[1]. The period of a conservative system vibrating in a constrained type about a position of stable equilibrium is stationary in value when the type is normal. We might prove this from the original equations of vibration, but it will be more convenient to employ the normal co-ordinates. The constraint, which may be supposed to be of such a character as to leave only one degree of freedom, is represented by taking the quantities ϕ in given ratios.

If we put

$$\phi_1 = A_1\theta, \quad \phi_2 = A_2\theta, \quad \&c. \ldots\ldots\ldots\ldots(1),$$

θ is a variable quantity, and A_1, A_2, &c. are given for a given constraint.

The expressions for T and V become

$$T = \{\tfrac{1}{2}a_1 A_1^2 + \tfrac{1}{2}a_2 A_2^2 + \ldots\ldots\} \, \theta'^2,$$

$$V = \{\tfrac{1}{2}c_1 A_1^2 + \tfrac{1}{2}c_2 A_2^2 + \ldots\ldots\} \, \theta^2,$$

whence, if θ varies as $\cos pt$,

$$p^2 = \frac{c_1 A_1^2 + c_2 A_2^2 + \ldots + c_m A_m^2}{a_1 A_1^2 + a_2 A_2^2 + \ldots + a_m A_m^2} \ldots\ldots\ldots\ldots(2).$$

This gives the period of the vibration of the constrained type; and it is evident that the period is stationary, when all but one of the coefficients A_1, A_2 ... vanish, that is to say, when the type coincides with one of those natural to the system, and no constraint is needed.

By means of this theorem we may prove that an increase in the mass of any part of a vibrating system is attended by a prolongation of all the natural periods, or at any rate that no period can be diminished. Suppose the increment of mass to be infinitesimal. After the alteration, the types of free vibration will in general be changed; but, by a suitable constraint, the system may

---

[1] *Proceedings of the Mathematical Society*, June 1873.

be made to retain any one of the former types. If this be done, it is certain that any vibration which involves a motion of the part whose mass has been increased will have its period prolonged. Only as a particular case (as, for example, when a load is placed at the node of a vibrating string) can the period remain unchanged. The theorem now allows us to assert that the removal of the constraint, and the consequent change of type, can only affect the period by a quantity of the second order; and that therefore in the limit the free period cannot be less than before the change. By integration we infer that a finite increase of mass must prolong the period of every vibration which involves a motion of the part affected, and that in no case can the period be diminished; but in order to see the correspondence of the two sets of periods, it may be necessary to suppose the alterations made by steps.

Conversely, the effect of a removal of part of the mass of a vibrating system must be to shorten the periods of all the free vibrations.

In like manner we may prove that if the system undergo such a change that the potential energy of a given configuration is diminished, while the kinetic energy of a given motion is unaltered, the periods of the free vibrations are all increased, and conversely. This proposition may sometimes be used for tracing the effects of a constraint; for if we suppose that the potential energy of any configuration violating the condition of constraint gradually increases, we shall approach a state of things in which the condition is observed with any desired degree of completeness. During each step of the process every free vibration becomes (in general) more rapid, and a number of the free periods (equal to the degrees of liberty lost) become infinitely small. The same practical result may be reached without altering the potential energy by supposing the *kinetic* energy of any *motion* violating the condition to increase without limit. In this case one or more periods become infinitely large, but the finite periods are ultimately the same as those arrived at when the potential energy is increased, although in one case the periods have been throughout increasing, and in the other diminishing. This example shews the necessity of making the alterations by steps; otherwise we should not understand the correspondence of the two sets of periods. Further illustrations will be given under the head of two degrees of freedom.

By means of the principle that the value of the free periods is stationary, we may easily calculate corrections due to any deviation in the system from theoretical simplicity. If we take as a hypothetical type of vibration that proper to the simple system, the period so found will differ from the truth by quantities depending on the squares of the irregularities. Several examples of such calculations will be given in the course of this work.

89. Another point of importance relating to the period of a system vibrating in an arbitrary type remains to be noticed. It appears from (2) § 88, that the period of the vibration corresponding to any hypothetical type is included between the greatest and least of those natural to the system. In the case of systems like strings and plates which are treated as capable of continuous deformation, there is no least natural period; but we may still assert that the period calculated from any hypothetical type cannot exceed that belonging to the gravest normal type. When therefore the object is to estimate the longest proper period of a system by means of calculations founded on an assumed type, we know *a priori* that the result will come out too small.

In the choice of a hypothetical type judgment must be used, the object being to approach the truth as nearly as can be done without too great a sacrifice of simplicity. Thus the type for a string heavily weighted at one point might suitably be taken from the extreme case of an infinite load, when the two parts of the string would be straight. As an example of a calculation of this kind, of which the result is known, we will take the case of a uniform string of length l, stretched with tension T_1, and inquire what the period would be on certain suppositions as to the type of vibration.

Taking the origin of x at the middle of the string, let the curve of vibration on the positive side be

$$y = \cos pt \left\{ 1 - \left(\frac{2x}{l} \right)^n \right\} \quad \ldots\ldots\ldots\ldots\ldots (1),$$

and on the negative side the image of this in the axis of y, n being not less than unity. This form satisfies the condition that y vanishes when $x = \pm \frac{1}{2}l$. We have now to form the expressions for T and V, and it will be sufficient to consider the positive half of the string only. Thus, ρ being the longitudinal density,

$$T = \tfrac{1}{2}\int_0^{\frac{l}{2}} \rho \dot{y}^2 dx = \frac{\rho\, n^2 l\, p^2 \sin^2 pt}{2\,(n+1)\,(2n+1)},$$

and

$$V = \tfrac{1}{2}T_1\int_0^{\frac{l}{2}} \left(\frac{dy}{dx}\right)^2 dx = \frac{n^2 T_1 \cos^2 pt}{(2n-1)\, l}.$$

Hence

$$p^2 = \frac{2\,(n+1)\,(2n+1)}{2n-1}\cdot\frac{T_1}{\rho l^2} \quad\dots\dots\dots\dots\dots (2).$$

If $n=1$, the string vibrates as if the mass were concentrated in its middle point, and

$$p^2 = \frac{12\,T_1}{\rho l^2}.$$

If $n=2$, the form is parabolic, and

$$p^2 = \frac{10\,T_1}{\rho l^2}.$$

The true value of p^2 for the gravest type is $\frac{\pi^2 T_1}{\rho l^2}$, so that the assumption of a parabolic form gives a period which is too small in the ratio $\pi : \sqrt{10}$ or $\cdot9936 : 1$. The minimum of p^2, as given by (2), occurs when $n = \frac{\sqrt{6}+1}{2} = 1\cdot72474$, and gives

$$p^2 = \cdot98990\,\frac{T_1}{\rho l^2}.$$

The period is now too small in the ratio

$$\pi : \sqrt{\cdot98990} = \cdot99851 : 1.$$

It will be seen that there is considerable latitude in the choice of a type, even the violent supposition that the string vibrates as two straight pieces giving a period less than ten per cent. in error. And whatever type we choose to take, the period calculated from it cannot be greater than the truth.

90. The rigorous determination of the periods and types of vibration of a given system is usually a matter of great difficulty, arising from the fact that the functions necessary to express the modes of vibration of most continuous bodies are not as yet recognised in analysis. It is therefore often necessary to fall back on methods of approximation, referring the proposed system to some

other of a character more amenable to analysis, and calculating corrections depending on the supposition that the difference between the two systems is small. The problem of approximately simple systems is thus one of great importance, more especially as it is impossible in practice actually to realise the simple forms about which we can most easily reason.

Let us suppose then that the vibrations of a simple system are thoroughly known, and that it is required to investigate those of a system derived from it by introducing small variations in the mechanical functions. If ϕ_1, ϕ_2, &c. be the normal co-ordinates of the original system,

$$T = \tfrac{1}{2} a_1 \dot{\phi}_1^2 + \tfrac{1}{2} a_2 \dot{\phi}_2^2 + \dots ,$$
$$V = \tfrac{1}{2} c_1 \phi_1^2 + \tfrac{1}{2} c_2 \phi_2^2 + \dots ,$$

and for the varied system, referred to the same co-ordinates, which are now only approximately normal,

$$\left.\begin{aligned} T + \delta T &= \tfrac{1}{2} (a_1 + \delta a_{11}) \dot{\phi}_1^2 + \dots + \delta a_{12} \dot{\phi}_1 \dot{\phi}_2 + \dots \\ V + \delta V &= \tfrac{1}{2} (c_1 + \delta c_{11}) \phi_1^2 + \dots + \delta c_{12} \phi_1 \phi_2 + \dots \end{aligned}\right\} \dots\dots (1),$$

in which δa_{11}, δa_{12}, δc_{11}, δc_{12}, &c. are to be regarded as small quantities. In certain cases new co-ordinates may appear, but if so their coefficients must be small. From (1) we obtain for the Lagrangian equations of motion,

$$\left.\begin{aligned} \overline{(a_1 + \delta a_{11}} D^2 + c_1 + \delta c_{11}) \, \phi_1 + (\delta a_{12} D^2 + \delta c_{12}) \, \phi_2 \\ + (\delta a_{13} D^2 + \delta c_{13}) \, \phi_3 + \dots = 0 \\ (\delta a_{21} D^2 + \delta c_{21}) \, \phi_1 + \overline{(a_2 + \delta a_{22}} D^2 + c_2 + \delta c_{22}) \, \phi_2 \\ + (\delta a_{23} D^2 + \delta c_{23}) \, \phi_3 + \dots = 0 \end{aligned}\right\} \dots\dots (2).$$

In the original system the fundamental types of vibration are those which correspond to the variation of but a single co-ordinate at a time. Let us fix our attention on one of them, involving say a variation of ϕ_r, while all the remaining co-ordinates vanish. The change in the system will in general entail an alteration in the fundamental or normal types; but under the circumstances contemplated the alteration is small. The new normal type is expressed by the synchronous variation of the other co-ordinates in addition to ϕ_r; but the ratio of any other ϕ_s to ϕ_r is small. When these ratios are known, the normal mode of the altered system will be determined.

Since the whole motion is simple harmonic, we may suppose that each co-ordinate varies as $\cos p_r t$, and substitute in the differential equations $-p_r^2$ for D^2. In the sth equation ϕ_s occurs with the finite coefficient

$$- a_s p_r^2 - \delta a_{ss} p_r^2 + c_s + \delta c_{ss}.$$

The coefficient of ϕ_r is

$$- \delta a_{rs} p_r^2 + \delta c_{rs}.$$

The other terms are to be neglected in a first approximation, since both the co-ordinate (relatively to ϕ_r) and its coefficient are small quantities. Hence

$$\phi_s : \phi_r = - \frac{\delta c_{rs} - p_r^2 \delta a_{rs}}{c_s - p_r^2 a_s} \quad \dotfill (3).$$

Now

$$- a_s p_s^2 + c_s = 0,$$

and thus

$$\phi_s : \phi_r = \frac{p_r^2 \delta a_{rs} - \delta c_{rs}}{a_s (p_s^2 - p_r^2)} \quad \dotfill (4),$$

the required result.

If the kinetic energy alone undergo variation,

$$\phi_s : \phi_r = \frac{p_r^2}{p_s^2 - p_r^2} \frac{\delta a_{rs}}{a_s} \quad \dotfill (5).$$

The corrected value of the period is determined by the rth equation of (2), not hitherto used. We may write it,

$$\phi_r \{- p_r^2 a_r - p_r^2 \delta a_{rr} + c_r + \delta c_{rr}\} + \Sigma \, \phi_s \, (- p_r^2 \delta a_{rs} + \delta c_{rs}) = 0.$$

Substituting for $\phi_s : \phi_r$ from (4), we get

$$p_r^2 = \frac{c_r + \delta c_{rr}}{a_r + \delta a_{rr}} - \Sigma \, \frac{(\delta c_{rs} - p_r^2 \delta a_{rs})^2}{a_s a_r \, (p_s^2 - p_r^2)} \quad \dotfill (6).$$

The first term gives the value of p_r^2 calculated without allowance for the change of type, and is sufficient, as we have already proved, when the square of the alteration in the system may be neglected. The terms included under the symbol Σ, in which the summation extends to all values of s other than r, give the correction due to the change of type and are of the second order. Since a_s and a_r are positive, the sign of any term depends upon that of $p_s^2 - p_r^2$. If $p_s^2 > p_r^2$, that is, if the mode s be more acute than the mode r, the correction is negative, and makes the calculated note graver than before; but if the mode s be the graver, the correction raises the note. If r refer to the gravest mode of the system, the whole correction is negative; and if r refer to the acutest mode, the whole correction is positive, as we have already seen by another method.

91. As an example of the use of these formulæ, we may take the case of a stretched string, whose longitudinal density ρ is not quite constant. If x be measured from one end, and y be the transverse displacement, the configuration at any time t will be expressed by

$$y = \phi_1 \sin \frac{\pi x}{l} + \phi_2 \sin \frac{2\pi x}{l} + \phi_3 \sin \frac{3\pi x}{l} + \ldots\ldots\ldots (1),$$

l being the length of the string. ϕ_1, ϕ_2, ... are the normal co-ordinates for $\rho = $ constant, and though here ρ is not strictly constant, the configuration of the system may still be expressed by means of the same quantities. Since the potential energy of any configuration is the same as if $\rho = $ constant, $\delta V = 0$. For the kinetic energy we have

$$T + \delta T = \tfrac{1}{2} \int_0^l \rho \left(\dot{\phi}_1 \sin \frac{\pi x}{l} + \dot{\phi}_2 \sin \frac{2\pi x}{l} + \ldots \right)^2 dx$$

$$= \tfrac{1}{2} \dot{\phi}_1{}^2 \int_0^l \rho \sin^2 \frac{\pi x}{l} dx + \tfrac{1}{2} \dot{\phi}_2{}^2 \int_0^l \rho \sin^2 \frac{2\pi x}{l} dx + \ldots$$

$$+ \dot{\phi}_1 \dot{\phi}_2 \int_0^l \rho \sin \frac{\pi x}{l} \sin \frac{2\pi x}{l} dx + \ldots$$

If ρ were constant, the products of the velocities would disappear, since ϕ_1, ϕ_2, &c. are, on that supposition, the normal co-ordinates. As it is, the integral coefficients, though not actually evanescent, are small quantities. Let $\rho = \rho_0 + \delta\rho$; then in our previous notation

$$a_r = \tfrac{1}{2} l\rho_0, \quad \delta a_{rr} = \int_0^l \delta\rho \sin^2 \frac{r\pi x}{l} dx, \quad \delta a_{rs} = \int_0^l \delta\rho \sin \frac{r\pi x}{l} \sin \frac{s\pi x}{l} dx.$$

Thus the type of vibration is expressed by

$$\phi_s : \phi_r = \frac{p_r{}^2}{p_s{}^2 - p_r{}^2} \cdot \frac{2}{l\rho_0} \int_0^l \delta\rho \sin \frac{r\pi x}{l} \sin \frac{s\pi x}{l} dx;$$

or, since

$$p_r{}^2 : p_s{}^2 = r^2 : s^2,$$

$$\phi_s : \phi_r = \frac{r^2}{s^2 - r^2} \int_0^l \frac{2\delta\rho}{l\rho_0} \sin \frac{r\pi x}{l} \sin \frac{s\pi x}{l} dx \ldots\ldots\ldots (2).$$

Let us apply this result to calculate the displacement of the nodal point of the second mode ($r = 2$), which would be in the middle, if the string were uniform. In the neighbourhood of this point, if $x = \frac{1}{2}\,l + \delta x$, the approximate value of y is

$$y = \phi_1 \sin\frac{\pi}{2} + \phi_2 \sin\frac{2\pi}{2} + \phi_3 \sin\frac{3\pi}{2} + \dots$$

$$+ \delta x \left\{ \frac{\pi}{l}\,\phi_1 \cos\frac{\pi}{2} + \frac{2\pi}{l}\cdot\phi_2 \cos\frac{2\pi}{2} + \dots \right\}$$

$$= \phi_1 - \phi_3 + \phi_5 - \dots + \frac{\pi}{l}\,\delta x\,\{-2\phi_2 + 2\phi_4 + \dots\}.$$

Hence when $y = 0$,

$$\delta x = \frac{l}{2\pi\phi_2}\,\{\phi_1 - \phi_3 + \phi_5 - \dots\} \dots\dots\dots\dots (3)$$

approximately, where

$$\phi_s : \phi_2 = \frac{4}{s^2 - 4}\int_0^l \frac{2\delta\rho}{l\rho_0}\sin\frac{2\pi x}{l}\sin\frac{s\pi x}{l}\,dx\dots\dots\dots(4).$$

To shew the application of these formulæ, we may suppose the irregularity to consist in a small load of mass $\rho_0\lambda$ situated at $x = \frac{1}{4}\,l$, though the result might be obtained much more easily directly. We have

$$\delta x = \frac{-2\lambda}{\pi\sqrt{2}}\left\{ \frac{2}{1^2 - 4} - \frac{2}{3^2 - 4} - \frac{2}{5^2 - 4} + \frac{2}{7^2 - 4} + \dots\dots\dots \right\},$$

from which the value of δx may be calculated by approximation. The real value of δx is, however, very simple. The series within brackets may be written

$$1 + \frac{1}{3} - \frac{1}{5} - \frac{1}{7} + \frac{1}{9} + \frac{1}{11} - \&c.$$

which is equal to

$$\int_0^1 \frac{1 + x^2}{1 + x^4}\,dx.$$

The value of the definite integral is

$$\pi \div 4\sin\frac{\pi}{4}\,*,$$

and thus

$$\delta x = -\frac{2\lambda}{\pi\sqrt{2}}\cdot\frac{\pi\sqrt{2}}{4} = -\frac{\lambda}{2},$$

Todhunter's *Int. Calc.* § 255.

as may also be readily proved by equating the periods of vibration of the two parts of the string, that of the loaded part being
calculated approximately on the assumption of unchanged type.

As an example of the formula (6) § 90 for the period, we
may take the case of a string carrying a small load $\rho_0\lambda$ at its
middle point. We have

$$a_r = \tfrac{1}{2} l\rho_0, \quad \delta a_{rr} = \rho_0\lambda \sin^2 \frac{r\pi}{2}, \quad \delta a_{rs} = \rho_0\lambda \sin \frac{r\pi}{2} \sin \frac{s\pi}{2},$$

and thus, if P_r be the value corresponding to $\lambda = 0$, we get when
r is even, $p_r = P_r$, and when r is odd,

$$p_r^2 = P_r^2 \left\{ \frac{1}{1 + \dfrac{2\lambda}{l}} - \Sigma \frac{4r^2}{s^2 - r^2} \frac{\lambda^2}{l^2} \right\} \quad \dots\dots\dots\dots (5),$$

where the summation is to be extended to all the odd values
of s other than r. If $r = 1$,

$$p_1^2 = P_1^2 \left\{ 1 - \frac{2\lambda}{l} + \frac{4\lambda^2}{l^2} - \Sigma \frac{4}{s^2 - 1} \frac{\lambda^2}{l^2} \right\}.$$

Now

$$2\Sigma \frac{1}{s^2 - 1} = \Sigma \frac{1}{s - 1} - \Sigma \frac{1}{s + 1},$$

in which the values of s are 3, 5, 7, 9 ... Accordingly

$$\Sigma \frac{1}{s^2 - 1} = \frac{1}{4},$$

and

$$p_1^2 = P_1^2 \left\{ 1 - \frac{2\lambda}{l} + \frac{3\lambda^2}{l^2} + \dots\dots \right\} \quad \dots\dots\dots\dots (6),$$

giving the pitch of the gravest tone accurately as far as the
square of the ratio $\lambda : l$.

In the general case the value of p_r^2, correct as far as the
first order in $\delta\rho$, will be

$$p_r^2 = P_r^2 \left\{ 1 - \frac{\delta a_{rr}}{a_r} \right\}$$

$$= P_r^2 \left\{ 1 - \frac{2}{l} \int_0^l \frac{\delta\rho}{\rho_0} \sin^2 \frac{r\pi x}{l} \, dx \right\} \quad \dots\dots\dots\dots (7).$$

92.   The theory of vibrations throws great light on expansions
of arbitrary functions in series of other functions of specified
types. The best known example of such expansions is that
generally called after Fourier, in which an arbitrary periodic

function is resolved into a series of harmonics, whose periods are submultiples of that of the given function. It is well known that the difficulty of the question is confined to the proof of the *possibility* of the expansion; if this be assumed, the determination of the coefficients is easy enough. What I wish now to draw attention to is, that in this, and an immense variety of similar cases, the possibility of the expansion may be inferred from physical considerations.

To fix our ideas, let us consider the small vibrations of a uniform string stretched between fixed points. We know from the general theory that the whole motion, whatever it may be, can be analysed into a series of component motions, each represented by a harmonic function of the time, and capable of existing by itself. If we can discover these normal types, we shall be in a position to represent the most general vibration possible by combining them, assigning to each an arbitrary amplitude and phase.

Assuming that a motion is harmonic with respect to time, we get to determine the type an equation of the form

$$\frac{d^2y}{dx^2} + k^2y = 0,$$

whence it appears that the normal functions are

$$y = \sin\frac{\pi x}{l}, \qquad y = \sin\frac{2\pi x}{l}, \qquad y = \sin\frac{3\pi x}{l}, \quad \&c.$$

We infer that the most general position which the string can assume is capable of representation by a series of the form

$$A_1 \sin\frac{\pi x}{l} + A_2 \sin\frac{2\pi x}{l} + A_3 \sin\frac{3\pi x}{l} + \ldots\ldots,$$

which is a particular case of Fourier's theorem. There would be no difficulty in proving the theorem in its most general form.

So far the string has been supposed uniform. But we have only to introduce a variable density, or even a single load at any point of the string, in order to alter completely the expansion whose possibility may be inferred from the dynamical theory. It is unnecessary to dwell here on this subject, as we shall have further examples in the chapters on the vibrations of particular systems, such as bars, membranes, and confined masses of air.

93. The determination of the coefficients to suit arbitrary initial conditions may always be readily effected by the fundamental property of the normal functions, and it may be convenient to sketch the process here for systems like strings, bars, membranes, plates, &c. in which there is only one dependent variable ζ to be considered. If u_1, u_2 ... be the normal functions, and ϕ_1, ϕ_2 ... the corresponding co-ordinates,

$$\zeta = \phi_1 u_1 + \phi_2 u_2 + \phi_3 u_3 + \dots\dots\dots\dots\dots\dots (1).$$

The equations of free motion are

$$\ddot{\phi}_1 + n_1{}^2 \phi_1 = 0, \quad \ddot{\phi}_2 + n_2{}^2 \phi_2 = 0, \text{ \&c. } \dots\dots\dots\dots (2),$$

of which the solutions are

$$\begin{aligned} \phi_1 &= A_1 \sin n_1 t + B_1 \cos n_1 t \\ \phi_2 &= A_2 \sin n_2 t + B_2 \cos n_2 t \\ &\dots\dots\dots\dots\dots\dots\dots\dots \end{aligned} \Bigg\} \dots\dots\dots\dots (3).$$

The initial values of ζ and $\dot{\zeta}$ are therefore

$$\left. \begin{aligned} \zeta_0 &= B_1 u_1 + B_2 u_2 + B_3 u_3 + \dots \\ \dot{\zeta}_0 &= n_1 A_1 u_1 + n_2 A_2 u_2 + n_3 A_3 u_3 + \dots \end{aligned} \right\} \dots\dots\dots\dots (4),$$

and the problem is to determine A_1, A_2, ... B_1, B_2 ... so as to correspond with arbitrary values of ζ_0 and $\dot{\zeta}_0$.

If $\rho\, dx$ be the mass of the element dx, we have from (1)

$$T = \tfrac{1}{2} \int \rho\, \dot{\zeta}^2 dx$$

$$= \tfrac{1}{2} \dot{\phi}_1{}^2 \int \rho\, u_1{}^2 dx + \tfrac{1}{2} \dot{\phi}_2{}^2 \int \rho\, u_2{}^2 dx + \dots + \dot{\phi}_1 \dot{\phi}_2 \int \rho\, u_1 u_2 dx + \dots$$

But the expression for T in terms of $\dot{\phi}_1$, $\dot{\phi}_2$, &c. cannot contain the products of the normal generalized velocities, and therefore every integral of the form

$$\int \rho\, u_r u_s dx = 0 \dots\dots\dots\dots\dots\dots (5).$$

Hence to determine B_r we have only to multiply the first of equations (4) by ρu_r and integrate over the system. We thus obtain

$$B_r \int \rho\, u_r{}^2 dx = \int \rho\, u_r \zeta_0 dx \dots\dots\dots\dots (6).$$

Similarly,

$$n_r A_r \int \rho\, u_r{}^2 dx = \int \rho\, u_r \dot{\zeta}_0 dx \dots\dots\dots\dots (7).$$

The process is just the same whether the element dx be a line, area, or volume.

The conjugate property, expressed by (5), depends upon the fact that the functions u are normal. As soon as this is known by the solution of a differential equation or otherwise, we may infer the conjugate property without further proof, but the property itself is most intimately connected with the fundamental variational equation of motion § 94.

94. If V be the potential energy of deformation, ζ the displacement, and ρ the density of the (line, area, or volume) element dx, the equation of virtual velocities gives immediately

$$\delta V + \int \rho \, \ddot{\zeta} \, \delta \zeta \, dx = 0 \dots\dots\dots\dots\dots(1).$$

In this equation δV is a symmetrical function of ζ and $\delta \zeta$, as may be readily proved from the expression for V in terms of generalized co-ordinates. In fact if

$$V = \tfrac{1}{2} c_{11} \psi_1^{\ 2} + \dots + c_{12} \psi_1 \psi_2 + \dots$$

$$\delta V = c_{11} \psi_1 \delta \psi_1 + c_{22} \psi_2 \delta \psi_2 + \dots$$
$$+ c_{12} (\psi_1 \delta \psi_2 + \psi_2 \delta \psi_1) + \dots$$

Suppose now that ζ refers to the motion corresponding to a normal function u_r, so that $\ddot{\zeta} + n_r^2 \zeta = 0$, while $\delta \zeta$ is identified with another normal function u_s; then

$$\delta V = n_r^2 \int \rho \, u_r u_s dx.$$

Again, if we suppose, as we are equally entitled to do, that ζ varies as u_s and $\delta \zeta$ as u_r, we get for the same quantity δV,

$$\delta V = n_s^2 \int \rho \, u_r u_s dx ;$$

and therefore

$$(n_r^2 - n_s^2) \int \rho \, u_r u_s dx = 0 \dots\dots\dots\dots\dots(2),$$

from which the conjugate property follows, if the motions represented respectively by u_r and u_s have different periods.

A good example of the connection of the two methods of treatment will be found in the chapter on the transverse vibrations of bars.

95. Professor Stokes[1] has drawn attention to a very general law connecting those parts of the free motion which depend on the initial *displacements* of a system not subject to frictional forces, with those which depend on the initial *velocities*. If a velocity of any type be communicated to a system at rest, and then after a small interval of time the opposite velocity be communicated, the effect in the limit will be to start the system without velocity, but with a displacement of the corresponding type. We may readily prove from this that in order to deduce the motion depending on initial displacements from that depending on the initial velocities, it is only necessary to differentiate with respect to the time, and to replace the arbitrary constants (or functions) which express the initial velocities by those which express the corresponding initial displacements.

Thus, if ϕ be any normal co-ordinate satisfying the equation

$$\ddot{\phi} + n^2\phi = 0,$$

the solution in terms of the initial values of ϕ and $\dot{\phi}$ is

$$\phi = \phi_0 \cos nt + \frac{1}{n}\dot{\phi}_0 \sin nt \ldots\ldots\ldots\ldots\ldots(1),$$

of which the first term may be obtained from the second by Stokes' rule.

[1] *Dynamical Theory of Diffraction, Cambridge Trans.* Vol. IX.

CHAPTER V.

VIBRATING SYSTEMS IN GENERAL.
CONTINUED.

96. WHEN dissipative forces act upon a system, the character of the motion is in general more complicated. If two only of the functions T, F, and V be finite, we may by a suitable linear transformation rid ourselves of the products of the co-ordinates, and obtain the normal types of motion. In the preceding chapter we have considered the case of $F = 0$. The same theory with obvious modifications will apply when $T = 0$, or $V = 0$, but these cases though of importance in other parts of Physics, such as Heat and Electricity, scarcely belong to our present subject.

The presence of friction will not interfere with the reduction of T and V to sums of squares; but the transformation proper for them will not in general suit also the requirements of F. The general equation can then only be reduced to the form

$$a_1\ddot{\phi}_1 + b_{11}\dot{\phi}_1 + b_{12}\dot{\phi}_2 + \dots + c_1\phi_1 = \Phi_1, \quad \&c\dots\dots (1),$$

and not to the simpler form applicable to a system of one degree of freedom, viz.

$$a_1\ddot{\phi}_1 + b_1\dot{\phi}_1 + c_1\phi_1 = \Phi_1, \quad \&c. \dots\dots\dots (2).$$

We may, however, choose which pair of functions we shall reduce, though in Acoustics the choice would almost always fall on T and V.

97. There is, however, a not unimportant class of cases in which the reduction of all three functions may be effected; and the theory then assumes an exceptional simplicity. Under this head the most important are probably those when F is of the same form as T or V. The first case occurs frequently, in books at any rate, when the motion of each part of the system is resisted by a retarding force, proportional both to the mass and velocity of the

part. The same exceptional reduction is possible when F is a linear function of T and V, or when T is itself of the same form as V. In any of these cases, the equations of motion are of the same form as for a system of one degree of freedom, and the theory possesses certain peculiarities which make it worthy of separate consideration.

The equations of motion are obtained at once from T, F and V:—

$$\left. \begin{array}{l} a_1 \ddot{\phi}_1 + b_1 \dot{\phi}_1 + c_1 \phi_1 = \Phi_1, \\ a_2 \ddot{\phi}_2 + b_2 \dot{\phi}_2 + c_2 \phi_2 = \Phi_2 \end{array} \right\} \text{ &c.} \quad \dots\dots\dots\dots (1),$$

in which the co-ordinates are separated.

For the free vibrations we have only to put $\Phi_1 = 0$, &c., and the solution is of the form

$$\phi = e^{-\frac{1}{2}\kappa t} \left\{ \dot{\phi}_0 \frac{\sin n't}{n'} + \phi_0 \left(\cos n't + \frac{\kappa}{2n'} \sin n't \right) \right\} \dots\dots (2),$$

where $\qquad \kappa = \dfrac{b}{a}, \quad n^2 = \dfrac{c}{a}, \quad n' = \sqrt{(n^2 - \tfrac{1}{4}\kappa^2)},$

and ϕ_0 and $\dot{\phi}_0$ are the initial values of ϕ and $\dot{\phi}$.

The whole motion may therefore be analysed into component motions, each of which corresponds to the variation of but one normal co-ordinate at a time. And the vibration in each of these modes is altogether similar to that of a system with only one degree of liberty. After a certain time, greater or less according to the amount of dissipation, the free vibrations become insignificant, and the system returns sensibly to rest.

Simultaneously with the free vibrations, but in perfect independence of them, there may exist forced vibrations depending on the quantities Φ. Precisely as in the case of one degree of freedom, the solution of

$$a\ddot{\phi} + b\dot{\phi} + c\phi = \Phi \dots\dots\dots\dots\dots\dots (3),$$

may be written

$$\phi = \frac{1}{n'} \int_0^t e^{-\frac{1}{2}\kappa(t-t')} \sin n'(t-t') \, \Phi \, dt' \dots\dots\dots (4),$$

where as above

$$\kappa = b \div a, \quad n^2 = c \div a, \quad n' = \sqrt{(n^2 - \tfrac{1}{4}\kappa^2)}.$$

To obtain the complete expression for ϕ we must add to the right-hand member of (4), which makes the initial values of ϕ and $\dot{\phi}$ vanish, the terms given in (2) which represent the residue at time t of the initial values ϕ_0 and $\dot{\phi}_0$. If there be no friction, the value of ϕ in (4) reduces to

$$\phi = \frac{1}{n}\int_0^t \sin n\,(t - t')\,\Phi\,dt' \dots\dots\dots\dots\dots (5).$$

98. The complete independence of the normal co-ordinates leads to an interesting theorem concerning the relation of the subsequent motion to the initial disturbance. For if the forces which act upon the system be of such a character that they do no work on the displacement indicated by $\delta\phi_1$, then $\Phi_1 = 0$. No such forces, however long continued, can produce any effect on the motion ϕ_1. If it exist, they cannot destroy it; if it do not exist, they cannot generate it. The most important application of the theorem is when the forces applied to the system act at a node of the normal component ϕ_1, that is, at a point which the component vibration in question does not tend to set in motion. Two extreme cases of such forces may be specially noted, (1) when the force is an impulse, starting the system from rest, (2) when it has acted so long that the system is again at rest under its influence in a disturbed position. So soon as the force ceases, natural vibrations set in, and in the absence of friction would continue for an indefinite time. We infer that whatever in other respects their character may be, they contain no component of the type ϕ_1. This conclusion is limited to cases where T, F, V admit of simultaneous reduction, including of course the case of no friction.

99. The formulæ quoted in § 97 are applicable to any kind of force, but it will often happen that we have to deal only with the effects of impressed forces of the harmonic type, and we may then advantageously employ the more special formulæ applicable to such forces. In using normal co-ordinates, we have first to calculate the forces Φ_1, Φ_2, &c. corresponding to each period, and thence deduce the values of the co-ordinates themselves. If among the natural periods (calculated without allowance for friction) there be any nearly agreeing in magnitude with the period of an impressed force, the corresponding component vibrations will be abnormally large, unless indeed the force itself be greatly attenuated in the preliminary resolution. Suppose, for example, that a transverse force of harmonic type and given period acts at a single point of a stretched string. All the normal modes of vibration will, in general, be excited, not however in their own proper periods, but

in the period of the impressed force; but any normal component, which has a node at the point of application will not be excited. The magnitude of each component thus depends on two things: (1) on the situation of its nodes with respect to the point at which the force is applied, and (2) on the degree of agreement between its own proper period and that of the force. It is important to remember that in response to a simple harmonic force, the system will vibrate in general in *all* its modes, although in particular cases it may sometimes be sufficient to attend to only one of them as being of paramount importance.

100. When the periods of the forces operating are very long relatively to the free periods of the system, an equilibrium theory is sometimes adequate, but in such a case the solution could generally be found more easily without the use of the normal co-ordinates. Bernoulli's theory of the Tides is of this class, and proceeds on the assumption that the free periods of the masses of water found on the globe are small relatively to the periods of the operative forces, in which case the inertia of the water might be left out of account. As a matter of fact this supposition is only very roughly and partially applicable, and we are consequently still in the dark on many important points relating to the tides. The principal forces have a semi-diurnal period, which is not sufficiently long in relation to the natural periods concerned, to allow of the inertia of the water being neglected. But if the rotation of the earth had been much slower, the equilibrium theory of the tides might have been adequate.

A corrected equilibrium theory is sometimes useful, when the period of the impressed force is sufficiently long in comparison with most of the natural periods of a system, but not so in the case of one or two of them. It will be sufficient to take the case where there is no friction. In the equation

$$a\ddot{\phi} + c\phi = \Phi, \quad \text{or} \quad \ddot{\phi} + n^2\phi = \frac{1}{a}\Phi,$$

suppose that the impressed force varies as $\cos pt$. Then

$$\phi = \Phi \div a\,(n^2 - p^2) \dots\dots\dots\dots\dots (1).$$

The equilibrium theory neglects p^2 in comparison with n^2, and takes

$$\phi = \Phi \div an^2 \dots\dots\dots\dots\dots\dots (2).$$

Suppose now that this course is justifiable, except in respect of the single normal co-ordinate ϕ_1. We have then only to add to the result of the equilibrium theory, the difference between the true and the there assumed value of ϕ_1, viz.

$$\phi_1 = \frac{\Phi_1}{a_1(n_1{}^2 - p^2)} - \frac{\Phi_1}{a_1 n_1{}^2} = \frac{p^2}{n_1{}^2 - p^2} \cdot \frac{\Phi_1}{a} \dots\dots\dots\dots(3).$$

The other extreme case ought also to be noticed. If the forced vibrations be extremely rapid, they may become nearly independent of the potential energy of the system. Instead of neglecting p^2 in comparison with n^2, we have then to neglect n^2 in comparison with p^2, which gives

$$\phi = -\Phi \div a p^2 \dots\dots\dots\dots\dots(4).$$

If there be one or two co-ordinates to which this treatment is not applicable, we may supplement the result, calculated on the hypothesis that V is altogether negligible, with corrections for these particular co-ordinates.

101.   Before passing on to the general theory of the vibrations of systems subject to dissipation, it may be well to point out some peculiarities of the free vibrations of continuous systems, started by a force applied at a single point. On the suppositions and notations of § 93, the configuration at any time is determined by

$$\zeta = \phi_1 u_1 + \phi_2 u_2 + \phi_3 u_3 + \dots\dots\dots\dots\dots(1),$$

where the normal co-ordinates satisfy equations of the form

$$a_r \ddot{\phi}_r + c_r \phi_r = \Phi_r \dots\dots\dots\dots\dots(2).$$

Suppose now that the system is held at rest by a force applied at the point Q. The value of Φ_r is determined by the consideration that $\Phi_r \delta \phi_r$ represents the work done upon the system by the impressed forces during a hypothetical displacement $\delta \zeta = \delta \phi_r u_r$, that is

$$\delta \phi_r \int Z u_r \, dx \, ;$$

thus

$$\Phi_r = \int Z u_r dx = u_r(Q) \int Z dx \, ;$$

so that initially by (2)

$$c_r \phi_r = u_r(Q) \int Z dx \dots\dots\dots\dots(3).$$

If the system be let go from this configuration at $t = 0$, we have at any subsequent time t,

$$\phi_r = \cos n_r t \, \frac{u_r(Q) \int Z dx}{c_r}$$

$$= \cos n_r t \, \frac{u_r(Q) \int Z dx}{n_r^2 \int \rho \, u_r^2 dx} \quad \dots\dots\dots\dots\dots (4),$$

and at the point P

$$\zeta = \Sigma \cos n_r t \, \frac{u_r(P) \, u_r(Q) \int Z dx}{n_r^2 \int \rho \, u_r^2 dx} \quad \dots\dots\dots (5).$$

At particular points $u_r(P)$ and $u_r(Q)$ vanish, but on the whole

$$u_r(P) \, u_r(Q) \div \int \rho \, u_r^2 dx$$

neither converges, nor diverges, with r. The series for ζ therefore converges with n_r^{-2}.

Again, suppose that the system is started by an impulse from the configuration of equilibrium. In this case initially

$$a_r \dot{\phi}_r = \int \Phi_r dt = u_r(Q) \int Z_1 dx,$$

whence at time t

$$\phi_r = \frac{\sin n_r t}{a_r n_r} \cdot u_r(Q) \cdot \int Z_1 dx$$

$$= \frac{\sin n_r t \cdot u_r(Q)}{n_r \int \rho u_r^2 dx} \int Z_1 dx \dots\dots\dots\dots (6).$$

This gives

$$\zeta = \Sigma \sin n_r t \, \frac{u_r(P) \, u_r(Q) \int Z_1 dx}{n_r \int \rho u_r^2 dx} \quad \dots\dots\dots\dots (7),$$

shewing that in this case the series converges with n_r^{-1}, that is more slowly than in the previous case.

In both cases it may be observed that the value of ζ is symmetrical with respect to P and Q, proving that the displacement at time t for the point P when the force or impulse is applied at Q, is the same as it would be at Q if the force or impulse had been applied at P. This is an example of a very general reciprocal theorem, which we shall consider at length presently.

As a third case we may suppose the body to start from rest as deformed by a force *uniformly distributed*, over its length, area, or volume. We readily find

$$\zeta = \Sigma \cos n_r t \, \frac{u_r(P) \cdot Z \cdot \int u_r dx}{n_r^2 \int \rho u_r^2 dx} \dots\dots\dots\dots (8).$$

The series for ζ will be more convergent than when the force is concentrated in a single point.

In exactly the same way we may treat the case of a continuous body whose motion is subject to dissipation, provided that the three functions T, F, V be simultaneously reducible, but it is not necessary to write down the formulæ.

102. If the three mechanical functions T, F and V of any system be not simultaneously reducible, the natural vibrations (as has already been observed) are more complicated in their character. When, however, the dissipation is small, the method of reduction is still useful; and this· class of cases besides being of some importance in itself will form a good introduction to the more general theory. We suppose then that T and V are expressed as sums of squares

$$\left. \begin{array}{l} T = \tfrac{1}{2}\, a_1 \dot{\phi}_1^{\,2} + \tfrac{1}{2}\, a_2 \dot{\phi}_2^{\,2} + \dots \\ V = \tfrac{1}{2}\, c_1 \phi_1^{\,2} + \tfrac{1}{2}\, c_2 \phi_2^{\,2} + \dots \end{array} \right\} \dots\dots\dots\dots (1),$$

while F still appears in the more general form

$$F = \tfrac{1}{2}\, b_{11} \dot{\phi}_1^{\,2} + \tfrac{1}{2}\, b_{22} \dot{\phi}_2^{\,2} + \dots + b_{12} \dot{\phi}_1 \dot{\phi}_2 + \ \dots\dots\dots (2).$$

The equations of motion are accordingly

$$\left. \begin{array}{l} a_1 \ddot{\phi}_1 + b_{11} \dot{\phi}_1 + b_{12} \dot{\phi}_2 + b_{13} \dot{\phi}_3 + \dots + c_1 \phi_1 = 0 \\ a_2 \ddot{\phi}_2 + b_{21} \dot{\phi} + b_{22} \dot{\phi}_2 + b_{23} \dot{\phi}_3 + \dots + c_2 \phi_2 = 0 \\ \dots\dots\dots\dots\dots\dots\dots\dots\dots\dots\dots\dots\dots\dots\dots \end{array} \right\} \dots\dots\dots (3),$$

in which the coefficients b_{11}, b_{12}, &c. are to be treated as small. If there were no friction, the above system of equations would be satisfied by supposing one co-ordinate ϕ_r to vary suitably, while the other co-ordinates vanish. In the actual case there will be a corresponding solution in which the value of any other co-ordinate ϕ will be small relatively to ϕ_r.

Hence, if we omit terms of the second order, the rth equation becomes,

$$a_r\ddot{\phi}_r + b_{rr}\dot{\phi}_r + c_r\phi_r = 0 \dots\dots\dots\dots\dots(4),$$

from which we infer that ϕ_r varies approximately as if there were no change due to friction in the type of vibration. If ϕ_r vary as $e^{p_r t}$, we obtain to determine p_r

$$a_r p_r^2 + b_{rr} p_r + c_r = 0 \dots\dots\dots\dots\dots (5).$$

The roots of this equation are complex, but the real part is small in comparison with the imaginary part.

From the sth equation, if we introduce the supposition that all the co-ordinates vary as $e^{p_r t}$, we get

$$(p_r^2 a_s + c_s)\,\phi_s + b_{rs} p_r \phi_r = 0,$$

terms of the second order being omitted; whence

$$\phi_s : \phi_r = -\frac{b_{rs} p_r}{p_r^2 a_s + c_s} = \frac{b_{rs} p_r}{a_s (p_s^2 - p_r^2)} \dots\dots(6).$$

This equation determines approximately the altered type of vibration. Since the chief part of p_r is imaginary, we see that the co-ordinates ϕ_s are approximately in the same phase, *but that that phase differs by a quarter period from the phase of ϕ_r.* Hence when the function F does not reduce to a sum of squares, the character of the elementary modes of vibration is less simple than otherwise, and the various parts of the system are no longer simultaneously in the same phase.

We proved above that, when the friction is small, the value of p_r may be calculated approximately without allowance for the change of type; but by means of (6) we may obtain a still closer approximation, in which the squares of the small quantities are retained. The rth equation (3) gives

$$a_r p_r^2 + c_r + b_{rr} p_r + \Sigma\,\frac{p_r^2 b_{rs}^2}{a_s (p_s^2 - p_r^2)} = 0 \dots\dots\dots\dots (7).$$

The leading part of the terms included under Σ being real, the correction has no effect on the real part of p_r on which the rate of decay depends.

103. We now return to the consideration of the general equations of § 84.

If ψ_1, ψ_2, &c. be the co-ordinates and Ψ_1, Ψ_2, &c. the forces, we have

$$\left. \begin{array}{l} e_{11}\psi_1 + e_{12}\psi_2 + \ldots = \Psi_1 \\ e_{21}\psi_1 + e_{22}\psi_2 + \ldots = \Psi_2 \ \&c. \end{array} \right\} \ \ldots\ldots\ldots\ldots\ldots(1),$$

where

$$e_{rs} = a_{rs}D^2 + b_{rs}D + c_{rs}\ldots\ldots\ldots\ldots\ldots\ldots\ldots(2).$$

For the free vibrations Ψ_1, &c. vanish. If ∇ be the determinant

$$\nabla = \begin{vmatrix} e_{11}, & e_{12}, & \ldots \\ e_{21}, & e_{22}, & \ldots \\ \ldots\ldots\ldots\ldots \end{vmatrix} \ \ldots\ldots\ldots\ldots\ldots\ldots\ldots(3),$$

the result of eliminating from (1) all the co-ordinates but one, is

$$\nabla\psi = 0 \ \ldots\ldots\ldots\ldots\ldots\ldots\ldots (4).$$

Since ∇ now contains odd powers of D, the $2m$ roots of the equation $\nabla = 0$ no longer occur in equal positive and negative pairs, but contain a real as well as an imaginary part. The complete integral may however still be written

$$\psi = Ae^{\mu_1 t} + A'e^{\mu_1' t} + Be^{\mu_2 t} + B'e^{\mu_2' t} + \ldots\ldots\ldots(5),$$

where the pairs of conjugate roots are μ_1, μ_1'; μ_2, μ_2'; &c. Corresponding to each root, there is a particular solution such as

$$\psi_1 = A_1 e^{\mu_1 t}, \quad \psi_2 = A_2 e^{\mu_1 t}, \quad \psi_3 = A_3 e^{\mu_1 t}, \text{ &c.},$$

in which the *ratios* $A_1 : A_2 : A_3 \ldots$ are determined by the equations of motion, and only the absolute value remains arbitrary. In the present case however (where ∇ contains odd powers of D) these ratios are not in general real, and therefore the variations of the co-ordinates ψ_1, ψ_2, &c. are not synchronous in phase. If we put $\mu_1 = \alpha_1 + i\beta_1$, $\mu_1' = \alpha_1 - i\beta_1$, &c., we see that none of the quantities α can be positive, since in that case the energy of the motion would increase with the time, as we know it cannot do.

Enough has now been said on the subject of the free vibrations of a system in general. Any further illustration that it may require will be afforded by the discussion of the case of two degrees of freedom, § 112, and by the vibrations of strings and other special bodies with which we shall soon be occupied. We resume the equations (1) with the view of investigating further the nature of *forced vibrations*.

104. In order to eliminate from the equations all the co-ordinates but one (ψ_1), operate on them in succession with the minor determinants

$$\frac{d\nabla}{de_{11}}, \quad \frac{d\nabla}{de_{21}}, \quad \frac{d\nabla}{de_{31}}, \&c.,$$

and add the results together; and in like manner for the other co-ordinates. We thus obtain as the equivalent of the original system of equations

$$\left.\begin{array}{l} \nabla\psi_1 = \dfrac{d\nabla}{de_{11}}\,\Psi_1 + \dfrac{d\nabla}{de_{21}}\,\Psi_2 + \dfrac{d\nabla}{de_{31}}\,\Psi_3 + \ldots \\[2mm] \nabla\psi_2 = \dfrac{d\nabla}{de_{12}}\,\Psi_1 + \dfrac{d\nabla}{de_{22}}\,\Psi_2 + \dfrac{d\nabla}{de_{32}}\,\Psi_3 + \ldots \\[2mm] \nabla\psi_3 = \dfrac{d\nabla}{de_{13}}\,\Psi_1 + \dfrac{d\nabla}{de_{23}}\,\Psi_3 + \dfrac{d\nabla}{de_{33}}\,\Psi_3 + \ldots \end{array}\right\} \ldots\ldots (1),$$

in which the differentiations of ∇ are to be made without re-cognition of the equality subsisting between e_{rs} and e_{sr}.

The forces Ψ_1, Ψ_2, &c. are any whatever, subject, of course, to the condition of not producing so great a displacement or motion that the squares of the small quantities become sensible. If, as is often the case, the forces operating be made up of two parts, one constant with respect to time, and the other periodic, it is convenient to separate in imagination the two classes of effects produced. The effect due to the constant forces is exactly the same as if they acted alone, and is found by the solution of a statical problem. It will therefore generally be sufficient to suppose the forces periodic, the effects of any constant forces, such as gravity, being merely to alter the configuration about which the vibrations proper are executed. We may thus without any real loss of generality confine ourselves to periodic, and therefore by Fourier's theorem to harmonic forces.

We might therefore assume as expressions for Ψ_1, &c. circular functions of the time; but, as we shall have frequent occasion to recognise in the course of this work, it is usually more con-venient to employ an imaginary exponential function, such as $E\,e^{ipt}$, where E is a constant which may be complex. When the corresponding symbolical solution is obtained, its real and imaginary parts may be separated, and belong respectively to the real and imaginary parts of the data. In this way the analysis gains considerably in brevity, inasmuch as differentiations and alterations of phase are expressed by merely modifying the complex coefficient without changing the form of the function. We therefore write

$$\Psi_1 = E_1 e^{ipt}, \quad \Psi_2 = E_2 e^{ipt}, \quad \&c.$$

The minor determinants of the type $\dfrac{d\nabla}{de_{rs}}$ are rational integral functions of the symbol D, and operate on Ψ_1, &c. according to the law

$$f(D) e^{ipt} = f(ip) e^{ipt} \dotfill (2).$$

Our equations therefore assume the form

$$\nabla\psi_1 = A_1 e^{ipt}, \quad \nabla\psi_2 = A_2 e^{ipt}, \quad \&c. \dotfill (3),$$

where A_1, A_2, &c. are certain complex constants. And the symbolical solutions are

$$\psi_1 = A_1 \nabla^{-1} e^{ipt}, \quad \&c.,$$

or by (2),

$$\psi_1 = A_1 \frac{e^{ipt}}{\nabla(ip)}, \quad \&c. \dotfill (4),$$

where $\nabla(ip)$ denotes the result of substituting ip for D in ∇.

Consider first the case of a system exempt from friction. ∇ and its differential coefficients are then *even* functions of D, so that $\nabla(ip)$ is real. Throwing away the imaginary part of the solution, writing $R_1 e^{i\theta_1}$ for A_1, &c. we have

$$\psi_1 = \frac{R_1}{\nabla(ip)} \cos(pt + \theta_1), \quad \&c. \dotfill (5).$$

If we suppose that the forces Ψ_1, &c. (in the case of more than one generalized component) have all the same phase, they may be expressed by

$$E_1 \cos(pt + \alpha), \quad E_2 \cos(pt + \alpha), \quad \&c. ;$$

and then, as is easily seen, the co-ordinates themselves agree in phase with the forces:

$$\psi_1 = \frac{R_1}{\nabla(ip)} \cos(pt + \alpha) \dotfill (6).$$

The amplitudes of the vibrations depend among other things on the magnitude of $\nabla(ip)$. Now, if the period of the forces be the same as one of those belonging to the free vibrations, $\nabla(ip) = 0$, and the amplitude becomes infinite. This is, of course, just the case in which it is essential to introduce the consideration of friction, from which no natural system is really exempt.

If there be friction, $\nabla(ip)$ is complex; but it may be divided into two parts—one real and the other purely imaginary, of which the latter depends entirely on the friction. Thus, if we put

$$\nabla(ip) = \nabla_1(ip) + ip\,\nabla_2(ip) \dots\dots\dots (7),$$

∇_1, ∇_2 are even functions of ip, and therefore real. If as before $A_1 \doteq R_1 e^{i\theta_1}$, our solution takes the form

$$\psi_1 = \frac{R_1 e^{i\theta_1} e^{i\gamma} e^{ipt}}{\{|\nabla_1(ip)|^2 + p^2|\nabla_2(ip)|^2\}^{\frac{1}{2}}},$$

or, on throwing away the imaginary part,

$$\psi_1 = \frac{R_1 \cos(pt + \theta_1 + \gamma)}{\{|\nabla_1(ip)|^2 + p^2|\nabla_2(ip)|^2\}^{\frac{1}{2}}} \dots\dots\dots (8),$$

where

$$\tan\gamma = -\frac{p\,\nabla_2(ip)}{\nabla_1(ip)} \dots\dots\dots (9).$$

We have said that $\nabla_2(ip)$ depends entirely on the friction; but it is not true, on the other hand, that $\nabla_1(ip)$ is exactly the same, as if there had been no friction. However, this is approximately the case, if the friction be small; because any part of $\nabla(ip)$, which depends on the first power of the coefficients of friction, is necessarily imaginary. Whenever there is a coincidence between the period of the force and that of one of the free vibrations, $\nabla_1(ip)$ vanishes, and we have $\tan\gamma = -\infty$, and therefore

$$\psi_1 = \frac{R_1 \sin(pt + \theta_1)}{p\,\nabla_2(ip)} \dots\dots\dots(10),$$

indicating a vibration of large amplitude, only limited by the friction.

On the hypothesis of small friction, θ is in general small, and so also is γ, except in case of approximate equality of periods. With certain exceptions, therefore, the motion has nearly the same (or opposite) phase with the force that excites it.

When a force expressed by a harmonic term acts on a system, the resulting motion is everywhere harmonic, and retains the original period, provided always that the squares of the displacements and velocities may be neglected. This important principle was enunciated by Laplace and applied by him to the theory of the tides. Its great generality was also recognised by Sir John Herschel, to whom we owe a formal demonstration of its truth[1].

If the force be not a harmonic function of the time, the types of vibration in different parts of the system are in general different from each other and from that of the force. The harmonic functions are thus the only ones which preserve their type unchanged, which, as was remarked in the Introduction, is a strong reason for anticipating that they correspond to simple tones.

105. We now turn to a somewhat different kind of forced vibration, where, instead of given *forces* as hitherto, given inexorable *motions* are prescribed.

If we suppose that the co-ordinates ψ_1, ψ_2, ... ψ_r are given functions of the time, while the forces of the remaining types Ψ_{r+1}, Ψ_{r+2}, ... Ψ_m vanish, the equations of motion divide themselves into two groups, viz.

$$\left. \begin{aligned} e_{11}\,\psi_1 + e_{12}\psi_2 + \ldots + e_{1m}\psi_m &= \Psi_1 \\ e_{21}\,\psi_1 + e_{22}\psi_2 + \ldots + e_{2m}\psi_m &= \Psi_2 \\ \cdots\cdots\cdots\cdots\cdots\cdots\cdots\cdots\cdots \\ e_{r1}\psi_1 + e_{r2}\psi_2 + \ldots + e_{rm}\psi_m &= \Psi_r \end{aligned} \right\} \ldots\ldots\ldots\ldots (1);$$

and

$$\left. \begin{aligned} e_{r+1,1}\,\psi_1 + e_{r+1,2}\psi_2 + \ldots + e_{r+1,m}\psi_m &= 0 \\ \cdots\cdots\cdots\cdots\cdots\cdots\cdots\cdots\cdots\cdots\cdots \\ e_{m1}\,\psi_1 + e_{m2}\,\psi_2 + \ldots + e_{mm}\,\psi_m &= 0 \end{aligned} \right\} \ldots\ldots (2).$$

In each of the $m - r$ equations of the latter group, the first r terms are known explicit functions of the time, and have the same effect as known forces acting on the system. The equations of this group are therefore sufficient to determine the unknown quantities; after which, if required, the forces necessary to maintain the prescribed motion may be determined from the first group. It is obvious that there is no essential difference between the two classes of problems of forced vibrations.

106. The motion of a system devoid of friction and executing simple harmonic vibrations in consequence of prescribed variations of some of the coordinates, possesses a peculiarity parallel to those considered in §§ 74, 79. Let

$$\psi_1 = A_1 \cos pt, \quad \psi_2 = A_2 \cos pt, \quad \&c.$$

in which the quantities $A_1 \ldots A_r$ are regarded as given, while the

---

[1] *Encyc. Metrop.* art. 323. Also *Outlines of Astronomy*, § 650.

remaining ones are arbitrary. We have from the expressions for T and V, § 82,

$$2\,(T + V) = \tfrac{1}{2}\,(c_{11} + p^2 a_{11})\,A_1{}^2 + \ldots + (c_{12} + p^2 a_{12})\,A_1 A_2 + \ldots$$
$$+ \{\tfrac{1}{2}\,(c_{11} - p^2 a_{11})\,A_1{}^2 + \ldots + (c_{12} - p^2 a_{12})\,A_1 A_2 + \ldots\}\cos 2pt,$$

from which we see that the equations of motion express the condition that E, the variable part of $T + V$, which is proportional to

$$\tfrac{1}{2}\,(c_{11} - p^2 a_{11})\,A_1{}^2 + \ldots + (c_{12} - p^2 a_{12})\,A_1 A_2 + \ldots \quad\ldots\ldots (1),$$

shall be stationary in value, for all variations of the quantities $A_{r+1} \ldots A_m$. Let p'^2 be the value of p^2 natural to the system when vibrating under the restraint defined by the ratios

$$A_1 : A_2 \ldots A_r : A_{r+1} : \ldots A_m;$$

then

$$p'^2 = \{\tfrac{1}{2} c_{11} A_1{}^2 + \ldots + c_{12} A_1 A_2 + \ldots\} \div \{\tfrac{1}{2} a_{11} A_1{}^2 + \ldots + a_{12} A_1 A_2 + \ldots\},$$

so that

$$E = (p'^2 - p^2)\,\{\tfrac{1}{2} a_{11} A_1{}^2 + \ldots + a_{12} A_1 A_2 + \ldots\} \quad\ldots\ldots\ldots(2).$$

From this we see that if p^2 be certainly less than p'^2; that is, if the prescribed period be greater than any of those natural to the system under the partial constraint represented by

$$A_1 : A_2 \ldots A_r,$$

then E is necessarily positive, and the stationary value—there can be but one—is an absolute minimum. For a similar reason, if the prescribed period be *less* than any of those natural to the partially constrained system, E is an absolute maximum algebraically, but arithmetically an absolute minimum. But when p^2 lies within the range of possible values of p'^2, E may be positive or negative, and the actual value is not the greatest or least possible. Whenever a natural vibration is consistent with the imposed conditions, that will be the vibration assumed. The variable part of $T + V$ is then zero.

For convenience of treatment we have considered apart the two great classes of forced vibrations and free vibrations; but there is, of course, nothing to prevent their coexistence. After the lapse of a sufficient interval of time, the free vibrations always disappear, however small the friction may be. The case of absolutely no friction is purely ideal.

There is one caution, however, which may not be superfluous in respect to the case where given *motions* are forced on the system. Suppose, as before, that the co-ordinates ψ_1, ψ_2,...ψ_r are given. Then the free vibrations, whose existence or non-existence is a matter of indifference so far as the forced motion is concerned, must be understood to be such as the system is capable of, when the co-ordinates ψ_1...ψ_r *are not allowed to vary from zero.* In order to prevent their varying, forces of the corresponding types must be introduced; so that from one point of view the motion in question may be regarded as forced. But the applied forces are merely of the nature of a constraint; and their effect is the same as a limitation on the freedom of the motion.

107. Very remarkable reciprocal relations exist between the forces and motions of different types, which may be regarded as extensions of the corresponding theorems for systems in which only V or T has to be considered (§ 72 and §§ 77, 78). If we suppose that all the component forces, except two—Ψ_1 and Ψ_2—are zero, we obtain from § 104,

$$\left. \begin{aligned} \nabla\psi_1 &= \frac{d\nabla}{de_{11}}\Psi_1 + \frac{d\nabla}{de_{21}}\Psi_2 \\ \nabla\psi_2 &= \frac{d\nabla}{de_{12}}\Psi_1 + \frac{d\nabla}{de_{22}}\Psi_2 \end{aligned} \right\} \quad \dots\dots\dots\dots (1).$$

We now consider two cases of motion for the same system; first when Ψ_2 vanishes, and secondly (with dashed letters) when Ψ_1' vanishes. If $\Psi_2 = 0$,

$$\psi_2 = \nabla^{-1}\frac{d\nabla}{de_{12}}\Psi_1 \dots\dots\dots\dots\dots (2).$$

Similarly, if $\Psi_1' = 0$,

$$\psi_1' = \nabla^{-1}\frac{d\nabla}{de_{21}}\Psi_2' \dots\dots\dots\dots\dots (3).$$

In these equations ∇ and its differential coefficients are rational integral functions of the symbol D; and since in every case $e_{rs} = e_{sr}$, ∇ is a symmetrical determinant, and therefore

$$\frac{d\nabla}{de_{rs}} = \frac{d\nabla}{de_{sr}} \dots\dots\dots\dots\dots\dots (4).$$

Hence we see that if a force Ψ_1 act on the system, the co-ordinate ψ_2 is related to it in the same way as the co-ordinate ψ_1' is related to the force Ψ_2', when this latter force is supposed to act alone.

In addition to the motion here contemplated, there may be free vibrations dependent on a disturbance already existing at the moment subsequent to which all new sources of disturbance are included in Ψ; but these vibrations are themselves the effect of forces which acted previously. However small the dissipation may be, there must be an interval of time after which free vibrations die out, and beyond which it is unnecessary to go in taking account of the forces which have acted on a system. If therefore we include under Ψ forces of sufficient remoteness, there are no independent vibrations to be considered, and in this way the theorem may be extended to cases which would not at first sight appear to come within its scope. Suppose, for example, that the system is at rest in its position of equilibrium, and then begins to be aĉted on by a force of the first type, gradually increasing in magnitude from zero to a finite value Ψ_1, at which point it ceases to increase. If now at a given epoch of time the force be suddenly destroyed and remain zero ever afterwards, free vibrations of the system will set in, and continue until destroyed by friction. At any time t subsequent to the given epoch, the co-ordinate ψ_2 has a value dependent upon t proportional to Ψ_1. The theorem allows us to assert that this value ψ_2 bears the same relation to Ψ_1 as ψ_1' would at the same moment have borne to Ψ_2', if the original cause of the vibrations had been a force of the second type increasing gradually from zero to Ψ_2', and then suddenly vanishing at the given epoch of time. We have already had an example of this in § 101, and a like result obtains when the cause of the original disturbance is an impulse, or, as in the problem of the pianoforte-string, a variable force of finite though short duration. In these applications of our theorem we obtain results relating to free vibrations, considered as the residual effect of forces whose actual operation may have been long before.

108. In an important class of cases the forces Ψ_1 and Ψ_2' are harmonic, and of the same period. We may represent them by $A_1 e^{ipt}$, $A_2' e^{ipt}$, where A_1 and A_2' may be assumed to be *real*, if the forces be in the same phase at the moments compared. The results may then be written

$$\left. \begin{aligned} \psi_2 &= A_1 \frac{d \log \nabla (ip)}{de_{12}} e^{ipt} \\ \psi_1' &= A_2' \frac{d \log \nabla (ip)}{de_{21}} e^{ipt} \end{aligned} \right\} \quad \text{................ (1),}$$

where ip is written for D. Thus,

$$A_2' \psi_2 = A_1 \psi_1' \quad \text{........................ (2)}$$

Since the ratio $A_1 : A_2'$ is by hypothesis real, the same is true of the ratio $\psi_1' : \psi_2$; which signifies that the motions represented by those symbols are in the same phase. Passing to real quantities we may state the theorem thus:—

If a force $\Psi_1 = A_1 \cos pt$, *acting on the system give rise to the motion* $\psi_2 = \theta A_1 \cos (pt - \epsilon)$; *then will a force* $\Psi_2' = A_2' \cos pt$ *produce the motion* $\psi_1' = \theta A_2' \cos (pt - \epsilon)$.

If there be no friction, ϵ will be zero.

If $A_1 = A_2'$, then $\psi_1' = \psi_2$. But it must be remembered that the forces Ψ_1 and Ψ_2' are not necessarily comparable, any more than the co-ordinates of corresponding types, one of which for example may represent a linear and another an angular displacement.

The reciprocal theorem may be stated in several ways, but before proceeding to these we will give another investigation, not requiring a knowledge of determinants.

If $\Psi_1, \Psi_2,... \psi_1, \psi_2,...$ and $\Psi_1', \Psi_2',... \psi_1', \psi_2',...$ be two sets of forces and corresponding displacements, the equations of motion, § 103, give

$$\Psi_1\psi_1' + \Psi_2\psi_2' + ... = \psi_1' (e_{11}\psi_1 + e_{12}\psi_2 + e_{13}\psi_2 + ...)$$
$$+ \psi_2' (e_{21}\psi_1 + e_{22}\psi_2 + e_{23}\psi_3 + ...) + ...$$

Now, if all the forces vary as e^{ipt}, the effect of a symbolic operator such as e_{rs} on any of the quantities ψ is merely to multiply that quantity by the constant found by substituting ip for D in e_{rs}. Supposing this substitution made, and having regard to the relations $e_{rs} = e_{sr}$, we may write

$$\Psi_1\psi_1' + \Psi_2\psi_2' + ... = e_{11}\psi_1\psi_1' + e_{22}\psi_2\psi_2' + ...$$
$$+ e_{12} (\psi_1'\psi_2 + \psi_2'\psi_1) + ... \quad\quad ......(3).$$

Hence by the symmetry

$$\Psi_1\psi_1' + \Psi_2\psi_2' + ... = \Psi_1'\psi_1 + \Psi_2'\psi_2 + ... (4),$$

which is the expression of the reciprocal relation.

109. In the applications that we are about to make it will be supposed throughout that the forces of all types but two (which we may as well take as the first and second) are zero. Thus

$$\Psi_1\psi_1' + \Psi_2\psi_2' = \Psi_1'\psi_1 + \Psi_2'\psi_2 .... (1).$$

The consequences of this equation may be exhibited in three different ways. In the first we suppose that

$$\Psi_2 = 0, \quad \Psi_1' = 0,$$

whence

$$\psi_2 : \Psi_1 = \psi_1' : \Psi_2' \quad\dots\dots\dots\dots (2),$$

shewing, as before, that the relation of ψ_2 to Ψ_1 in the first case when $\Psi_2 = 0$ is the same as the relation of ψ_1' to Ψ_2' in the second case, when $\Psi_1' = 0$, the identity of relationship extending to phase as well as amplitude.

A few examples may promote the comprehension of a law, whose extreme generality is not unlikely to convey an impression of vagueness.

If P and Q be two points of a horizontal bar supported in any manner (e.g. with one end clamped and the other free), a given harmonic transverse force applied at P will give at any moment the same vertical deflection at Q as would have been found at P, had the force acted at Q.

If we take angular instead of linear displacements, the theorem will run :—A given harmonic *couple* at P will give the same *rotation* at Q as the couple at Q would give at P.

Or if one displacement be linear and the other angular, the result may be stated thus :—Suppose for the first case that a harmonic couple acts at P, and for the second that a vertical force of the same period and phase acts at Q, then the linear displacement at Q in the first case has at every moment the same phase as the rotatory displacement at P in the second, and the amplitudes of the two displacements are so related that the maximum couple at P would do the same work in acting over the maximum rotation at P due to the force at Q, as the maximum force at Q would do in acting through the maximum displacement at Q due to the couple at P. In this case the statement is more complicated, as the forces, being of different kinds, cannot be taken equal.

If we suppose the period of the forces to be excessively long, the momentary position of the system tends to coincide with that in which it would be maintained at rest by the then acting forces, and the equilibrium theory becomes applicable. Our theorem then reduces to the statical one proved in § 72.

As a second example, suppose that in a space occupied by air, and either wholly, or partly, confined by solid boundaries, there are two spheres A and B, whose centres have one degree of freedom. Then a periodic force acting on A will produce the same motion in B, as if the parts were interchanged; and this, whatever membranes, strings, forks on resonance cases, or other bodies capable of being set into vibration, may be present in their neighbourhood.

Or, if A and B denote two points of a solid elastic body of any shape, a force parallel to OX, acting at A, will produce the same motion of the point B parallel to OY as an equal force parallel to OY acting at B would produce in the point A, parallel to OX.

Or, again, let A and B be two points of a space occupied by air, between which are situated obstacles of any kind. Then a sound originating at A is perceived at B with the same intensity as that with which an equal sound originating at B would be perceived at A.[1] The obstacle, for instance, might consist of a rigid wall pierced with one or more holes. This example corresponds to the optical law that if by any combination of reflecting or refracting surfaces one point can be seen from a second, the second can also be seen from the first. In Acoustics the sound shadows are usually only partial in consequence of the not insignificant value of the wave-length in comparison with the dimensions of ordinary obstacles: and the reciprocal relation is of considerable interest.

A further example may be taken from electricity. Let there be two circuits of insulated wire A and B, and in their neighbourhood any combination of wire-circuits or solid conductors in communication with condensers. A periodic electro-motive force in the circuit A will give rise to the same current in B as would be excited in A if the electro-motive force operated in B.

Our last example will be taken from the theory of conduction and radiation of heat, Newton's law of cooling being assumed as a basis. The temperature at any point A of a conducting and radiating system due to a steady (or harmonic) source of heat at B is the same as the temperature at B due to an equal source at A. Moreover, if at any time the source at B be removed, the whole subsequent course of temperature at A will be the same as it would be at B if the parts of B and A were interchanged.

---

[1] Helmholtz, *Crelle*, Bd. LVII. The sounds must be such as in the absence of obstacles would diffuse themselves equally in all directions.

110. The second way of stating the reciprocal theorem is arrived at by taking in (1) of § 109,

$$\psi_1 = 0, \quad \psi_2' = 0;$$

whence $\qquad \Psi_1 \psi_1' = \Psi_2' \psi_2$ ....(1),

or $\qquad \Psi_1 : \psi_2 = \Psi_2' : \psi_1'$ ....(2),

shewing that the relation of Ψ_1 to ψ_2 in the first case, when $\psi_1 = 0$, is the same as the relation of Ψ_2' to ψ_1' in the second case, when $\psi_2' = 0$.

Thus in the example of the rod, if the point P be held at rest while a given vibration is imposed upon Q (by a force there applied), the reaction at P is the same both in amplitude and phase as it would be at Q if that point were held at rest and the given vibration were imposed upon P.

So if A and B be two electric circuits in the neighbourhood of any number of others, $C, D, \ldots$ whether closed or terminating in condensers, and a given periodic current be excited in A by the necessary electro-motive force, the induced electro-motive force in B is the same as it would be in A, if the parts of A and B were interchanged.

The third form of statement is obtained by putting in (1) of § 109,

$$\Psi_1 = 0, \quad \psi_2' = 0;$$

whence $\qquad \Psi_1' \psi_1 + \Psi_2' \psi_2 = 0$ .... (3),

or $\qquad \psi_1 : \psi_2 = - \Psi_2' : \Psi_1'$ ....(4),

proving that the ratio of ψ_1 to ψ_2 in the first case, when Ψ_2 acts alone, is the negative of the ratio of Ψ_2' to Ψ_1' in the second case, when the forces are so related as to keep ψ_2' equal to zero.

Thus if the point P of the rod be held at rest while a periodic force acts at Q, the reaction at P bears the same numerical ratio to the force at Q as the displacement at Q would bear to the displacement at P, if the rod were caused to vibrate by a force applied at P.

111. The reciprocal theorem has been proved for all systems in which the frictional forces can be represented by the function F, but it is susceptible of a further and an important generalization. We have indeed proved the existence of the function F for a large class of cases where the motion is resisted by forces proportional to the absolute or relative velocities, but there are

other sources of dissipation not to be brought under this head, whose effects it is equally important to include; for example, the dissipation due to the conduction or radiation of heat. Now although it be true that the forces in these cases are not *for all possible motions* in a constant ratio to the velocities or displacements, yet in any actual case of periodic motion (τ) they are necessarily periodic, and therefore, whatever their phase, expressible by a sum of two terms, one proportional to the displacement (absolute or relative) and the other proportional to the velocity of the part of the system affected. If the coefficients be the same, not necessarily for all motions whatever, *but for all motions of the period* τ, the function F exists in the only sense required for our present purpose. In fact since it is exclusively with motions of period τ that the theorem is concerned, it is plainly a matter of indifference whether the functions T, F, V are dependent upon τ or not. Thus extended, the theorem is perhaps sufficiently general to cover the whole field of dissipative forces.

It is important to remember that the Principle of Reciprocity is limited to systems which vibrate about a configuration of *equilibrium*, and is therefore not to be applied without reservation to such a problem as that presented by the transmission of sonorous waves through the atmosphere when disturbed by wind. The vibrations must also be of such a character that the square of the motion can be neglected throughout; otherwise our demonstration would not hold good. Other apparent exceptions depend on a misunderstanding of the principle itself. Care must be taken to observe a proper correspondence between the forces and displacements, the rule being that the action of the force over the displacement is to represent *work done*. Thus *couples* correspond to *rotations*, *pressures* to increments of *volume*, and so on.

112. In Chapter III. we considered the vibrations of a system with one degree of freedom. The remainder of the present Chapter will be devoted to some details of the case where the degrees of freedom are two.

If x and y denote the two co-ordinates, the expressions for T and V are of the form

$$\left.\begin{aligned}2T &= L\dot{x}^2 + 2M\dot{x}\dot{y} + N\dot{y}^2\\ 2V &= Ax^2 + 2Bxy + Cy^2\end{aligned}\right\}\dotfill(1);$$

so that, in the absence of friction, the equations of motion are

$$L\ddot{x} + M\ddot{y} + Ax + By = X \atop M\ddot{x} + N\ddot{y} + Bx + Cy = Y \Big\} \quad\quad\quad (2).$$

When there are no impressed forces, we have for the natural vibrations

$$(LD^2 + A)x + (MD^2 + B)y = 0 \atop (MD^2 + B)x + (ND^2 + C)y = 0 \Big\} \quad\quad (3),$$

D being the symbol of differentiation with respect to time.

If a solution of (3) be $x = l e^{\lambda t}$, $y = m e^{\lambda t}$, λ^2 is one of the roots of

$$(L\lambda^2 + A)(N\lambda^2 + C) - (M\lambda^2 + B)^2 = 0 \quad\quad (4),$$

or

$$\lambda^4(LN - M^2) + \lambda^2(LC + NA - 2MB) + AC - B^2 = 0 \quad (5).$$

The constants L, M, N; A, B, C, are not entirely arbitrary. Since T and V are essentially positive, the following inequalities must be satisfied:—

$$LN > M^2, \quad AC > B^2 \quad\quad\quad (6).$$

Moreover, L, N, A, C must themselves be positive.

We proceed to examine the effect of these restrictions on the roots of (5).

In the first place the three coefficients in the equation are positive. For the first and third, this is obvious from (6). The coefficient of λ^2

$$= (\sqrt{LC} - \sqrt{NA})^2 + 2\sqrt{LNAC} - 2MB,$$

in which, as is seen from (6), $\sqrt{LNAC}$ is necessarily greater than MB. We conclude that the values of λ^2, if real, are both negative.

It remains to prove that the roots are in fact real. The condition to be satisfied is that the following quantity be not negative:—

$$(LC + NA - 2MB)^2 - 4(LN - M^2)(AC - B^2).$$

After reduction this may be brought into the form

$$4(\sqrt{LN}.B - \sqrt{AC}.M)^2$$
$$+ (\sqrt{LC} - \sqrt{NA})^2 \{(\sqrt{LC} - \sqrt{NA})^2 + 4(\sqrt{LNAC} - MB)\},$$

which shews that the condition is satisfied, since $\sqrt{LNAC} - MB$ is positive. This is the analytical proof that the values of λ^2 are both real and negative; a fact that might have been anticipated without any analysis from the physical constitution of the system, whose vibrations they serve to express.

The two values of λ^2 are different, unless *both*

$$\left. \begin{array}{c} \sqrt{LN}.B - \sqrt{AC}.M = 0 \\ \sqrt{LC} - \sqrt{NA} = 0 \end{array} \right\},$$

which require that

$$L : M : N = A : B : C \dots\dots\dots\dots(7).$$

The common spherical pendulum is an example of this case.

By means of a suitable force Y the co-ordinate y may be prevented from varying. The system then loses one degree of freedom, and the period corresponding to the remaining one is in general different from either of those possible before the introduction of Y. Suppose that the types of the motions obtained by thus preventing in turn the variation of y and x are respectively $e^{\mu_1 t}$, $e^{\mu_2 t}$. Then μ_1^2, μ_2^2 are the roots of the equation

$$(L\lambda^2 + A)(N\lambda^2 + C) = 0,$$

being that obtained from (4) by suppressing M and B. Hence (4) may itself be put into the form

$$LN(\lambda^2 - \mu_1^2)(\lambda^2 - \mu_2^2) = (M\lambda^2 + B)^2 \dots\dots\dots\dots(8),$$

which shews at once that neither of the roots of λ^2 can be intermediate in value between μ_1^2 and μ_2^2. A little further examination will prove that one of the roots is greater than both the quantities μ_1^2, μ_2^2, and the other less than both. For if we put

$$f(\lambda^2) = LN(\lambda^2 - \mu_1^2)(\lambda^2 - \mu_2^2) - (M\lambda^2 + B)^2,$$

we see that when λ^2 is very small, f is positive $(AC - B^2)$; when λ^2 decreases (algebraically) to μ_1^2, f changes sign and becomes negative. Between 0 and μ_1^2 there is therefore a root; and also by similar reasoning between μ_2^2 and $-\infty$. We conclude that the tones obtained by subjecting the system to the two kinds of constraint in question are both intermediate in pitch between the tones given by the natural vibrations of the system. In particular cases μ_1^2, μ_2^2 may be equal, and then

$$\lambda^2 = \frac{\sqrt{LN}\,\mu^2 \pm B}{\sqrt{LN} \mp M} = \frac{-\sqrt{AC} \pm B}{\sqrt{LN} \mp M} \dots\dots\dots\dots(9).$$

This proposition may be generalized. *Any* kind of constraint which leaves the system still in possession of one degree of freedom may be regarded as the imposition of a forced relation between the co-ordinates, such as

$$\alpha x + \beta y = 0 \dots\dots\dots\dots(10).$$

Now if $\alpha x + \beta y$, and any other homogeneous linear function of x and y, be taken as new variables, the same argument proves that the single period possible to the system after the introduction of the constraint, is intermediate in value between those two in which the natural vibrations were previously performed. Conversely, the two periods which become possible when a constraint is removed, lie one on each side of the original period.

If the values of λ^2 be equal, which can only happen when

$$L : M : N = A : B : C,$$

the introduction of a constraint has no effect on the period; for instance, the limitation of a spherical pendulum to one vertical plane.

113. As a simple example of a system with two degrees of freedom, we may take a stretched string of length l, itself without inertia, but carrying two equal masses m at distances a and b from one end (Fig. 17). Tension $= T_1$.

Fig. 17.

If x and y denote the displacements,

$$2T = m\left(\dot{x}^2 + \dot{y}^2\right),$$

$$2V = T_1\left\{\frac{x^2}{a} + \frac{(x-y)^2}{b-a} + \frac{y^2}{l-b}\right\}$$

Since T and V are not of the same form, it follows that the two periods of vibration are in every case unequal.

If the loads be symmetrically attached, the character of the two component vibrations is evident. In the first, which will have the longer period, the two weights move together, so that x and y remain equal throughout the vibration. In the second x and y are numerically equal, but opposed in sign. The middle point of the string then remains at rest, and the two masses are always to be found on a straight line passing through it. In the first case $x - y = 0$, and in the second $x + y = 0$; so that $x - y$, and $x + y$ are the new variables which must be assumed in order to reduce the functions T and V simultaneously to a sum of squares.

For example, if the masses be so attached as to divide the string into three equal parts,

$$\left.\begin{aligned}
2T &= \frac{m}{2} \left\{ (\dot{x} + \dot{y})^2 + (\dot{x} - \dot{y})^2 \right\} \\
2V &= \frac{3T_1}{2l} \left\{ (x + y)^2 + 3(x - y)^2 \right\}
\end{aligned}\right\} \quad \ldots\ldots\ldots\ldots(1),$$

from which we obtain as the complete solution,

$$\left.\begin{aligned}
x + y &= A \cos\left(\sqrt{\frac{3T_1}{lm}} \cdot t + \alpha \right) \\
x - y &= B \cos\left(\sqrt{\frac{9T_1}{lm}} \cdot t + \beta \right)
\end{aligned}\right\} \quad \ldots\ldots\ldots\ldots(2),$$

where, as usual, the constants A, α, B, β are to be determined by the initial circumstances.

114.   When the two natural periods of a system are nearly equal, the phenomenon of intermittent vibration sometimes presents itself in a very curious manner. In order to illustrate this, we may recur to the string loaded, we will now suppose, with two equal masses at distances from its ends equal to one-fourth of the length. If the middle point of the string were absolutely fixed, the two similar systems on either side of it would be completely independent, or, if the whole be considered as one system, the two periods of vibration would be equal. We now suppose that instead of being absolutely fixed, the middle point is attached to springs, or other machinery, destitute of inertia, so that it is capable of yielding *slightly*. The reservation as to inertia is to avoid the introduction of a third degree of freedom.

From the symmetry it is evident that the fundamental vibrations of the system are those represented by $x + y$ and $x - y$. Their periods are slightly different, because, on account of the yielding of the centre, the potential energy of a displacement when x and y are equal, is less than that of a displacement when x and y are opposite; whereas the kinetic energies are the same for the two kinds of vibration. In the solution

$$\left.\begin{aligned}
x + y &= A \cos\left(n_1 t + \alpha \right) \\
x - y &= B \cos\left(n_2 t + \beta \right)
\end{aligned}\right\} \quad \ldots\ldots\ldots\ldots(1),$$

we are therefore to regard n_1 and n_2 as nearly, but not quite, equal. Now let us suppose that initially x and $\dot{x}$ vanish. The conditions are

$$\left.\begin{aligned}
A \cos\alpha + \quad B \cos\beta &= 0 \\
n_1 A \sin\alpha + n_2 B \sin\beta &= 0
\end{aligned}\right\},$$

which give approximately
$$A + B = 0, \quad \alpha = \beta.$$

Thus
$$x = A \sin \frac{n_2 - n_1}{2} t \; \sin \left(\frac{n_1 + n_2}{2} t + \alpha \right) \Bigg\}$$
$$y = A \cos \frac{n_2 - n_1}{2} t \; \cos \left(\frac{n_1 + n_2}{2} t + \alpha \right) \Bigg\} \quad \ldots\ldots\ldots\ldots(2).$$

The value of the co-ordinate x is here approximately expressed by a harmonic term, whose amplitude, being proportional to $\sin \frac{n_2 - n_1}{2} t$, is a slowly varying harmonic function of the time. The vibrations of the co-ordinates are therefore intermittent, and so adjusted that each amplitude vanishes at the moment that the other is at its maximum.

This phenomenon may be prettily shewn by a tuning fork of very low pitch, heavily weighted at the ends, and firmly held by screwing the stalk into a massive support. When the fork vibrates in the normal manner, the rigidity, or want of rigidity, of the stalk does not come into play; but if the displacements of the two prongs be in the same direction, the slight yielding of the stalk entails a small change of period. If the fork be excited by striking one prong, the vibrations are intermittent, and appear to transfer themselves backwards and forwards between the prongs. Unless, however, the support be very firm, the abnormal vibration, which involves a motion of the centre of inertia, is soon dissipated; and then, of course, the vibration appears to become steady. If the fork be merely held in the hand, the phenomenon of intermittence cannot be obtained at all.

115. The stretched string with two attached masses may be used to illustrate some general principles. For example, the period of the vibration which remains possible when one mass is held at rest, is intermediate between the two free periods. Any increase in either load depresses the pitch of both the natural vibrations, and conversely. If the new load be situated at a point of the string not coinciding with the places where the other loads are attached, nor with the node of one of the two previously possible free vibrations (the other has no node), the effect is still to prolong both the periods already present. With regard to the third finite period, which becomes possible for the first time after the addition of the new load, it must be regarded as derived from

one of infinitely small magnitude, of which an indefinite number may be supposed to form part of the system. It is instructive to trace the effect of the introduction of a new load and its gradual increase from zero to infinity, but for this purpose it will be simpler to take the case where there is but one other. At the commencement there is one finite period τ_1, and another of infinitesimal magnitude τ_2. As the load increases τ_2 becomes finite, and both τ_1 and τ_2 continually increase. Let us now consider what happens when the load becomes very great. One of the periods is necessarily large and capable of growing beyond all limit. The other must approach a fixed finite limit. The first belongs to a motion in which the larger mass vibrates nearly as if the other were absent; the second is the period of the vibration of the smaller mass, taking place much as if the larger were fixed. Now since τ_1 and τ_2 can never be equal, τ_1 must be always the greater; and we infer, that as the load becomes continually larger, it is τ_1 that increases indefinitely, and τ_2 that approaches a finite limit.

We now pass to the consideration of forced vibrations.

116. The general equations for a system of two degrees of freedom including friction are

$$\left. \begin{array}{l} (LD^2 + \alpha D + A)\, x + (MD^2 + \beta D + B)\, y = X \\ (MD^2 + \beta D + B)\, x + (ND^2 + \gamma D + C)\, y = Y \end{array} \right\} \ \ldots\ldots\ (1).$$

In what follows we shall suppose that $Y = 0$, and that $X = e^{ipt}$. The solution for y is

$$y = - \frac{(B - p^2 M + i\beta p)\, e^{ipt}}{(A - p^2 L + i\alpha p)\,(C - p^2 N + i\gamma p) - (B - p^2 M + i\beta p)^2} \ \ldots\ (2).$$

If the connection between x and y be of a loose character, the constants M, β, B are small, so that the term $(B - p^2 M + i\beta p)^2$ in the denominator may in general be neglected. When this is permissible, the co-ordinate y is the same as if x had been prevented from varying, and a force Y had been introduced whose magnitude is independent of N, γ, and C. But if, in consequence of an approximate isochronism between the force and one of the motions which become possible when x or y is constrained to be zero, either $A - p^2 L + i\alpha p$, or $C - p^2 N + i\gamma p$ be small, then the term in the denominator containing the coefficients of mutual influence must be retained, being no longer *relatively* unimportant; and the solution is accordingly of a more complicated character.

Symmetry shews that if we had assumed $X=0$, $Y=e^{ipt}$, we should have found the same value for x as now obtains for y. This is the Reciprocal Theorem of § 108 applied to a system capable of two independent motions. The string and two loads may again be referred to as an example.

117. So far for an imposed force. We shall next suppose that it is a *motion* of one co-ordinate ($x=e^{ipt}$) that is prescribed, while $Y=0$; and for greater simplicity we shall confine ourselves to the case where $\beta=0$. The value of y is

$$y=-\frac{(B-Mp^2)\,e^{ipt}}{C-Np^2+i\gamma p} \quad\text{...............(1).}$$

Let us now inquire into the reaction of this motion on x. We have

$$(MD^2+B)\,y=-\frac{(B-Mp^2)^2\,e^{ipt}}{C-Np^2+i\gamma p} \quad\text{...............(2).}$$

If the real and imaginary parts of the coefficient of e^{ipt} be respectively A' and $i\,\alpha'p$, we may put

$$(MD^2+B)\,y=A'x+\alpha'\dot{x} \quad\text{...............(3),}$$

and

$$A'=-\frac{(B-Mp^2)^2\,(C-Np^2)}{(C-Np^2)^2+\gamma^2p^2} \quad\text{...............(4),}$$

$$\alpha'=\frac{(B-Mp^2)^2\gamma}{(C-Np^2)^2+\gamma^2p^2} \quad\text{...............(5).}$$

It appears that the effect of the reaction of y (over and above what would be caused by holding $y=0$) is represented by changing A into $A+A'$, and α into $\alpha+\alpha'$, where A' and α' have the above values, and is therefore equivalent to the effect of an alteration in the coefficients of spring and friction. These alterations, however, are not constants, *but functions of the period of the motion contemplated*, whose character we now proceed to consider.

Let n be the value of p corresponding to the natural frictionless period of y (x being maintained at zero); so that $C-n^2N=0$. Then

$$\left.\begin{array}{l} A'=(B-Mp^2)^2\dfrac{N(p^2-n^2)}{N^2(p^2-n^2)^2+\gamma^2p^2} \\[2mm] \alpha'=(B-Mp^2)^2\dfrac{\gamma}{N^2(p^2-n^2)^2+\gamma^2p^2} \end{array}\right\}\quad\text{.........(6).}$$

In most cases with which we are practically concerned γ is small, and interest centres mainly on values of p not much differing from n. We shall accordingly leave out of account the variations of the positive factor $(B - Mp^2)^2$, and in the small term $\gamma^2 p^2$, substitute for p its approximate value n. When p is not nearly equal to n, the term in question is of no importance.

As might be anticipated from the general principle of work, α' is always positive. Its maximum value occurs when $p = n$ nearly, and is then proportional to $\dfrac{1}{\gamma n^2}$, which varies *inversely* with γ. This might not have been expected on a superficial view of the matter, for it seems rather a paradox that, the greater the friction, the less should be its result. But it must be remembered that γ is only the *coefficient* of friction, and that when γ is small the maximum motion is so much increased that the whole work spent against friction is greater than if γ were more considerable.

But the point of most interest is the dependence of A' on p. If p be less than n, A' is·negative. As p passes through the value n, A' vanishes, and changes sign. When A' is negative, the influence of y is to diminish the recovering power of the vibration x, and we see that this happens when the forced vibration is slower than that natural to y. The tendency of the vibration y is thus to retard the vibration x, if the latter be already the slower, but to accelerate it, if it be already the more rapid, only vanishing in the critical case of perfect isochronism. The attempt to make x vibrate at the rate determined by n is beset with a peculiar difficulty, analogous to that met with in balancing a heavy body with the centre of gravity above the support. On whichever side a slight departure from precision of adjustment may occur the influence of the dependent vibration is always to increase the error. Examples of the instability of pitch accompanying a strong resonance will come across us hereafter; but undoubtedly the most interesting application of the results of this section is to the explanation of the anomalous refraction, by substances possessing a very marked selective absorption, of the two kinds of light situated (in a normal spectrum) immediately on either side of the absorption band[1]. It was observed by Christiansen and Kundt, the discoverers of this remarkable phenomenon, that media of the kind in question (for example, *fuchsine* in alcoholic solution) refract the ray immediately *below* the absorption-band abnormally *in excess*, and that *above* it *in defect*. If we suppose, as on other grounds it would be natural to do, that the intense absorption is

[1] *Phil. Mag.*, May, 1872. Also Sellmeier, *Pogg. Ann.* t. cxliii. p. 272.

the result of an agreement between the vibrations of the kind of light affected, and some vibration proper to the molecules of the absorbing agent, our theory would indicate that for light of somewhat greater period the effect must be the same as a relaxation of the natural elasticity of the ether, manifesting itself by a slower propagation and increased refraction. On the other side of the absorption-band its influence must be in the opposite direction.

In order to trace the law of connection between A' and p, take, for brevity, $\gamma n = a$, $N(p^2 - n^2) = x$, so that

$$A' \propto \frac{x}{x^2 + a^2}.$$

When the sign of x is changed, A' is reversed with it, but preserves its numerical value. When $x = 0$, or $\pm \infty$, A' vanishes.

Fig. 18.

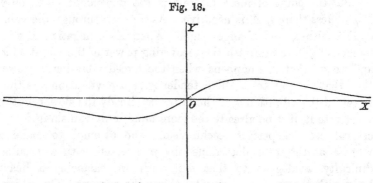

Hence the origin is on the representative curve (Fig. 18), and the axis of x is an asymptote. The maximum and minimum values of A' occur when x is respectively equal to $+a$, or $-a$; and then

$$\frac{x}{x^2 + a^2} = \pm \frac{1}{2a}.$$

The corresponding values of p are given by

$$p^2 = n^2 \pm \frac{\gamma n}{N} \quad\text{...........................} (7).$$

Hence, the smaller the value of a or γ, the greater will be the maximum alteration of A, and the corresponding value of p will approach nearer and nearer to n. It may be well to repeat, that in the optical application a diminished γ is attended by an *increased* maximum absorption. When the adjustment of periods is such as to favour A' as much as possible, the corresponding value of a' is one half of *its* maximum.

CHAPTER VI.

118. AMONG vibrating bodies there are none that occupy a more prominent position than Stretched Strings. From the earliest times they have been employed for musical purposes, and in the present day they still form the essential parts of such important instruments as the pianoforte and the violin. To the mathematician they must always possess a peculiar interest as the battle-field on which were fought out the controversies of D'Alembert, Euler, Bernoulli and Lagrange, relating to the nature of the solutions of partial differential equations. To the student of Acoustics they are doubly important. In consequence of the comparative simplicity of their theory, they are the ground on which difficult or doubtful questions, such as those relating to the nature of simple tones, can be most advantageously faced; while in the form of a Monochord or Sonometer, they afford the most generally available means for the comparison of pitch.

The 'string' of Acoustics is a perfectly uniform and flexible filament of solid matter stretched between two fixed points—in fact an ideal body, never actually realized in practice, though closely approximated to by most of the strings employed in music. We shall afterwards see how to take account of any small deviations from complete flexibility and uniformity.

The vibrations of a string may be divided into two distinct classes, which are practically independent of one another, if the amplitudes do not exceed certain limits. In the first class the displacements and motions of the particles are *longitudinal*, so that the string always retains its straightness. The potential energy of a displacement depends, not on the whole tension, but on the *changes* of tension which occur in the various parts of the string, due to the increased or diminished extension. In order to calculate it we must know the relation between the extension of a string and the stretching force. The approximate law (given by Hooke) may be expressed by saying that the extension varies as the tension, so that if l and l' denote the natural and the stretched lengths of a string, and T the tension,

$$\frac{l' - l}{l} = \frac{T}{E} \quad \dots\dots\dots\dots\dots\dots\dots\dots\dots(1),$$

where E is a constant, depending on the material and the section, which may be interpreted to mean the tension that would be necessary to stretch the string to twice its natural length, if the law applied to so great extensions, which, in general, it is far from doing.

119. The vibrations of the second kind are *transverse;* that is to say, the particles of the string move sensibly in planes perpendicular to the line of the string. In this case the potential energy of a displacement depends upon the general tension, and the small variations of tension accompanying the additional stretching due to the displacement may be left out of account. It is here assumed that the stretching due to the motion may be neglected in comparison with that to which the string is already subject in its position of equilibrium. Once assured of the fulfilment of this condition, we do not, in the investigation of transverse vibrations, require to know anything further of the law of extension.

The most general vibration of the transverse, or lateral, kind may be resolved, as we shall presently prove, into two sets of component normal vibrations, executed in perpendicular planes. Since it is only in the initial circumstances that there can be any distinction, pertinent to the question, between one plane and another, it is sufficient for most purposes to regard the motion as entirely confined to a single plane passing through the line of the string.

In treating of the theory of strings it is usual to commence with two particular solutions of the partial differential equation, representing the transmission of waves in the positive and negative directions, and to combine these in such a manner as to suit the case of a finite string, whose ends are maintained at rest; neither of the solutions taken by itself being consistent with the existence of *nodes,* or places of permanent rest. This aspect of the question is very important, and we shall fully consider it; but it seems scarcely desirable to found the solution in the first instance on a property so peculiar to a *uniform* string as the undisturbed transmission of waves. We will proceed by the more general method of assuming (in conformity with what was proved in the last chapter) that the motion may be resolved into normal components of the harmonic type, and determining their periods and character by the special conditions of the system.

Towards carrying out this design the first step would naturally be the investigation of the partial differential equation, to which the motion of a continuous string is subject. But in order to throw light on a point, which it is most important to understand clearly,—the connection between finite and infinite freedom, and the passage corresponding thereto between arbitrary constants and arbitrary functions, we will commence by following a somewhat different course.

120. In Chapter III. it was pointed out that the fundamental vibration of a string would not be entirely altered in character, if the mass were concentrated at the middle point. Following out this idea, we see that if the whole string were divided into a number of small parts and the mass of each concentrated at its centre, we might by sufficiently multiplying the number of parts arrive at a system, still of finite freedom, but capable of representing the continuous string with any desired accuracy, so far at least as the lower component vibrations are concerned. If the analytical solution for any number of divisions can be obtained, its limit will give the result corresponding to a uniform string. This is the method followed by Lagrange.

Let l be the length, ρl the whole mass of the string, so that ρ denotes the mass per unit length, T_1 the tension.

Fig. 19.

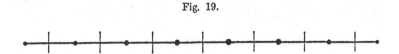

The length of the string is divided into $m+1$ equal parts (a), so that

$$(m+1)\,a = l \dots\dots\dots\dots\dots\dots\dots\dots(1).$$

At the m points of division equal masses (μ) are supposed concentrated, which are the representatives of the mass of the portions (a) of the string, which they severally bisect. The mass of each terminal portion of length $\frac{1}{2}a$ is supposed to be concentrated at the final points. On this understanding, we have

$$(m+1)\,\mu = \rho l \dotfill (2).$$

We proceed to investigate the vibrations of a string, itself devoid of inertia, but loaded at each of m points equidistant (a) from themselves and from the ends, with a mass μ.

If $\psi_1, \psi_2 \ldots\ldots \psi_{m+2}$ denote the lateral displacements of the loaded points, including the initial and final points, we have the following expressions for T and V,

$$T = \tfrac{1}{2}\,\mu\,\{\dot\psi_1^{\,2} + \dot\psi_2^{\,2} + \ldots + \dot\psi_{m+1}^{\,2} + \dot\psi_{m+2}^{\,2}\} \dotfill (3),$$

$$V = \frac{T_1}{2a}\,\{(\psi_2 - \psi_1)^2 + (\psi_3 - \psi_2)^2 + \ldots + (\psi_{m+2} - \psi_{m+1})^2\} \ldots (4),$$

with the conditions that ψ_1 and ψ_{m+2} vanish. These give by Lagrange's Method the m equations of motion,

$$\left.\begin{array}{l} B\psi_1 + A\psi_2 + B\psi_3 \ = 0 \\ B\psi_2 + A\psi_3 + B\psi_4 \ = 0 \\ B\psi_3 + A\psi_4 + B\psi_5 \ = 0 \\ \cdots\cdots\cdots\cdots\cdots\cdots \\ B\psi_m + A\psi_{m+1} + B\psi_{m+2} = 0 \end{array}\right\} \dotfill (5),$$

where
$$A = \mu D^2 + \frac{2T_1}{a}, \qquad B = -\frac{T_1}{a} \dotfill (6).$$

Supposing now that the vibration under consideration is one of normal type, we assume that $\psi_1,\ \psi_2$, &c. are all proportional to $\cos(nt - \epsilon)$, where n remains to be determined. A and B may then be regarded as constants, with a substitution of $-n^2$ for D^2.

If for the sake of brevity we put

$$C = A \div B = -2 + \frac{\mu a n^2}{T_1} \dotfill (7),$$

the determinantal equation, which gives the values of n^2, assumes the form

$$\left|\begin{array}{c} C,\ 1,\ 0,\ 0,\ 0...... \\ 1,\ C,\ 1,\ 0,\ 0...... \\ 0,\ 1,\ C,\ 1,\ 0...... \\ 0,\ 0,\ 1,\ C,\ 1...... \\ 0,\ 0,\ 0,\ 1,\ C...... \\ .... \\ .... \end{array}\right| \begin{array}{l} m\ rows \\ \\ \\ =0..............(8). \end{array}$$

From this equation the values of the roots might be found. It may be proved that, if $C = 2\cos\theta$, the determinant is equivalent to $\sin(m+1)\theta \div \sin\theta$; but we shall attain our object with greater ease directly from (5) by acting on a hint derived from the known results relating to a continuous string, and assuming for trial a particular type of vibration. Thus let a solution be

$$\psi_r = P\ \sin(r-1)\beta\ \cos(nt-\epsilon)..............(9),$$

a form which secures that $\psi_1 = 0$. In order that ψ_{m+2} may vanish,

$$(m+1)\beta = s\pi.....................(10),$$

where s is an integer. Substituting the assumed values of ψ in the equations (5), we find that they are satisfied, provided that

$$2B\cos\beta + A = 0 ....(11);$$

so that the value of n in terms of β is

$$n = 2\sin\frac{\beta}{2}\sqrt{\frac{T_1}{\mu a}} ....(12).$$

A normal vibration is thus represented by

$$\psi_r = P_s \sin\frac{(r-1)s\pi}{m+1}\cos(n_s t - \epsilon_s) ....(13),$$

where

$$n_s = 2\sqrt{\frac{T_1}{\mu a}}\ \sin\frac{s\pi}{2(m+1)} ....(14),$$

and P_s, ϵ_s denote arbitrary constants independent of the general constitution of the system. The m admissible values of n are found from (14) by ascribing to s in succession the values 1, 2, 3...m, and are all different. If we take $s = m+1$, ψ_r vanishes, so that this does not correspond to a possible vibration. Greater values of s give only the same periods over again. If $m+1$ be even, one of the values of n—that, namely, corresponding to

9—2

$s = \frac{1}{2}(m+1)$,—is the same as would be found in the case of only a single load ($m = 1$). The interpretation is obvious. In the kind of vibration considered every alternate particle remains at rest, so that the intermediate ones really move as though they were attached to the centres of strings of length $2a$, fastened at the ends.

The most general solution is found by putting together all the possible particular solutions of normal type

$$\psi_r = \sum_{s=1}^{s=m} P_s \sin \frac{(r-1)s\pi}{m+1} \cos(n_s t - \epsilon_s) \dots\dots(15),$$

and, by ascribing suitable values to the arbitrary constants, can be identified with the vibration resulting from arbitrary initial circumstances.

Let x denote the distance of the particle r from the end of the string, so that $(r-1)a = x$; then by substituting for μ and a from (1) and (2), our solution may be written,

$$\psi(x) = P_s \sin s\frac{\pi x}{l} \cos(n_s t - \epsilon_s)\dots\dots\dots(16),$$

$$n_s = \frac{2(m+1)}{l}\sqrt{\frac{T_1}{\rho}} \sin \frac{s\pi}{2(m+1)}\dots\dots\dots(17).$$

In order to pass to the case of a continuous string, we have only to put m infinite. The first equation retains its form, and specifies the displacement at any point x. The limiting form of the second is simply

$$n = \frac{s\pi}{l}\sqrt{\frac{T_1}{\rho}}\dots\dots\dots\dots\dots(18),$$

whence for the periodic time,

$$\tau = \frac{2\pi}{n} = \frac{2l}{s}\sqrt{\frac{\rho}{T_1}}\dots\dots\dots\dots(19).$$

The periods of the component tones are thus aliquot parts of that of the gravest of the series, found by putting $s = 1$. The whole motion is in all cases periodic; and the period is $2l\sqrt{\frac{\rho}{T_1}}$. This statement, however, must not be understood as excluding a shorter period; for in particular cases any number of the lower components may be absent. All that is asserted is that the above-mentioned interval of time is *sufficient* to bring about a complete recurrence. We defer for the present any further discussion of the important formula (19), but it is interesting to observe the approach to a limit in (17), as m is made successively greater and greater. For this purpose it will be sufficient to take the gravest tone for which $s = 1$, and accordingly to trace the variation of

$$\frac{2(m+1)}{\pi} \sin \frac{\pi}{2(m+1)}.$$

The following are a series of simultaneous values of the function and variable :—

m	1	2	3	4	9	19	39
$\dfrac{2(m+1)}{\pi} \sin \dfrac{\pi}{2(m+1)}$	·9003	·9549	·9745	·9836	·9959	·9990	·9997

It will be seen that for very moderate values of m the limit is closely approached. Since m is the number of (moveable) loads, the case $m = 1$ corresponds to the problem investigated in Chapter III., but in comparing the results we must remember that we there supposed the *whole* mass of the string to be concentrated at the centre. In the present case the load at the centre is only half as great; the remainder being supposed concentrated at the ends, where it is without effect.

From the fact that our solution is general, it follows that any initial form of the string can be represented by

$$\psi(x) = \sum_{s=1}^{s=\infty} (P \cos \epsilon)_s \sin s \frac{\pi x}{l} \quad \dots\dots\dots\dots (20).$$

And, since any form possible for the string at all may be regarded as initial, we infer that any finite single valued function of x, which vanishes at $x = 0$ and $x = l$, can be expanded within those limits in a series of sines of $\frac{\pi x}{l}$ and its multiples,—which is a case of Fourier's theorem. We shall presently shew how the more general form can be deduced.

121. We might now determine the constants for a continuous string by integration as in § 93, but it is instructive to solve the problem first in the general case (m finite), and afterwards to proceed to the limit. The initial conditions are

$$\psi(a) = A_1 \sin \frac{\pi a}{l} + A_2 \sin 2\frac{\pi a}{l} + \dots + A_m \sin m\frac{\pi a}{l},$$

$$\psi(2a) = A_1 \sin 2\frac{\pi a}{l} + A_2 \sin 4\frac{\pi a}{l} + \dots + A_m \sin 2m\frac{\pi a}{l},$$

$$\dots\dots\dots\dots\dots\dots\dots\dots\dots\dots\dots\dots\dots\dots\dots\dots\dots\dots\dots\dots$$

$$\psi(ma) = A_1 \sin m\frac{\pi a}{l} + A_2 \sin 2m\frac{\pi a}{l} + \dots + A_m \sin mm\frac{\pi a}{l};$$

where, for brevity, $A_s = P_s \cos \epsilon_s$, and $\psi(a)$, $\psi(2a) \dots\dots \psi(ma)$ are the initial displacements of the m particles.

To determine any constant A_s, multiply the first equation by $\sin s\frac{\pi a}{l}$, the second by $\sin 2s\frac{\pi a}{l}$, &c., and add the results. Then, by Trigonometry, the coefficients of all the constants, except A_s, vanish, while that of $A_s = \frac{1}{2}(m+1)$[1]. Hence

$$A_s = \frac{2}{m+1} \Sigma_{r=1}^{r=m} \psi(ra) \sin rs\frac{\pi a}{l} \dots\dots\dots\dots(1).$$

We need not stay here to write down the values of B_s (equal to $P_s \sin \epsilon_s$) as depending on the initial velocities. When a becomes infinitely small, ra under the sign of summation ranges by infinitesimal steps from zero to l. At the same time $\frac{1}{m+1} = \frac{a}{l}$, so that writing $ra = x$, $a = dx$, we have ultimately

$$A_s = \frac{2}{l}\int_0^l \psi(x) \sin\left(\frac{s\pi x}{l}\right) dx \dots\dots\dots\dots(2),$$

expressing A_s in terms of the initial displacements.

122. We will now investigate independently the partial differential equation governing the transverse motion of a perfectly flexible string, on the suppositions (1) that the magnitude of the tension may be considered constant, (2) that the square of the inclination of any part of the string to its initial direction may be neglected. As before, ρ denotes the linear density at any point, and T_1 is the constant tension. Let rectangular co-ordinates be taken parallel, and perpendicular to the string, so that x gives the equilibrium and x, y, z the displaced position of any particle at time t. The forces acting on the element dx are the tensions at its two ends, and any impressed forces $Y\rho\, dx$, $Z\rho\, dx$. By D'Alembert's Prin-

[1] Todhunter's *Int. Calc.*, p. 267.

ciple these form an equilibrating system with the reactions against acceleration, $-\rho \dfrac{d^2y}{dt^2}$, $-\rho \dfrac{d^2z}{dt^2}$. At the point x the components of tension are

$$T_1 \frac{dy}{dx}, \quad T_1 \frac{dz}{dx},$$

if the squares of $\dfrac{dy}{dx}$, $\dfrac{dz}{dx}$ be neglected; so that the forces acting on the element dx arising out of the tension are

$$T_1 \frac{d}{dx}\left(\frac{dy}{dx}\right) dx, \quad T_1 \frac{d}{dx}\left(\frac{dz}{dx}\right) dx.$$

Hence for the equations of motion,

$$\left.\begin{aligned} \frac{d^2y}{dt^2} &= \frac{T_1}{\rho} \frac{d^2y}{dx^2} + Y \\ \frac{d^2z}{dt^2} &= \frac{T_1}{\rho} \frac{d^2z}{dx^2} + Z \end{aligned}\right\} \dots\dots\dots\dots\dots\dots\dots(1),$$

from which it appears that the dependent variables y and z are altogether independent of one another.

The student should compare these equations with the corresponding equations of finite differences in § 120. The latter may be written

$$\mu \frac{d^2}{dt^2} \psi(x) = \frac{T_1}{a} \{\psi(x-a) + \psi(x+a) - 2\psi(x)\}.$$

Now in the limit, when a becomes infinitely small,

$$\psi(x-a) + \psi(x+a) - 2\psi(x) = \psi''(x)\, a^2,$$

while $\mu = \rho a$; and the equation assumes ultimately the form

$$\frac{d^2}{dt^2} \psi(x) = \frac{T_1}{\rho} \frac{d^2}{dx^2} \psi(x),$$

agreeing with (1).

In like manner the limiting forms of (3) and (4) of § 120 are

$$T = \tfrac{1}{2} \int \rho \left(\frac{dy}{dt}\right)^2 dx \dots\dots\dots\dots\dots\dots(2),$$

$$V = \tfrac{1}{2} T_1 \int \left(\frac{dy}{dx}\right)^2 dx \dots\dots\dots\dots\dots\dots(3),$$

which may also be proved directly.

The first is obvious from the definition of T. To prove the second, it is sufficient to notice that the potential energy in any configuration is the work required to produce the necessary stretching against the tension T_1. Reckoning from the configuration of equilibrium, we have

$$V = T_1 \int \left(\frac{ds}{dx} - 1 \right) dx ;$$

and, so far as the third power of $\frac{dy}{dx}$,

$$\frac{ds}{dx} - 1 = \tfrac{1}{2} \left(\frac{dy}{dx} \right)^2.$$

123.   In most of the applications that we shall have to make, the density ρ is constant, there are no impressed forces, and the motion may be supposed to take place in one plane. We may then conveniently write

$$\frac{T_1}{\rho} = a^2 \ \dotfill \ (1),$$

and the differential equation is expressed by

$$\frac{d^2y}{d(at)^2} = \frac{d^2y}{dx^2} \ \dotfill (2).$$

If we now assume that y varies as $\cos mat$, our equation becomes

$$\frac{d^2y}{dx^2} + m^2 y = 0 \ \dotfill (3),$$

of which the most general solution is

$$y = (A \sin mx + C \cos mx) \cos mat \dotfill (4).$$

This, however, is not the most general harmonic motion of the period in question. In order to obtain the latter, we must assume

$$y = y_1 \cos mat + y_2 \sin mat \ \dotfill (5),$$

where y_1, y_2 are functions of x, not necessarily the same. On substitution in (2) it appears that y_1 and y_2 are subject to equations of the form (3), so that finally

$$\left. \begin{array}{l} y = (A \sin mx + C \cos mx) \cos mat \\ \quad + (B \sin mx + D \cos mx) \sin mat \end{array} \right\} \ \dotfill (6),$$

an expression containing four arbitrary constants. For any continuous length of string satisfying without interruption the differential equation, this is the most general solution possible, under the condition that the motion at every point shall be simple harmonic. But whenever the string forms part of a system vibrating freely and without dissipation, we know from former chapters that all parts are simultaneously in the same phase, which requires that

$$A : B = C : D \quad\text{.........................(7)};$$

and then the most general vibration of simple harmonic type is

$$y = \{\alpha \sin mx + \beta \cos mx\} \cos (mat - \epsilon) \quad\text{......... (8)}.$$

124. The most simple as well as the most important problem connected with our present subject is the investigation of the free vibrations of a finite string of length l held fast at both its ends. If we take the origin of x at one end, the terminal conditions are that when $x = 0$, and when $x = l$, y vanishes for all values of t. The first requires that in (6) of § 123

$$C = 0, \quad D = 0 \quad\text{............................(1)};$$

and the second that

$$\sin ml = 0 \quad\text{............................(2)},$$

or that $ml = s\pi$, where s is an integer. We learn that the only harmonic vibrations possible are such as make

$$m = \frac{s\pi}{l} \quad\text{...............................(3)},$$

and then

$$y = \sin \frac{s\pi x}{l} \left(A \cos \frac{s\pi at}{l} + B \sin \frac{s\pi at}{l} \right) \text{............(4)}.$$

Now we know *a priori* that whatever the motion may be, it can be represented as a sum of simple harmonic vibrations, and we therefore conclude that the most general solution for a string, fixed at 0 and l, is

$$y = \sum_{s=1}^{s=\infty} \sin \frac{s\pi x}{l} \left(A_s \cos \frac{s\pi at}{l} + B_s \sin \frac{s\pi at}{l} \right) \text{.........(5)}.$$

The slowest vibration is that corresponding to $s = 1$. Its period (τ_1) is given by

$$\tau_1 = \frac{2l}{a} = 2l \sqrt{\frac{\rho}{T_1}} \quad\text{.....................(6)}.$$

The other components have periods which are aliquot parts of τ_1:—

$$\tau_s = \tau_1 \div s \quad\text{...........................(7)};$$

so that, as has been already stated, the whole motion is under all circumstances periodic in the time τ_1. The sound emitted constitutes in general a musical *note*, according to our definition of that term, whose pitch is fixed by τ_1, the period of its gravest component. It may happen, however, in special cases that the gravest vibration is absent, and yet that the whole motion is not periodic in any shorter time. This condition of things occurs, if $A_1^2 + B_1^2$ vanish, while, for example, $A_2^2 + B_2^2$ and $A_3^2 + B_3^2$ are finite. In such cases the sound could hardly be called a note; but it usually happens in practice that, when the gravest tone is absent, some other takes its place in the character of fundamental, and the sound still constitutes a note in the ordinary sense, though, of course, of elevated pitch. A simple case is when all the odd components beginning with the first are missing. The whole motion is then periodic in the time $\frac{1}{2}\tau_1$, and if the second component be present, the sound presents nothing unusual.

The pitch of the note yielded by a string (6), and the character of the fundamental vibration, were first investigated on mechanical principles by Brook Taylor in 1715; but it is to Daniel Bernoulli (1755) that we owe the general solution contained in (5). He obtained it, as we have done, by the synthesis of particular solutions, permissible in accordance with his Principle of the Co-existence of Small Motions. In his time the generality of the result so arrived at was open to question; in fact, it was the opinion of Euler, and also, strangely enough, of Lagrange[1], that the series of sines in (5) was not capable of representing an arbitrary function; and Bernoulli's argument on the other side, drawn from the infinite number of the disposable constants, was certainly inadequate[2].

Most of the laws embodied in Taylor's formula (6) had been discovered experimentally long before (1636) by Mersenne. They may be stated thus:—

---

[1] See Riemann's *Partielle Differential Gleichungen*, § 78.

[2] Dr Young, in his memoir of 1800, seems to have understood this matter quite correctly. He says, "At the same time, as M. Bernoulli has justly observed, since every figure may be infinitely approximated, by considering its ordinates as composed of the ordinates of an infinite number of trochoids of different magnitudes, it may be demonstrated that all these constituent curves would revert to their initial state, in the same time that a similar chord bent into a trochoidal curve would perform a single vibration: and this is in some respects a convenient and compendious method of considering the problem."

(1) For a given string and a given tension, the time varies as the length.

This is the fundamental principle of the monochord, and appears to have been understood by the ancients[1].

(2) When the length of the string is given, the time varies inversely as the square root of the tension.

(3) Strings of the same length and tension vibrate in times, which are proportional to the square roots of the linear density.

These important results may all be obtained by the method of dimensions, if it be assumed that τ depends only on l, ρ, and T_1.

For, if the units of length, time and mass be denoted respectively by $[L]$, $[T_1]$, $[M]$, the dimensions of these symbols are given by

$$l = [L], \quad \rho = [ML^{-1}], \quad T_1 = [MLT^{-2}],$$

and thus (see § 52) the only combination of them capable of representing a time is $T_1^{-\frac{1}{2}} \cdot \rho^{\frac{1}{2}} \cdot l$. The only thing left undetermined is the numerical factor.

125. Mersenne's laws are exemplified in all stringed instruments. In playing the violin different notes are obtained from the same string by shortening its efficient length. In tuning the violin or the pianoforte, an adjustment of pitch is effected with a constant length by varying the tension; but it must be remembered that ρ is not quite invariable.

To secure a prescribed pitch with a string of given material, it is requisite that one relation only be satisfied between the length, the thickness, and the tension; but in practice there is usually no great latitude. The length is often limited by considerations of convenience, and its curtailment cannot always be compensated by an increase of thickness, because, if the tension be not increased proportionally to the section, there is a loss of flexibility, while if the tension be so increased, nothing is effected towards lowering the pitch. The difficulty is avoided in the lower strings of the pianoforte and violin by the addition of a coil of fine wire, whose effect is to impart inertia without too much impairing flexibility.

---

[1] Aristotle "knew that a pipe or a chord of double length produced a sound of which the vibrations occupied a double time; and that the properties of concords depended on the proportions of the times occupied by the vibrations of the separate sounds."—Young's *Lectures on Natural Philosophy*, Vol. I. p. 404.

For quantitative investigations into the laws of strings, the sonometer is employed. By means of a weight hanging over a pulley, a catgut, or a metallic wire, is stretched across two bridges mounted on a resonance case. A moveable bridge, whose position is estimated by a scale running parallel to the wire, gives the means of shortening the efficient portion of the wire to any desired extent. The vibrations may be excited by plucking, as in the harp, or with a bow (well supplied with rosin), as in the violin.

If the moveable bridge be placed half-way between the fixed ones, the note is raised an octave; when the string is reduced to one-third, the note obtained is the twelfth.

By means of the law of lengths, Mersenne determined for the first time the frequencies of known musical notes. He adjusted the length of a string until its note was one of assured position in the musical scale, and then prolonged it under the same tension until the vibrations were slow enough to be counted.

For experimental purposes it is convenient to have two, or more, strings mounted side by side, and to vary in turn their lengths, their masses, and the tensions to which they are subjected. Thus in order that two strings of equal length may yield the interval of the octave, their tensions must be in the ratio of 1 : 4, if the masses be the same; or, if the tensions be the same, the masses must be in the reciprocal ratio.

The sonometer is very useful for the numerical determination of pitch. By varying the tension, the string is tuned to unison with a fork, or other standard of known frequency, and then by adjustment of the moveable bridge, the length of the string is determined, which vibrates in unison with any note proposed for measurement. The law of lengths then gives the means of effecting the desired comparison of frequencies.

Another application by Scheibler to the determination of absolute pitch is important. The principle is the same as that explained in Chapter III., and the method depends on deducing the absolute pitch of two notes from a knowledge of both the *ratio* and the *difference* of their frequencies. The lengths of the sonometer string when in unison with a fork, and when giving with it four beats per second, are carefully measured. The ratio of the lengths is the inverse ratio of the frequencies, and the difference of the frequencies is four. From these data the absolute pitch of the fork can be calculated.

The pitch of a string may be calculated also by Taylor's formula from the mechanical elements of the system, but great precautions are necessary to secure accuracy. The tension is produced by a weight, whose mass (expressed with the same unit as ρ) may be called P; so that $T_1 = gP$, where $g = 32\cdot2$, if the units of length and time be the foot and the second. In order to secure that the whole tension acts on the vibrating segment, no bridge must be interposed, a condition only to be satisfied by suspending the string vertically. After the weight is attached, a portion of the string is isolated by clamping it firmly at two points, and the length is measured. The mass of the unit of length ρ refers to the stretched state of the string, and may be found indirectly by observing the elongation due to a tension of the same order of magnitude as T_1, and calculating what would be produced by T_1 according to Hooke's law, and by weighing a known length of the string in its normal state. After the clamps have been secured great care is required to avoid fluctuations of temperature, which would seriously influence the tension. In this way Seebeck obtained very accurate results.

126. When a string vibrates in its gravest normal mode, the excursion is at any moment proportional to $\sin\dfrac{\pi x}{l}$, increasing numerically from either end towards the centre; no intermediate point of the string remains permanently at rest. But it is otherwise in the case of the higher normal components. Thus, if the vibration be of the mode expressed by

$$y = \sin\frac{s\pi x}{l}\left(A_s\cos\frac{s\pi at}{l} + B_s\sin\frac{s\pi at}{l}\right),$$

the excursion is proportional to $\sin\dfrac{s\pi x}{l}$, which vanishes at $s-1$ points, dividing the string into s equal parts. These points of no motion are called nodes, and may evidently be touched or held fast without in any way disturbing the vibration. The production of 'harmonics' by lightly touching the string at the points of aliquot division is a well-known resource of the violinist. All component modes are excluded which have not a node at the point touched; so that, as regards pitch, the effect is the same as if the string were securely fastened there.

127. The constants, which occur in the general value of y, § 124, depend on the special circumstances of the vibration, and may be expressed in terms of the initial values of y and $\dot{y}$.

Putting $t = 0$, we find

$$y_0 = \Sigma_{s=1}^{s=\infty} A_s \sin \frac{s\pi x}{l}; \quad \dot{y}_0 = \frac{\pi a}{l} \Sigma_{s=1}^{s=\infty} s B_s \sin \frac{s\pi x}{l} \ldots\ldots(1).$$

Multiplying by $\sin \frac{s\pi x}{l}$, and integrating from 0 to l, we obtain

$$A_s = \frac{2}{l} \int_0^l y_0 \sin \frac{s\pi x}{l} dx; \quad B_s = \frac{2}{\pi a s} \int_0^l \dot{y}_0 \sin \frac{s\pi x}{l} dx\ldots..(2).$$

These results exemplify Stokes' law, § 95; for that part of y, which depends on the initial velocities, is

$$y = \Sigma_{s=1}^{s=\infty} \frac{2}{\pi a s} \sin \frac{s\pi x}{l} \sin \frac{s\pi a t}{l} \int_0^l \dot{y}_0 \sin \frac{s\pi x}{l} dx,$$

and from this the part depending on initial displacements may be inferred, by differentiating with respect to the time, and substituting y_0 for $\dot{y}_0$.

When the condition of the string at some one moment is thoroughly known, these formulæ allow us to calculate the motion for all subsequent time. For example, let the string be initially at rest, and so displaced that it forms two sides of a triangle. Then $B_s = 0$; and

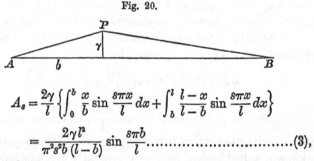

Fig. 20.

$$A_s = \frac{2\gamma}{l} \left\{ \int_0^b \frac{x}{b} \sin \frac{s\pi x}{l} dx + \int_b^l \frac{l-x}{l-b} \sin \frac{s\pi x}{l} dx \right\}$$

$$= \frac{2\gamma l^2}{\pi^2 s^2 b (l-b)} \sin \frac{s\pi b}{l} \ldots\ldots\ldots\ldots\ldots\ldots\ldots\ldots\ldots\ldots(3),$$

on integration.

We see that A_s vanishes, if $\sin \frac{s\pi b}{l} = 0$, that is, if there be a node of the component in question situated at P. A more comprehensive view of the subject will be afforded by another mode of solution to be given presently.

128. In the expression for y the coefficients of $\sin\frac{s\pi x}{l}$ are the normal co-ordinates of Chapters IV. and V. We will denote them therefore by ϕ_s, so that the configuration and motion of the system at any instant are defined by the values of ϕ_s and $\dot\phi_s$ according to the equations

$$\left.\begin{array}{l} y = \phi_1\sin\frac{\pi x}{l} + \phi_2\sin\frac{2\pi x}{l} + \dots + \phi_s\sin\frac{s\pi x}{l} + \dots \\ \dot y = \dot\phi_1\sin\frac{\pi x}{l} + \dot\phi_2\sin\frac{2\pi x}{l} + \dots + \dot\phi_s\sin\frac{s\pi x}{l} + \dots \end{array}\right\}\dots(1).$$

We proceed to form the expressions for T and V, and thence to deduce the normal equations of vibration.

For the kinetic energy,

$$T = \tfrac12\rho\int_0^l \dot y^2 dx = \tfrac12\rho\int_0^l \left\{\Sigma_{s=1}^{s=\infty}\dot\phi_s\sin\frac{s\pi x}{l}\right\}^2 dx$$

$$= \tfrac12\rho\int_0^l \Sigma_{s=1}^{s=\infty}\dot\phi_s^2\sin^2\frac{s\pi x}{l}\,dx,$$

the product of every pair of terms vanishing by the general property of normal co-ordinates. Hence

$$T = \tfrac14\rho l\,\Sigma_{s=1}^{s=\infty}\dot\phi_s^2 \dots(2).$$

In like manner,

$$V = \tfrac12 T_1\int_0^l\left(\frac{dy}{dx}\right)^2 dx = \tfrac12 T_1\int_0^l\left\{\Sigma_{s=1}^{s=\infty}\phi_s\frac{s\pi}{l}\cos\frac{s\pi x}{l}\right\}^2 dx$$

$$= \tfrac14 T_1 l.\Sigma_{s=1}^{s=\infty}\frac{s^2\pi^2}{l^2}\phi_s^2 \dots(3).$$

These expressions do not presuppose any particular motion, either natural, or otherwise; but we may apply them to calculate the whole energy of a string vibrating naturally, as follows:—If M be the whole mass of the string (ρl), and its equivalent $(a^2\rho)$ be substituted for T_1, we find for the sum of the energies,

$$T+V = \tfrac14 M.\Sigma_{s=1}^{s=\infty}\left\{\dot\phi_s^2 + \frac{s^2\pi^2 a^2}{l^2}\phi_s^2\right\}\dots(4),$$

or, in terms of A_s and B_s of § 126,

$$T+V = \pi^2 M.\Sigma_{s=1}^{s=\infty}\frac{A_s^2+B_s^2}{\tau_s^2}\dots(5).$$

If the motion be not confined to the plane of xy, we have merely to add the energy of the vibrations in the perpendicular plane.

Lagrange's method gives immediately the equation of motion

$$\ddot{\phi}_s + \left(\frac{s\pi a}{l}\right)^2 \phi_s = \frac{2}{l\rho} \Phi_s \dots\dots\dots\dots\dots (6),$$

which has been already considered in § 66. If ϕ_0 and $\dot{\phi}_0$ be the initial values of ϕ and $\dot{\phi}$, the general solution is

$$\phi = \dot{\phi}_0 \frac{\sin nt}{n} + \phi_0 \cos nt$$

$$+ \frac{2}{l\rho n} \int_0^t \sin n\,(t - t')\, \Phi\, dt' \dots\dots\dots\dots \dots(7),$$

where n is written for $\frac{s\pi a}{l}$.

By definition Φ_s is such that $\Phi_s\, \delta\phi_s$ represents the work done by the impressed forces on the displacement $\delta\phi_s$. Hence, if the force acting at time t on an element of the string $\rho\, dx$ be $\rho Y dx$,

$$\Phi_s = \int_0^l \rho Y \sin\frac{s\pi x}{l}\, dx \dots\dots\dots\dots\dots\dots(8).$$

In these equations ϕ_s is a linear quantity, as we see from (1); and Φ_s is therefore a force of the ordinary kind.

129. In the applications that we have to make, the only impressed force will be supposed to act in the immediate neighbourhood of one point $x = b$, and may usually be reckoned as a whole, so that

$$\Phi_s = \sin\frac{s\pi b}{l} \int \rho Y dx \dots\dots\dots\dots\dots\dots(1).$$

If the point of application of the force coincide with a node of the mode (s), $\Phi_s = 0$, and we learn that the force is altogether without influence on the component in question. This principle is of great importance; it shews, for example, that if a string be at rest in its position of equilibrium, no force applied at its centre, whether in the form of plucking, striking, or bowing, can generate any of the even normal components[1]. If after the operation of the force, its point of application be damped, as by touching it

---

[1] The observation that a harmonic is not generated, when one of its nodal points is plucked, is due to Young.

with the finger, all motion must forthwith cease; for those components which have not a node at the point in question are stopped by the damping, and those which have, are absent from the beginning[1]. More generally, by damping any point of a sounding string, we stop all the component vibrations which have not, and leave entirely unaffected those which have a node at the point touched.

The case of a string pulled aside at one point and afterwards let go from rest may be regarded as included in the preceding statements. The complete solution may be obtained thus. Let the motion commence at the time $t = 0$; from which moment $\Phi_s = 0$. The value of ϕ_s at time t is

$$\phi_s = (\phi_s)_0 \cos nt + \frac{1}{n} (\dot{\phi}_s)_0 \sin nt \dots\dots\dots\dots(2),$$

where $(\phi_s)_0$, $(\dot{\phi}_s)_0$ denote the initial values of the quantities affected with the suffix s. Now in the problem in hand $(\dot{\phi}_s)_0 = 0$, and $(\phi_s)_0$ is determined by

$$n^2 (\phi_s)_0 = \frac{2}{l\rho} \Phi_s = \frac{2}{l\rho} Y' \sin \frac{s\pi b}{l} \dots\dots\dots\dots(3),$$

if Y' denote the force with which the string is held aside at the point b. Hence at time t

$$\phi_s = \frac{2}{l\rho n^2} Y' \sin \frac{s\pi b}{l} \cos nt \dots\dots\dots\dots(4),$$

and by (1) of § 128

$$y = \frac{2}{l\rho} Y' . \Sigma_{s=1}^{s=\infty} \sin \frac{s\pi b}{l} \sin \frac{s\pi x}{l} \frac{\cos nt}{n^2} \dots\dots\dots(5),$$

where $n = s\pi a : l$.

The symmetry of the expression (5) in x and b is an example of the principle of § 107.

The problem of determining the subsequent motion of a string set into vibration by an impulse acting at the point b, may be treated in a similar manner. Integrating (6) of § 128 over the duration of the impulse, we find ultimately, with the same notation as before,

$$(\dot{\phi}_s)_0 = \frac{2}{l\rho} \sin \frac{s\pi b}{l} Y_1,$$

---

[1] A like result ensues when the point which is damped is at the same distance from one end of the string as the point of excitation is from the other end.

if $\int Y' dt$ be denoted by Y_1. At the same time $(\phi_s)_0 = 0$, so that by (2) at time t

$$y = \frac{2 Y_1}{l\rho} \sum_{s=1}^{s=\infty} \sin\frac{s\pi b}{l} \sin\frac{s\pi x}{l} \frac{\sin nt}{n} \dots\dots\dots(6).$$

The series of component vibrations is less convergent for a struck than for a plucked string, as the preceding expressions shew. The reason is that in the latter case the initial value of y is continuous, and only $\frac{dy}{dx}$ discontinuous, while in the former it is $\dot{y}$ itself that makes a sudden spring. See §§ 32, 101.

The problem of a string set in motion by an impulse may also be solved by the general formulæ (7) and (8) of § 128. The force finds the string at rest at $t = 0$, and acts for an infinitely short time from $t = 0$ to $t = \tau'$. Thus $(\phi_s)_0$ and $(\dot{\phi}_s)_0$ vanish, and (7) of § 128 reduces to

$$\phi_s = \frac{2}{l\rho n} \sin nt \int_0^{\tau'} \Phi_s dt',$$

while by (8) of § 128

$$\int_0^{\tau'} \Phi_s dt' = \sin\frac{s\pi b}{l} \int_0^{\tau'} Y' dt' = \sin\frac{s\pi b}{l} Y_1.$$

Hence, as before,

$$\phi_s = \frac{2}{l\rho n} Y_1 \sin\frac{s\pi b}{l} \sin nt \dots\dots\dots\dots(7).$$

Hitherto we have supposed the disturbing force to be concentrated at a single point. If it be distributed over a distance β on either side of b, we have only to integrate the expressions (6) and (7) with respect to b, substituting, for example, in (7) in place of $Y_1 \sin\frac{s\pi b}{l}$,

$$\int_{b-\beta}^{b+\beta} Y_1' \sin\frac{s\pi b}{l} db.$$

If Y_1' be constant between the limits, this reduces to

$$Y_1' \frac{2l}{s\pi} \sin\frac{s\pi\beta}{l} \sin\frac{s\pi b}{l} \dots\dots\dots\dots(8).$$

The principal effect of the distribution of the force is to render the series for y more convergent.

130. The problem which will next engage our attention is that of the pianoforte wire. The cause of the vibration is here the blow of a hammer, which is projected against the string, and after the impact rebounds. But we should not be justified in assuming, as in the last section, that the mutual action occupies so short a time that its duration may be neglected. Measured by the standards of ordinary life the duration of the contact is indeed very small, but here the proper comparison is with the natural periods of the string. Now the hammers used to strike the wires of a pianoforte are covered with several layers of cloth for the express purpose of making them more yielding, with the effect of prolonging the contact. The rigorous treatment of the problem would be difficult, and the solution, when obtained, probably too complicated to be of use; but by introducing a certain simplification Helmholtz has obtained a solution representing all the essential features of the case. He remarks that since the actual yielding of the string must be slight in comparison with that of the covering of the hammer, the law of the force called into play during the contact must be nearly the same as if the string were absolutely fixed, in which case the force would vary very nearly as a circular function. We shall therefore suppose that at the time $t = 0$, when there are neither velocities nor displacements, a force $F \sin pt$ begins to act on the string at $x = b$, and continues through half a period of the circular function, that is, until $t = \pi \div p$, after which the string is once more free. The magnitude of p will depend on the mass and elasticity of the hammer, but not to any great extent on the velocity with which it strikes the string.

The required solution is at once obtained by substituting for Φ_s in the general formula (7) of § 128 its value given by

$$\Phi_s = F \sin \frac{s\pi b}{l} \sin pt' \dots\dots\dots\dots\dots (1),$$

the range of the integration being from 0 to $\dfrac{\pi}{p}$. We find $\left(t > \dfrac{\pi}{p}\right)$

$$\phi_s = \frac{2F}{ln\rho} \sin \frac{s\pi b}{l} \int_0^{\frac{\pi}{p}} \sin n\,(t-t') \sin pt'\, dt'$$

$$= \frac{4p \cos \dfrac{n\pi}{2p}}{l\rho n\,(p^2 - n^2)} \cdot F \sin \frac{s\pi b}{l} \cdot \sin n\left(t - \frac{\pi}{2p}\right) \dots\dots\dots (2),$$

and the final solution for y becomes, if we substitute for n and ρ their values,

$$y = \frac{4apl^2F}{\pi T_1} \sum_{s=1}^{s=\infty} \frac{\cos \frac{s\pi^2 a}{2pl} \cdot \sin \frac{s\pi b}{l}}{s\left(l^2p^2 - s^2a^2\pi^2\right)} \sin \frac{s\pi x}{l} \sin \frac{s\pi a}{l}\left(t - \frac{\pi}{2p}\right)..(3).$$

We see that all components vanish which have a node at the point of excitement, but this conclusion does not depend on any particular law of force. The interest of the present solution lies in the information that may be elicited from it as to the dependence of the resulting vibrations on the duration of contact. If we denote the ratio of this quantity to the fundamental period of the string by ν, so that $\nu = \pi a : 2pl$, the expression for the amplitude of the component s is

$$\frac{8Fl}{\pi^2 T_1} \cdot \frac{\nu \cos (s\pi\nu)}{s\left(1 - 4s^2\nu^2\right)} \sin \frac{s\pi b}{l} .... (4).$$

We fall back on the case of an impulse by putting $\nu = 0$, and

$$Y_1 = \int_0^{\frac{\pi}{p}} F \sin pt \, dt = \frac{2F}{p}.$$

When ν is finite, those components disappear, whose periods are $\frac{2}{3}$, $\frac{2}{5}$, $\frac{2}{7}$, ... of the duration of contact; and when s is very great, the series converges with s^{-3}. Some allowance must also be made for the finite breadth of the hammer, the effect of which will also be to favour the convergence of the series.

The laws of the vibration of strings may be verified, at least in their main features, by optical methods of observation—either with the vibration-microscope, or by a tracing point recording the character of the vibration on a revolving drum. This character depends on two things,—the mode of excitement, and the point whose motion is selected for observation. Those components do not appear which have nodes either at the point of excitement, or at the point of observation. The former are not generated, and the latter do not manifest themselves. Thus the simplest motion is obtained by plucking the string at the centre, and observing one of the points of trisection, or *vice versa*. In this case the first harmonic which contaminates the purity of the principal vibration is the fifth component, whose intensity is usually insufficient to produce much disturbance. In a future chapter we shall compare the results of the dynamical theory with aural observation, but rather with the view of discovering and testing the laws of hearing, than of confirming the theory itself.

131. The case of a periodic force is included in the general solution of § 128, but we prefer to follow a somewhat different method, in order to make an extension in another direction. We have hitherto taken no account of dissipative forces, but we will now suppose that the motion of each element of the string is resisted by a force proportional to its velocity. The partial differential equation becomes

$$\frac{d^2y}{dt^2} + \kappa \frac{dy}{dt} = a^2 \frac{d^2y}{dx^2} + Y \dots\dots\dots\dots (1),$$

by means of which the subject may be treated. But it is still simpler to avail ourselves of the results of the last chapter, remarking that in the present case the friction-function F is of the same form as T. In fact

$$F = \tfrac{1}{4}\kappa l . \Sigma_{s=1}^{s=\infty} \dot{\phi}_s{}^2 \dots\dots\dots\dots\dots (2),$$

where $\phi_1, \phi_2, \dots$ are the normal co-ordinates, by means of which T and V are reduced to sums of squares. The equations of motion are therefore simply

$$\ddot{\phi}_s + \kappa \dot{\phi}_s + n^2\phi_s = \frac{2}{l\rho} \Phi_s \dots\dots\dots\dots\dots (3),$$

of the same form as obtains for systems with but one degree of freedom. It is only necessary to add to what was said in Chapter III., that since κ is independent of s, the natural vibrations subside in such a manner that the amplitudes maintain their relative values.

If a periodic force $F\cos pt$ act at a single point, we have

$$\Phi_s = F \sin \frac{s\pi b}{l} \cos pt \dots\dots\dots\dots\dots (4),$$

and § 46

$$\phi_s = \frac{2F \sin \epsilon}{l\rho\, p\kappa} \sin \frac{s\pi b}{l} \cos (pt - \epsilon) \dots\dots\dots\dots (5),$$

where

$$\tan \epsilon = \frac{p\kappa}{n^2 - p^2} \dots\dots\dots\dots\dots\dots (6).$$

If among the natural vibrations there be any one nearly isochronous with $\cos pt$, then a large vibration of that type will be forced, unless indeed the point of excitement should happen to

fall near a node. In the case of exact coincidence, the component
vibration in question vanishes; for no force applied at a node can
generate it, under the present law of friction, which however, it
may be remarked, is very special in character. If there be no
friction, $\kappa = 0$, and

$$l\rho\phi_s = \frac{2F}{n^2 - p^2} \sin \frac{s\pi b}{l} \cos pt \dots\dots\dots\dots(7),$$

which would make the vibration infinite, in the case of perfect
isochronism, unless $\sin \frac{s\pi b}{l} = 0$.

The value of y is here, as usual,

$$y = \phi_1 \sin \frac{\pi x}{l} + \phi_2 \sin \frac{2\pi x}{l} + \phi_3 \sin \frac{3\pi x}{l} + \dots\dots\dots(8).$$

132. The preceding solution is an example of the use of
normal co-ordinates in a problem of forced vibrations. It is of
course to free vibrations that they are more especially applicable,
and they may generally be used with advantage throughout,
whenever the system after the operation of various forces is
ultimately left to itself. Of this application we have already had
examples.

In the case of vibrations due to periodic forces, one advantage
of the use of normal co-ordinates is the facility of comparison with
the *equilibrium theory*, which it will be remembered is the theory
of the motion on the supposition that the inertia of the system
may be left out of account. If the value of the normal co-ordinate ϕ_s on the equilibrium theory be $A_s \cos pt$, then the actual
value will be given by the equation

$$\phi_s = \frac{n^2 A_s}{n^2 - p^2} \cos pt\dots\dots\dots\dots\dots(1),$$

so that, when the result of the equilibrium theory is known and
can readily be expressed in terms of the normal co-ordinates, the
true solution with the effects of inertia included can at once be
written down.

In the present instance, if a force $F \cos pt$ of very long period
act at the point b of the string, the result of the equilibrium
theory, in accordance with which the string would at any moment
consist of two straight portions, will be

$$l\rho\phi_s = \frac{2F}{n^2} \sin \frac{s\pi b}{l} \cos pt \dots\dots\dots\dots(2),$$

from which the actual result for all values of p is derived by simply writing $n^2 - p^2$ in place of n^2.

The value of y in this and similar cases may however be expressed in finite terms, and the difficulty of obtaining the finite expression is usually no greater than that of finding the form of the normal functions when the system is free. Thus in the equation of motion

$$\frac{d^2y}{dt^2} = a^2 \frac{d^2y}{dx^2} + Y,$$

suppose that Y varies as $\cos mat$. The forced vibration will then satisfy

$$\frac{d^2y}{dx^2} + m^2y = -\frac{1}{a^2} Y \dots\dots\dots\dots\dots(3).$$

If $Y = 0$, the investigation of the normal functions requires the solution of

$$\frac{d^2y}{dx^2} + m^2y = 0,$$

and a subsequent determination of m to suit the boundary conditions. In the problem of forced vibrations m is given, and we have only to supplement any particular solution of (3) with the complementary function containing two arbitrary constants. This function, apart from the value of m and the ratio of the constants, is of the same form as the normal functions; and all that remains to be effected is the determination of the two constants in accordance with the prescribed boundary conditions which the complete solution must satisfy. Similar considerations apply in the case of any continuous system.

133. If a periodic force be applied at a single point, there are two distinct problems to be considered; the first, when at the point $x = b$, a given periodic force acts; the second, when it is the actual motion of the point b that is obligatory. But it will be convenient to treat them together.

The usual differential equation

$$\frac{d^2y}{dt^2} + \kappa \frac{dy}{dt} = a^2 \frac{d^2y}{dx^2} \dots\dots\dots\dots\dots(1),$$

is satisfied over both the parts into which the string is divided at b, but is violated in crossing from one to the other.

In order to allow for a change in the arbitrary constants, we must therefore assume distinct expressions for y, and afterwards introduce the two conditions which must be satisfied at the point of junction. These are

(1) That there is no discontinuous change in the value of y;

(2) That the resultant of the tensions acting at b balances the impressed force.

Thus, if $F \cos pt$ be the force, the second condition gives

$$T_1 \Delta \left(\frac{dy}{dx}\right) + F \cos pt = 0 \dots\dots\dots\dots\dots\dots(2),$$

where $\Delta \left(\frac{dy}{dx}\right)$ denotes the alteration in the value of $\frac{dy}{dx}$ incurred in crossing the point $x = b$ in the positive direction.

We shall, however, find it advantageous to replace $\cos pt$ by the complex exponential e^{ipt}, and finally discard the imaginary part, when the symbolical solution is completed. On the assumption that y varies as e^{ipt}, the differential equation becomes

$$\frac{d^2y}{dx^2} + \lambda^2 y = 0\dots\dots\dots\dots\dots\dots\dots(3);$$

where λ^2 is the complex constant,

$$\lambda^2 = \frac{1}{a^2}(p^2 - i p\kappa)\dots\dots\dots\dots\dots\dots(4).$$

The most general solution of (3) consists of two terms, proportional respectively to $\sin \lambda x$, and $\cos \lambda x$; but the condition to be satisfied at $x = 0$, shews that the second does not occur here. Hence if γe^{ipt} be the value of y at $x = b$,

$$y = \gamma \frac{\sin \lambda x}{\sin \lambda b} \cdot e^{ipt} \dots\dots\dots\dots\dots\dots (5),$$

is the solution applying to the first part of the string from $x = 0$ to $x = b$. In like manner it is evident that for the second part we shall have

$$y = \gamma \frac{\sin \lambda (l - x)}{\sin \lambda (l - b)} e^{ipt} \dots\dots\dots\dots\dots\dots(6).$$

If γ be given, these equations constitute the symbolical solution of the problem; but if it be the force that be given, we require further to know the relation between it and γ.

Differentiation of (5) and (6) and substitution in the equation analogous to (2) gives

$$\gamma = \frac{F}{T_1} \frac{\sin \lambda b \, \sin \lambda (l-b)}{\lambda \sin \lambda l} \quad \dots\dots\dots\dots\dots (7).$$

Thus

$$y = \frac{F}{T_1} \frac{\sin \lambda x \, \sin \lambda (l-b)}{\lambda \sin \lambda l} e^{ipt}$$
$$\text{from } x = 0 \text{ to } x = b$$
$$y = \frac{F}{T_1} \frac{\sin \lambda (l-x) \, \sin \lambda b}{\lambda \sin \lambda l} e^{ipt}$$
$$\text{from } x = b \text{ to } x = l \qquad \dots(8)^1.$$

These equations exemplify the general law of reciprocity proved in the last chapter; for it appears that the motion at x due to the force at b is the same as would have been found at b, had the force acted at x.

In discussing the solution we will take first the case in which there is no friction. The coefficient κ is then zero; while λ is real, and equal to $p \div a$. The real part of the solution, corresponding to the force $F \cos pt$, is found by simply putting $\cos pt$ for e^{ipt} in (8), but it seems scarcely necessary to write the equations again for the sake of so small a change. The same remark applies to the forced motion given in terms of γ.

It appears that the motion becomes infinite in case the force is isochronous with one of the natural vibrations of the entire string, unless the point of application be a node; but in practice it is not easy to arrange that a string shall be subject to a force of given magnitude. Perhaps the best method would be to attach a small mass of iron, attracted periodically by an electro-magnet, whose coils are traversed by an intermittent current. But unless some means of compensation were devised, the mass would have to be very small in order to avoid its inertia introducing a new complication.

A better approximation may be obtained to the imposition of an obligatory motion. A massive fork of low pitch, excited by a bow or sustained in permanent operation by electro-magnetism, executes its vibrations in approximate independence of the re-actions of any light bodies which may be connected with it. In order therefore to subject any point of a string to an obligatory

¹ Donkin's *Acoustics*, p. 121.

transverse motion, it is only necessary to attach it to the extremity of one prong of such a fork, whose plane of vibration is perpendicular to the length of the string. This method of exhibiting the forced vibrations of a string appears to have been first used by Melde.

Another arrangement, better adapted for aural observation, has been employed by Helmholtz. The end of the stalk of a powerful tuning-fork, set into vibration with a bow, or otherwise, is pressed against the string. It is advisable to file the surface, which comes into contact with the string, into a suitable (saddle-shaped) form, the better to prevent slipping and jarring.

Referring to (5) we see that, if sin λb vanished, the motion (according to this equation) would become infinite, which may be taken to prove that in the case contemplated, the motion would really become great,—so great that corrections, previously insignificant, rise into importance. Now sin λb vanishes, when the force is isochronous with one of the natural vibrations of the first part of the string, supposed to be held fixed at 0 and b.

When a fork is placed on the string of a monochord, or other instrument properly provided with a sound-board, it is easy to find by trial the places of maximum resonance. A very slight displacement on either side entails a considerable falling off in the volume of the sound. The points thus determined divide the string into a number of equal parts, of such length that the natural note of any one of them (when fixed at both ends) is the same as the note of the fork, as may readily be verified. The important applications of resonance which Helmholtz has made to purify a simple tone from extraneous accompaniment will occupy our attention later.

134. Returning now to the general case where λ is complex, we have to extract the real parts from (5), (6), (8) of § 133. For this purpose the sines which occur as factors, must be reduced to the form Re^{ie}. Thus let

$$\sin \lambda x = R_x \, e^{i\epsilon_x} \dots\dots\dots\dots\dots\dots\dots(1),$$

with a like notation for the others. From (5) § 133 we shall thus obtain

$$y = \gamma \frac{R_x}{R_b} \cos(pt + \epsilon_x - \epsilon_b)\dots\dots\dots\dots\dots(2),$$

from $x = 0$ to $x = b$,

and from (6) § 133

$$y = \gamma \frac{R_{l-x}}{R_{l-b}} \cos (pt + \epsilon_{l-x} - \epsilon_{l-b}),$$

$$\text{from } x = b \text{ to } x = l,$$

corresponding to the obligatory motion $y = \gamma \cos pt$ at b.

By a similar process from (8) § 133, if

$$\lambda = \alpha + i\beta \dots\dots\dots\dots\dots\dots\dots (3),$$

we should obtain

$$y = \frac{F}{T_1} \frac{R_x \cdot R_{l-b}}{\sqrt{\alpha^2 + \beta^2} \cdot R_l} \cos \left(pt + \epsilon_x + \epsilon_{l-b} - \epsilon_l - \tan^{-1}\frac{\beta}{\alpha} \right)$$

$$\text{from } x = 0 \text{ to } x = b$$

$$y = \frac{F}{T_1} \frac{R_{l-x} \cdot R_b}{\sqrt{\alpha^2 + \beta^2} \cdot R_l} \cos \left(pt + \epsilon_{l-x} + \epsilon_b - \epsilon_l - \tan^{-1}\frac{\beta}{\alpha} \right)$$

$$\text{from } x = b \text{ to } x = l$$

$$\dots (4),$$

corresponding to the impressed force $F \cos pt$ at b. It remains to obtain the forms of R_x, ϵ_x, &c.

The values of α and β are determined by

$$\alpha^2 - \beta^2 = \frac{p^2}{a^2}, \quad 2\alpha\beta = -\frac{p\kappa}{a^2}\dots\dots\dots\dots\dots (5),$$

and $\quad \sin \lambda x = \sin \alpha x \cos i\beta x + \cos \alpha x \sin i\beta x$

$$= \sin \alpha x \frac{e^{\beta x} + e^{-\beta x}}{2} + i \cos \alpha x \frac{e^{\beta x} - e^{-\beta x}}{2},$$

so that

$$R_x^2 = \sin^2 \alpha x \left(\frac{e^{\beta x} + e^{-\beta x}}{2} \right)^2 + \cos^2 \alpha x \left(\frac{e^{\beta x} - e^{-\beta x}}{2} \right)^2 \dots (6),$$

$$\tan \epsilon_x = \frac{e^{\beta x} - e^{-\beta x}}{e^{\beta x} + e^{-\beta x}} \cot \alpha x \dots\dots\dots\dots\dots (7),$$

while

$$\sqrt{\alpha^2 + \beta^2} = \frac{1}{a} \sqrt[4]{p^4 + p^2\kappa^2}\dots\dots\dots\dots\dots (8).$$

This completes the solution.

If the friction be very small, the expressions may be simplified. For instance, in this case, to a sufficient approximation,

$$\alpha = \frac{p}{a}, \quad \beta = -\frac{\kappa}{2a}, \quad \sqrt{\alpha^2 + \beta^2} = \frac{p}{a},$$

$$\frac{e^{\beta x} + e^{-\beta x}}{2} = 1, \quad \frac{e^{\beta x} - e^{-\beta x}}{2} = -\frac{\kappa x}{2a};$$

so that corresponding to the obligatory motion at b $y = \gamma \cos pt$ the amplitude of the motion between $x = 0$ and $x = b$ is, approximately

$$\gamma \left\{ \frac{\sin^2 \dfrac{px}{a} + \dfrac{\kappa^2 x^2}{4a^2} \cos^2 \dfrac{px}{a}}{\sin^2 \dfrac{pb}{a} + \dfrac{\kappa^2 b^2}{4a^2} \cos^2 \dfrac{pb}{a}} \right\}^{\frac{1}{2}} \quad \dots\dots\dots\dots(9),$$

which becomes great, but not infinite, when $\sin \dfrac{pb}{a} = 0$, or the point of application is a node.

If the imposed force, or motion, be not expressed by a single harmonic term, it must first be resolved into such. The preceding solution may then be applied to each component separately, and the results added together. The extension to the case of more than one point of application of the impressed forces is also obvious. To obtain the most general solution satisfying the conditions, the expression for the natural vibrations must also be added; but these become reduced to insignificance after the motion has been in progress for a sufficient time.

The law of friction assumed in the preceding investigation is the only one whose results can be easily followed deductively, and it is sufficient to give a general idea of the effects of dissipative forces on the motion of a string. But in other respects the conclusions drawn from it possess a fictitious simplicity, depending on the fact that F—the friction function—is similar in form to T, which makes the normal co-ordinates independent of each other. In almost any other case (for example, when but a single point of the string is retarded by friction) there are no normal co-ordinates properly so called. There exist indeed elementary types of vibration into which the motion may be resolved, and which are perfectly independent, but these are essentially different in character from those with which we have been concerned hitherto, for the various parts of the system (as affected by one elementary vibration) are not simultaneously in the same phase. Special cases excepted, no linear transformation of the co-ordinates (with real coefficients) can reduce T, F, and V together to a sum of squares.

If we suppose that the string has no inertia, so that $T = 0$, F and V may then be reduced to sums of squares. This problem is of no acoustical importance, but it is interesting as being mathematically analogous to that of the conduction and radiation of heat in a bar whose ends are maintained at a constant temperature.

135. Thus far we have supposed that at two fixed points, $x = 0$ and $x = l$, the string is held at rest. Since absolute fixity cannot be attained in practice, it is not without interest to inquire in what manner the vibrations of a string are liable to be modified by a yielding of the points of attachment; and the problem will furnish occasion for one or two remarks of importance. For the sake of simplicity we shall suppose that the system is symmetrical with reference to the centre of the string, and that each extremity is attached to a mass M (treated as unextended in space), and is urged by a spring (μ) towards the position of equilibrium. If no frictional forces act, the motion is necessarily resolvable into normal vibrations. Assume

$$y = \{\alpha \sin mx + \beta \cos mx\} \cos (mat - \epsilon)\ldots\ldots\ldots(1).$$

The conditions at the ends are that

$$\left.\begin{array}{ll}\text{when} \quad x = 0, & M\ddot{y} + \mu y = \quad T_1 \dfrac{dy}{dx} \\[2mm] \text{when} \quad x = l, & M\ddot{y} + \mu y = - T_1 \dfrac{dy}{dx}\end{array}\right\} \ldots\ldots\ldots\ldots (2),$$

which give

$$\frac{\alpha}{\beta} = \frac{\beta \tan ml - \alpha}{\alpha \tan ml + \beta} = \frac{\mu - Ma^2 m^2}{mT_1} \ldots\ldots\ldots\ldots (3),$$

two equations, sufficient to determine m, and the ratio of β to α. Eliminating the latter ratio, we find

$$\tan ml = \frac{2\nu}{1 - \nu^2} \ldots\ldots\ldots\ldots\ldots\ldots (4),$$

if for brevity we write ν for $\dfrac{\mu - Ma^2 m^2}{mT_1}$.

Equation (3) has an infinite number of roots, which may be found by writing $\tan \theta$ for ν, so that $\tan ml = \tan 2\theta$, and the result of adding together *all* the corresponding particular solutions, each with its two arbitrary constants α and ϵ, is necessarily the most general solution of which the problem is capable, and is therefore adequate to represent the motion due to an arbitrary initial distribution of displacement and velocity. We infer that any function of x may be expanded between $x = 0$ and $x = l$ in a series of terms

$$\phi_1 (\nu_1 \sin m_1 x + \cos m_1 x) + \phi_2 (\nu_2 \sin m_2 x + \cos m_2 x) + \ldots\ldots (5),$$

m_1, m_2, &c. being the roots of (3) and ν_1, ν_2, &c. the corresponding values of ν. The quantities ϕ_1, ϕ_2, &c. are the *normal* co-ordinates of the system.

From the symmetry of the system it follows that in each normal vibration the value of y is numerically the same at points equally distant from the middle of the string, for example, at the two ends, where $x = 0$ and $x = l$. Hence $\nu_s \sin m_s l + \cos m_s l = \pm 1$, as may be proved also from (4).

The kinetic energy T of the whole motion is made up of the energy of the string, and that of the masses M. Thus

$$T = \tfrac{1}{2} \rho \int_0^l \{\Sigma \dot{\phi} \, (\nu \sin mx + \cos mx)\}^2 \, dx$$

$$+ \tfrac{1}{2} M \{\dot{\phi}_1 + \dot{\phi}_2 + \ldots\}^2 + \tfrac{1}{2} M \{\dot{\phi}_1 \, (\nu_1 \sin m_1 l + \cos m_1 l) + \ldots\}^2.$$

But by the characteristic property of normal co-ordinates, terms containing their products cannot be really present in the expression for T, so that

$$\rho \int_0^l (\nu_r \sin m_r x + \cos m_r x)(\nu_s \sin m_s x + \cos m_s x) \, dx$$

$$+ M + M (\nu_r \sin m_r l + \cos m_r l)(\nu_s \sin m_s l + \cos m_s l) = 0 \ldots \ldots (6),$$

if r and s be different.

This theorem suggests how to determine the arbitrary constants, so that the series (5) may represent an arbitrary function y. Take the expression

$$\rho \int_0^l y \, (\nu_s \sin m_s x + \cos m_s x) \, dx + My_0 + My_l (\nu_s \sin m_s l + \cos m_s l) \ldots (7),$$

and substitute in it the series (5) expressing y. The result is a series of terms of the type

$$\rho \int_0^l \phi_r \, (\nu_r \sin m_r x + \cos m_r x)(\nu_s \sin m_s x + \cos m_s x) \, dx$$

$$+ M\phi_r + M\phi_r \, (\nu_r \sin m_r l + \cos m_r l)(\nu_s \sin m_s l + \cos m_s l),$$

all of which vanish by (6), except the one for which $r = s$. Hence ϕ_s is equal to the expression (7), divided by

$$\rho \int_0^l (\nu_s \sin m_s x + \cos m_s x)^2 \, dx + M + M (\nu_s \sin m_s l + \cos m_s l)^2 \ldots (8),$$

and thus the coefficients of the series are determined. If $M = 0$, even although μ be finite, the process is of course much simpler, but the unrestricted problem is instructive. So much stress is often laid on special proofs of Fourier's and Laplace's series, that the student is apt to acquire too contracted a view of the nature of those important results of analysis.

We shall now shew how Fourier's theorem in its general form can be deduced from our present investigation. Let $M = 0$; then if $\mu = \infty$, the ends of the string are fast, and the equation determining m becomes $\tan ml = 0$, or $ml = s\pi$, as we know it must be. In this case the series for y becomes

$$y = A_1 \sin \frac{\pi x}{l} + A_2 \sin \frac{2\pi x}{l} + A_3 \sin \frac{3\pi x}{l} + \ldots\ldots\ldots \quad (9),$$

which must be general enough to represent any arbitrary functions of x, vanishing at 0 and l, between those limits. But now suppose that μ is zero, M still vanishing. The ends of the string may be supposed capable of sliding on two smooth rails perpendicular to its length, and the terminal condition is the vanishing of $\frac{dy}{dx}$. The equation in m is the *same as before*; and we learn that any function y' whose rates of variation vanish at $x = 0$ and $x = l$, can be expanded in a series

$$y' = B_1 \cos \frac{\pi x}{l} + B_2 \cos \frac{2\pi x}{l} + B_3 \cos \frac{3\pi x}{l} + \ldots\ldots \quad (10).$$

This series remains unaffected when the sign of x is changed, and the first series merely changes sign without altering its numerical magnitude. If therefore y' be an even function of x, (10) represents it from $-l$ to $+l$. And in the same way, if y be an odd function of x, (9) represents it between the same limits.

Now, whatever function of x $\phi(x)$ may be, it can be divided into two parts, one of which is even, and the other odd, thus:

$$\phi(x) = \frac{\phi(x) + \phi(-x)}{2} + \frac{\phi(x) - \phi(-x)}{2};$$

so that, if $\phi(x)$ be such that $\phi(-l) = \phi(+l)$ and $\phi'(-l) = \phi'(+l)$, it can be represented between the limits $\pm l$ by the mixed series

$$A_1 \sin \frac{\pi x}{l} + B_1 \cos \frac{\pi x}{l} + A_2 \sin \frac{2\pi x}{l} + B_2 \cos \frac{2\pi x}{l} + \ldots\ldots \quad (11).$$

This series is periodic, with the period $2l$. If therefore $\phi(x)$ possess the same property, no matter what in other respects its character may be, the series is its complete equivalent. This is Fourier's theorem[1].

We now proceed to examine the effects of a slight yielding of the supports, in the case of a string whose ends are approximately fixed. The quantity ν may be great, either through μ or through M. We shall confine ourselves to the two principal cases, (1) when μ is great and M vanishes, (2) when μ vanishes and M is great.

In the first case $\qquad \nu = \dfrac{\mu}{T_1 m}$,

and the equation in m is approximately

$$\tan ml = -\frac{2}{\nu} = -\frac{2T_1 m}{\mu}.$$

Assume $ml = s\pi + x$, where x is small; then

$$x = \tan x = -\frac{2T_1 . s\pi}{\mu l} \text{ approximately,}$$

and $\qquad ml = s\pi \left(1 - \frac{2T_1}{\mu l}\right) \dots\dots\dots\dots\dots\dots(12).$

To this order of approximation the tones do not cease to form a harmonic scale, but the pitch of the whole is slightly lowered. The effect of the yielding is in fact the same as that of an increase in the length of the string in the ratio $1 : 1 + \dfrac{2T_1}{\mu l}$, as might have been anticipated.

The result is otherwise if μ vanish, while M is great. Here

$$\nu = -\frac{Ma^2 m}{T_1},$$

and $\qquad \tan ml = \dfrac{2T_1}{Ma^2 m}$ approximately.

Hence

$$ml = s\pi + \frac{2T_1 l}{Ma^2 . s\pi} \quad \dots\dots\dots\dots\dots(13).$$

The effect is thus equivalent to a decrease in l in the ratio

$$1 : 1 - \frac{2T_1 l}{Ma^2 . s^2 \pi^2},$$

[1] The best 'system' for proving Fourier's theorem from dynamical considerations is an endless chain stretched round a smooth cylinder (§ 139), or a thin re-entrant column of air enclosed in a ring-shaped tube.

and consequently there is a rise in pitch, the rise being the greater the lower the component tone. It might be thought that any kind of yielding would depress the pitch of the string, but the preceding investigation shews that this is not the case. Whether the pitch will be raised or lowered, depends on the sign of ν, and this again depends on whether the natural note of the mass M urged by the spring μ is lower or higher than that of the component vibration in question.

136. The problem of an otherwise uniform string carrying a finite load M at $x = b$ can be solved by the formulæ investigated in § 133. For, if the force $F \cos pt$ be due to the reaction against acceleration of the mass M,

$$F = \gamma p^2 M \dots\dots\dots\dots\dots\dots\dots(1),$$

which combined with equation (7) of § 133 gives, to determine the possible values of λ (or $p : a$),

$$a^2 M \lambda \sin \lambda b \, \sin \lambda (l - b) = T_1 \sin \lambda l \dots\dots\dots(2).$$

The value of y for any normal vibration corresponding to λ is

$$\left. \begin{array}{l} y = P \sin \lambda x \, \sin \lambda (l - b) \cos (a \lambda t - \epsilon) \\ \qquad \text{from } x = 0 \text{ to } x = b \\ y = P \sin \lambda (l - x) \, \sin \lambda b \cos (a \lambda t - \epsilon) \\ \qquad \text{from } x = b \text{ to } x = l \end{array} \right\} \dots\dots\dots(3),$$

where P and ϵ are arbitrary constants.

It does not require analysis to prove that any normal components which have a node at the point of attachment are unaffected by the presence of the load. For instance, if a string be weighted at the centre, its component vibrations of even orders remain unchanged, while all the odd components are depressed in pitch. Advantage may sometimes be taken of this effect of a load, when it is desired for any purpose to disturb the harmonic relation of the component tones.

If M be very great, the gravest component is widely separated in pitch from all the others. We will take the case when the load is at the centre, so that $b = l - b = \frac{1}{2} l$. The equation in then becomes

$$\sin \frac{\lambda l}{2} \cdot \left\{ \frac{\lambda l}{2} \tan \frac{\lambda l}{2} - \frac{\rho l}{M} \right\} = 0 \dots\dots\dots\dots\dots(4),$$

where $\rho l : M$, denoting the ratio of the masses of the string and the load, is a small quantity which may be called α^2. The first root corresponding to the tone of lowest pitch occurs when $\frac{1}{2}\lambda l$ is small, and such that

$$(\tfrac{1}{2}\lambda l)^2 \{1 + \tfrac{1}{3}(\tfrac{1}{2}\lambda l)^2\} = \alpha^2 \text{ nearly,}$$

whence

$$\tfrac{1}{2}\lambda l = \alpha(1 - \tfrac{1}{6}\alpha^2),$$

and the periodic time is given by

$$\tau = \pi \sqrt{\frac{Ml}{T_1}} \left(1 + \frac{\rho l}{6M}\right) \quad\dots\dots\dots\dots\dots\dots(5).$$

The second term constitutes a correction to the rough value obtained in a previous chapter (§ 52), by neglecting the inertia of the string altogether. That it would be additive might have been expected, and indeed the formula as it stands may be obtained from the consideration that in the actual vibration the two parts of the string are nearly straight, and may be assumed to be exactly so in computing the kinetic and potential energies, without entailing any appreciable error in the calculated period. On this supposition the retention of the inertia of the string increases the kinetic energy corresponding to a given velocity of the load in the ratio of $M : M + \tfrac{1}{3}\rho l$, which leads to the above result. This method has indeed the advantage in one respect, as it might be applied when ρ is not uniform, or nearly uniform. All that is necessary is that the load M should be sufficiently predominant.

Fig. 21.

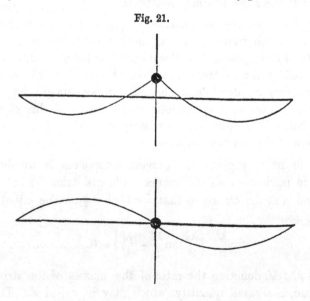

There is no other root of (4), until $\sin \frac{1}{2}\lambda l = 0$, which gives the second component of the string,—a vibration independent of the load. The roots after the first occur in closely contiguous pairs; for one set is given by $\frac{1}{2}\lambda l = s\pi$, and the other approximately by $\frac{1}{2}\lambda l = s\pi + \dfrac{\rho l}{s\pi M}$, in which the second term is small. The two types of vibration for $s = 1$ are shewn in the figure.

The general formula (2) may also be applied to find the effect of a small load on the pitch of the various components.

137. Actual strings and wires are not perfectly flexible. They oppose a certain resistance to bending, which may be divided into two parts, producing two distinct effects. The first is called viscosity, and shews itself by damping the vibrations. This part produces no sensible effect on the periods. The second is conservative in its character, and contributes to the potential energy of the system, with the effect of shortening the periods. A complete investigation cannot conveniently be given here, but the case which is most interesting in its application to musical instruments, admits of a sufficiently simple treatment.

When rigidity is taken into account, something more must be specified with respect to the terminal conditions than that y vanishes. Two cases may be particularly noted :—

(1)   When the ends are clamped, so that $\dfrac{dy}{dx} = 0$ at the ends.

(2)   When the terminal directions are perfectly free, in which case $\dfrac{d^2y}{dx^2} = 0$.

It is the latter which we propose now to consider.

If there were no rigidity, the type of vibration would be

$$y \propto \sin \frac{s\pi x}{l}, \quad \text{satisfying the second condition.}$$

The effect of the rigidity might be slightly to disturb the type; but whether such a result occur or not, the period calculated from the potential and kinetic energies on the supposition that the type remains unaltered is necessarily correct as far as the first order of small quantities (§ 88).

Now the potential energy due to the stiffness is expressed by

$$\delta V = \tfrac{1}{2}\kappa \int_0^l \left(\frac{d^2y}{dx^2}\right)^2 dx \dotfill (1),$$

where κ is a quantity depending on the nature of the material and on the form of the section in a manner that we are not now prepared to examine. The *form* of δV is evident, because the force required to bend any element ds is proportional to ds, and to the amount of bending already effected, that is to $ds \div \rho$. The whole work which must be done to produce a curvature $1 \div \rho$ in ds is therefore proportional to $ds \div \rho^2$; while to the approximation to which we work $ds = dx$, and $\dfrac{1}{\rho} = \dfrac{d^2y}{dx^2}$.

Thus, if $y = \phi \sin \dfrac{s\pi x}{l}$,

$$T = \tfrac{1}{4}\rho l \, \dot\phi^2; \quad V = \tfrac{1}{4}T_1 l \cdot \frac{s^2\pi^2}{l^2} \, \phi^2 \left(1 + \frac{\kappa}{T_1} \frac{s^2\pi^2}{l^2}\right),$$

and the period of ϕ is given by

$$\tau = \tau_0 \left(1 - \frac{\kappa}{2T_1} \frac{s^2\pi^2}{l^2}\right) \dots\dots\dots\dots\dots\dots(2),$$

if τ_0 denote what the period would become if the string were endowed with perfect flexibility. It appears that the effect of the stiffness increases rapidly with the order of the component vibrations, which cease to belong to a harmonic scale. However, in the strings employed in music, the tension is usually sufficient to reduce the influence of rigidity to insignificance.

The method of this section cannot be applied without modification to the other case of terminal condition, namely, when the ends are clamped. In their immediate neighbourhood the type of vibration must differ from that assumed by a perfectly flexible string by a quantity, which is no longer small, and whose square therefore cannot be neglected. We shall return to this subject, when treating of the transverse vibrations of rods.

138. There is one problem relating to the vibrations of strings which we have not yet considered, but which is of some practical interest, namely, the character of the motion of a violin (or cello) string under the action of the bow. In this problem the *modus operandi* of the bow is not sufficiently understood to allow us to follow exclusively the *a priori* method: the indications of theory must be supplemented by special observation. By a dexterous combination of evidence drawn from both sources Helmholtz has succeeded in determining the principal features of the case, but some of the details are still obscure.

Since the note of a good instrument, well handled, is musical, we infer that the vibrations are strictly periodic, or at least that strict periodicity is the ideal. Moreover—and this is very import- ant—the note elicited by the bow has nearly, or quite, the same pitch as the natural note of the string. The vibrations, although forced, are thus in some sense free. They are wholly dependent for their maintenance on the energy drawn from the bow, and yet the bow does not determine, or even sensibly modify, their periods. We are reminded of the self-acting electrical interrupter, whose motion is indeed forced in the technical sense, but has that kind of freedom which consists in determining (wholly, or in part) under what influences it shall come.

But it does not at once follow from the fact that the string vibrates with its natural periods, that it conforms to its natural types. If the coefficients of the Fourier expansion

$$y = \phi_1 \sin \frac{\pi x}{l} + \phi_2 \sin \frac{2\pi x}{l} + \ldots\ldots$$

be taken as the independent co-ordinates by which the configura- tion of the system is at any moment defined, we know that when there is no friction, or friction such that $F \propto T$, the natural vibra- tions are expressed by making each co-ordinate a *simple* harmonic (or quasi-harmonic) function of the time; while, for all that has hitherto appeared to the contrary, each co-ordinate in the present case might be *any* function of the time periodic in time τ. But a little examination will shew that the vibrations must be sensibly natural in their types as well as in their periods.

The force exercised by the bow at its point of application may be expressed by

$$Y = \Sigma A_r \cos\left(\frac{2r\pi t}{\tau} - \epsilon_r\right);$$

so that the equation of motion for the co-ordinate ϕ_s is

$$\ddot{\phi}_s + \kappa \dot{\phi}_s + \frac{s^2\pi^2 a^2}{l^2}\, \phi_s = \frac{2}{l\rho} \sin \frac{s\pi b}{l} . \Sigma A_r \cos\left(\frac{2r\pi t}{\tau} - \epsilon_r\right),$$

b being the point of application. Each of the component parts of Φ_s will give a corresponding term of its own period in the solu- tion, but the one whose period is the same as the natural period of ϕ_s will rise enormously in relative importance. Practically then, if the damping be small, we need only retain that part of ϕ_s

which depends on $A_s \cos\left(\dfrac{2s\pi t}{\tau} - \epsilon_s\right)$, that is to say, we may regard the vibrations as natural in their types.

Another material fact, supported by evidence drawn both from theory and aural observation, is this. All component vibrations are absent which have a node at the point of excitation. "In order, however, to extinguish these tones, it is necessary that the coincidence of the point of application of the bow with the node should be very *exact*. A very small deviation reproduces the missing tones with considerable strength[1]."

The remainder of the evidence on which Helmholtz' theory rests, was derived from direct observation with the vibration-microscope. As explained in Chapter II., this instrument affords a view of the curve representing the motion of the point under observation, as it would be seen traced on the surface of a transparent cylinder. In order to deduce the representative curve in its ordinary form, the imaginary cylinder must be conceived to be unrolled, or developed, into a plane.

The simplest results are obtained when the bow is applied at a node of one of the higher components, and the point observed is one of the other nodes of the same system. If the bow works fairly so as to draw out the fundamental tone clearly and strongly, the representative curve is that shewn in figure 22; where the abscissæ correspond to the time (AB being a complete period), and the ordinates represent the displacement. The remarkable

Fig. 22.

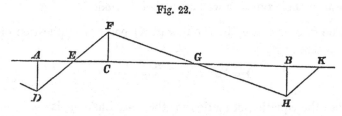

fact is disclosed that the whole period τ may be divided into two parts τ_0 and $\tau - \tau_0$, during each of which the velocity of the observed point is constant; but the velocities to and fro are in general unequal.

We have now to represent this curve by a series of harmonic terms. If the origin of time correspond to the point A, and

[1] Donkin's *Acoustics*, p. 131.

$AD = FC = \gamma$, Fourier's theorem gives

$$y = \frac{2\gamma\tau^2}{\pi^2\tau_0(\tau-\tau_0)} \sum_{s=1}^{s=\infty} \frac{1}{s^2} \sin\frac{s\pi\tau_0}{\tau} \sin\frac{2s\pi}{\tau}\left(t-\frac{\tau_0}{2}\right)......(1).$$

With respect to the value of τ_0, we know that all those components of y must vanish for which $\sin\frac{s\pi x_0}{l} = 0$ (x_0 being the point of observation), because under the circumstances of the case the bow cannot generate them. There is therefore reason to suppose that $\tau_0 : \tau = x_0 : l$; and in fact observation proves that $AC : CB$ (in the figure) is equal to the ratio of the two parts into which the string is divided by the point of observation.

Now the free vibrations of the string are represented in general by

$$y = \sum_{s=1}^{s=\infty} \sin\frac{s\pi x}{l}\left\{A_s\cos\frac{2s\pi t}{\tau} + B_s\sin\frac{2s\pi t}{\tau}\right\};$$

and this at the point $x = x_0$ must agree with (1). For convenience of comparison, we may write

$$A_s\cos\frac{2s\pi t}{\tau} + B_s\sin\frac{2s\pi t}{\tau} = C_s\cos\frac{2s\pi}{\tau}\left(t-\frac{\tau_0}{2}\right)$$
$$+ D_s\sin\frac{2s\pi}{\tau}\left(t-\frac{\tau_0}{2}\right),$$

and it then appears that $C_s = 0$.

We find also to determine D_s

$$\sin\frac{s\pi x_0}{l}\cdot D_s = \frac{2\gamma\tau^2}{\pi^2\tau_0(\tau-\tau_0)}\frac{1}{s^2}\sin\frac{s\pi x_0}{l},$$

whence

$$D_s = \frac{2\gamma\tau^2}{\pi^2\tau_0(\tau-\tau_0)}\frac{1}{s^2} \quad ....(2),$$

unless $\sin\frac{s\pi x_0}{l} = 0$.

In the case reserved, the comparison leaves D_s undetermined, but we know on other grounds that D_s then vanishes. However, for the sake of simplicity, we shall suppose for the present that D_s is always given by (2). If the point of application of the bow do not coincide with a node of any of the lower components, the error committed will be of no great consequence.

On this understanding the complete solution of the problem is

$$y = \frac{2\gamma\tau^2}{\pi^2\tau_0(\tau-\tau_0)} \sum_{s=1}^{s=\infty} \frac{1}{s^2}\sin\frac{s\pi x}{l}\sin\frac{2s\pi}{\tau}\left(t-\frac{\tau_0}{2}\right).................(3).$$

The amplitudes of the components are therefore proportional to s^{-2}. In the case of a plucked string we found for the corresponding function $s^{-2}\sin\dfrac{s\pi b}{l}$, which is somewhat similar. If the string be plucked at the middle, the even components vanish, but the odd ones follow the same law as obtains for a violin string. The equation (3) indicates that the string is always in the form of two straight lines meeting at an angle. In order more conveniently to shew this, let us change the origin of the time, and the constant multiplier, so that

$$y = \frac{8P}{\pi^2} \Sigma \frac{1}{s^2}\sin\frac{s\pi x}{l}\ \sin\frac{2s\pi t}{\tau} \ldots\ldots\ldots\ldots\ldots(4),$$

will be the equation expressing the form of the string at any time.

Now we know (§ 127) that the equation of the pair of lines proceeding from the fixed ends of the string, and meeting at a point whose co-ordinates are α, β, is

$$y = \frac{2\beta l^2}{\pi^2\alpha\,(l-\alpha)} \Sigma \frac{1}{s^2}\sin\frac{s\pi\alpha}{l}\ \sin\frac{s\pi x}{l}\,.$$

Thus at the time t, (4) represents such a pair of lines, meeting at the point whose co-ordinates are given by

$$\frac{\beta l^2}{\alpha\,(l-\alpha)} = \pm\,4P,$$

$$\sin\frac{s\pi\alpha}{l} = \pm\,\sin\frac{2s\pi t}{\tau}$$

These equations indicate that the projection on the axis of x of the point of intersection moves uniformly backwards and forwards between $x=0$ and $x=l$, and that the point of intersection itself is situated on one or other of two parabolic arcs, of which the equilibrium position of the string is a common chord.

Since the motion of the string as thus defined by that of the point of intersection of its two straight parts, has no especial relation to x_0 (the point of observation), it follows that, according to these equations, the same kind of motion might be observed at any other point. And this is approximately true. But the theoretical result, it will be remembered, was only obtained by assuming the presence in certain proportions of component vibrations having nodes at x_0, though in fact their absence is required by mechanical laws. The presence or absence of these components is a matter of indifference when a node is the point of observation, but not in any other case. When the node is departed from, the vibration curve shews a series of ripples, due to the absence of the components in question. Some further details will be found in Helmholtz and Donkin.

The sustaining power of the bow depends upon the fact that solid friction is less at moderate than at small velocities, so that when the part of the string acted upon is moving with the bow (not improbably at the same velocity), the mutual action is greater than when the string is moving in the opposite direction with a greater relative velocity. The accelerating effect in the first part of the motion is thus not entirely neutralised by the subsequent retardation, and an outstanding acceleration remains capable of maintaining the vibration in spite of other losses of energy. A curious effect of the same peculiarity of solid friction has been observed by Mr Froude, who found that the vibrations of a pendulum swinging from a shaft might be maintained or even increased by causing the shaft to rotate.

139. A string stretched on a smooth curved surface will in equilibrium lie along a geodesic line, and, subject to certain conditions of stability, will vibrate about this configuration, if displaced. The simplest case that can be proposed is when the surface is a cylinder of any form, and the equilibrium position of the string is perpendicular to the generating lines. The student will easily prove that the motion is independent of the curvature of the cylinder, and that the vibrations are in all essential respects the same as if the surface were developed into a plane. The case of an endless string, forming a necklace round the cylinder, is worthy of notice.

In order to illustrate the characteristic features of this class of problems, we will take the comparatively simple example of a string stretched on the surface of a smooth sphere, and lying, when in equilibrium, along a great circle. The co-ordinates to which it will be most convenient to refer the system are the latitude θ measured from the great circle as equator, and the longitude ϕ measured along it. If the radius of the sphere be a, we have

$$T = \frac{1}{2}\int \rho\,(a\dot{\theta})^2 a\,d\phi = \frac{a^3\rho}{2}\int \dot{\theta}^2 d\phi \dots\dots\dots\dots\dots(1).$$

The extension of the string is denoted by

$$\int (ds - a\,d\phi) = a \int \left(\frac{ds}{a\,d\phi} - 1\right) d\phi.$$

Now

$$ds^2 = (a\,d\theta)^2 + (a \cos\theta\,d\phi)^2 ;$$

so that

$$\frac{ds}{a\,d\phi} - 1 = \left\{\left(\frac{d\theta}{d\phi}\right)^2 + \cos^2\theta\right\}^{\frac{1}{2}} - 1 = \frac{1}{2}\left(\frac{d\theta}{d\phi}\right)^2 - \frac{\theta^2}{2}\text{, approximately.}$$

Thus

$$V = \tfrac{1}{2} aT_1 \int \left\{\left(\frac{d\theta}{d\phi}\right)^2 - \theta^2\right\} d\phi \dots\dots\dots\dots\dots(2) ;[1]$$

and

$$\delta V = aT_1 . \delta\theta \left[\frac{d\theta}{d\phi}\right]_0^l - aT_1 \int_0^l \delta\theta \left(\frac{d^2\theta}{d\phi^2} + \theta\right) d\phi.$$

If the ends be fixed,

$$\delta\theta \left[\frac{d\theta}{d\phi}\right]_0^l = 0,$$

and the equation of virtual velocities is

$$a^3\rho \int_0^l \ddot{\theta}\,\delta\theta\,d\phi - aT_1 \int_0^l \delta\theta \left(\frac{d^2\theta}{d\phi^2} + \theta\right) d\phi = 0,$$

whence, since $\delta\theta$ is arbitrary,

$$a^2\rho\,\ddot{\theta} = T_1 \left(\frac{d^2\theta}{d\phi^2} + \theta\right) \dots\dots\dots\dots\dots\dots(3).$$

This is the equation of motion.

If we assume $\theta \propto \cos pt$, we get

$$\frac{d^2\theta}{d\phi^2} + \theta + \frac{a^2\rho}{T_1}p^2\theta = 0 \dots\dots\dots\dots\dots(4),$$

of which the solution, subject to the condition that θ vanishes with ϕ, is

$$\theta = A \sin \left\{\frac{a^2\rho}{T_1}p^2 + 1\right\}^{\frac{1}{2}} \phi \,.\, \cos pt \dots\dots\dots\dots(5).$$

The remaining condition to be satisfied is that θ vanishes when $a\phi = l$, or $\phi = \alpha$, if $\alpha = l \div a$.

This gives

$$p^2 = \frac{T_1}{a^2\rho}\left(\frac{m^2\pi^2}{\alpha^2} - 1\right) = \frac{T_1}{\rho}\left(\frac{m^2\pi^2}{l^2} - \frac{1}{a^2}\right) \dots\dots\dots(6),$$

where m is an integer.

[1] Cambridge Mathematical Tripos Examination, 1876.

The normal functions are thus of the same form as for a straight string, viz.

$$\theta = A \sin \frac{m\pi\phi}{\alpha} \cos pt \quad \text{.....................(7),}$$

but the series of periods is different. The effect of the curvature is to make each tone graver than the corresponding tone of a straight string. If $\alpha > \pi$, one at least of the values of p^2 is negative, indicating that the corresponding modes are unstable. If $\alpha = \pi$, p_1 is zero, the string being of the same length in the displaced position, as when $\theta = 0$.

A similar method might be applied to calculate the motion of a string stretched round the equator of any surface of revolution.

140.   The approximate solution of the problem for a vibrating string of nearly but not quite uniform longitudinal density has been fully considered in Chapter IV. § 91, as a convenient example of the general theory of approximately simple systems. It will be sufficient here to repeat the result. If the density be $\rho_0 + \delta\rho$, the period τ_r of the rth component vibration is given by

$$\tau_r^2 = \frac{4l^2\rho_0}{r^2T_1} \left\{ 1 + \frac{2}{l} \int_0^l \frac{\delta\rho}{\rho_0} \sin^2 \frac{r\pi x}{l} \, dx \right\} \text{............ (1).}$$

If the irregularity take the form of a small load of mass m at the point $x = b$, the formula may be written

$$\tau_r^2 = \frac{4l^2\rho_0}{r^2T_1'} \left\{ 1 + \frac{2m}{l\rho_0} \sin^2 \frac{r\pi b}{l} \right\} \text{.................. (2).}$$

These values of τ^2 are correct as far as the first power of the small quantities $\delta\rho$ and m, and give the means of calculating a correction for such slight departures from uniformity as must always occur in practice.

As might be expected, the effect of a small load vanishes at nodes, and rises to a maximum at the points midway between consecutive nodes. When it is desired merely to make a rough estimate of the effective density of a nearly uniform string, the formula indicates that attention is to be given to the neighbourhood of loops rather than to that of nodes.

141.   The differential equation determining the motion of a string, whose longitudinal density ρ is variable, is

$$\rho \frac{d^2y}{dt^2} = T_1 \frac{d^2y}{dx^2} \text{..................... (1),}$$

from which, if we assume $y \propto \cos nt$, we obtain to determine the normal functions

$$\frac{d^2y}{dx^2} + \nu^2 \rho y = 0 \dots\dots\dots\dots\dots\dots (2),$$

where ν^2 is written for $n^2 \div T_1$. This equation is of the second order and linear, but has not hitherto been solved in finite terms. Considered as defining the curve assumed by the string in the normal mode under consideration, it determines the *curvature* at any point, and accordingly embodies a rule by which the curve can be constructed graphically. Thus in the application to a string fixed at both ends, if we start from either end at an arbitrary inclination, and with zero curvature, we are always directed by the equation with what curvature to proceed, and in this way we may trace out the entire curve.

If the assumed value of ν^2 be right, the curve will cross the axis of x at the required distance, and the law of vibration will be completely determined. If ν^2 be not known, different values may be tried until the curve ends rightly; a sufficient approximation to the value of ν^2 may usually be arrived at by a calculation founded on an assumed type (§§ 88, 90).

Whether the longitudinal density be uniform or not, the periodic time of any simple vibration varies *cœteris paribus* as the square root of the density and inversely as the square root of the tension under which the motion takes place.

The converse problem of determining the density, when the period and the type of vibration are given, is always soluble. For this purpose it is only necessary to substitute the given value of y, and of its second differential coefficient in equation (2). Unless the density be infinite, the extremities of a string are points of zero curvature.

When a given string is shortened, every component tone is raised in pitch. For the new state of things may be regarded as derived from the old by introduction, at the proposed point of fixture, of a spring (without inertia), whose stiffness is gradually increased without limit. At each step of the process the potential energy of a given deformation is augmented, and therefore (§ 88) the pitch of every tone is raised. In like manner an addition to the length of a string depresses the pitch, even though the added part be destitute of inertia.

142. Although a general integration of equation (2) of § 141 is beyond our powers, we may apply to the problem some of the many interesting properties of the solution of the linear equation of the second order, which have been demonstrated by MM. Sturm and Liouville[1]. It is impossible in this work to give anything like a complete account of their investigations; but a sketch, in which the leading features are included, may be found interesting, and will throw light on some points connected with the general theory of the vibrations of continuous bodies. I have not thought it necessary to adhere very closely to the methods adopted in the original memoirs.

At no point of the curve satisfying the equation

$$\frac{d^2y}{dx^2} + \nu^2 \rho\, y = 0 \dots\dots\dots\dots\dots\dots(1),$$

can both y and $\frac{dy}{dx}$ vanish together. By successive differentiations of (1) it is easy to prove that, if y and $\frac{dy}{dx}$ vanish simultaneously, all the higher differential coefficients $\frac{d^2y}{dx^2}$, $\frac{d^3y}{dx^3}$, &c. must also vanish at the same point, and therefore by Taylor's theorem the curve must coincide with the axis of x.

Whatever value be ascribed to ν^2, the curve satisfying (1) is *sinuous*, being concave throughout towards the axis of x, since ρ is everywhere positive. If at the origin y vanish, and $\frac{dy}{dx}$ be positive, the ordinate will remain positive for all values of x below a certain limit dependent on the value ascribed to ν^2. If ν^2 be very small, the curvature is slight, and the curve will remain on the positive side of the axis for a great distance. We have now to prove that as ν^2 increases, all the values of x which satisfy the equation $y = 0$ gradually diminish in magnitude.

Let y' be the ordinate of a second curve satisfying the equation

$$\frac{d^2y'}{dx^2} + \nu'^2 \rho\, y' = 0 \dots\dots\dots\dots\dots\dots (2),$$

as well as the condition that y' vanishes at the origin, and let us suppose that ν'^2 is somewhat greater than ν^2. Multiplying (2) by y,

<hr>

[1] The memoirs referred to in the text are contained in the first volume of Liouville's *Journal* (1836).

and (1) by y', subtracting, and integrating with respect to x between the limits 0 and x, we obtain, since y and y' both vanish with x,

$$y'\frac{dy}{dx} - y\frac{dy'}{dx} = (\nu'^2 - \nu^2)\int_0^x \rho\, y\, y'\, dx \,\ldots\ldots\ldots (3).$$

If we further suppose that x corresponds to a point at which y vanishes, and that the difference between ν'^2 and ν^2 is very small, we get ultimately

$$y'\frac{dy}{dx} = \delta\nu^2\int_0^x \rho\, y^2\, dx \,\ldots\ldots\ldots\ldots\ldots\ldots (4).$$

The right-hand member of (4) being essentially positive, we learn that y' and $\frac{dy}{dx}$ are of the same sign, and therefore that, whether $\frac{dy}{dx}$ be positive or negative, y' is already of the same sign as that to which y is changing, or in other words, the value of x for which y' vanishes is less than that for which y vanishes.

If we fix our attention on the portion of the curve lying between $x = 0$ and $x = l$, the ordinate continues positive throughout as the value of ν^2 increases, until a certain value is attained, which we will call ν_1^2. The function y is now identical in form with the first normal function u_1 of a string of density ρ fixed at 0 and l, and has no root except at those points. As ν^2 again increases, the first root moves inwards from $x = l$ until, when a second special value ν_2^2 is attained, the curve again crosses the axis at the point $x = l$, and then represents the second normal function u_2. This function has thus one internal root, and one only. In like manner corresponding to a higher value ν_3^2 we obtain the third normal function u_3 with two internal roots, and so on. The nth function u_n has thus exactly $n - 1$ internal roots, and since its first differential coefficient never vanishes simultaneously with the function, it changes sign each time a root is passed.

From equation (3) it appears that if u_r and u_s be two different normal functions,

$$\int_0^l \rho\, u_r\, u_s\, dx = 0 \,\ldots\ldots\ldots\ldots\ldots\ldots (5).$$

A beautiful theorem has been discovered by Sturm relating to the number of the roots of a function derived by addition from a finite number of normal functions. If u_m be the component of lowest order, and u_n the component of highest order, the function

$$f(x) = \phi_m u_m + \phi_{m+1} u_{m+1} + \ldots\ldots + \phi_n u_n \ldots\ldots\ldots (6),$$

where ϕ_m, ϕ_{m+1}, &c. are arbitrary coefficients, has *at least* $m-1$ internal roots, and *at most* $n-1$ internal roots. The extremities at $x = 0$ and at $x = l$ correspond of course to roots in all cases. The following demonstration bears some resemblance to that given by Liouville, but is considerably simpler, and, I believe, not less rigorous.

If we suppose that $f(x)$ has exactly μ internal roots (any number of which may be equal), the derived function $f'(x)$ cannot have less than $\mu + 1$ internal roots, since there must be at least one root of $f'(x)$ between each pair of consecutive roots of $f(x)$, and the whole number of roots of $f(x)$ concerned is $\mu + 2$. In like manner, we see that there must be at least μ roots of $f''(x)$, besides the extremities, which themselves necessarily correspond to roots; so that in passing from $f(x)$ to $f''(x)$ it is impossible that any roots can be lost. Now

$$f''(x) = \phi_m u_m'' + \phi_{m+1} u''_{m+1} + \ldots\ldots + \phi_n u_n''$$
$$= - \rho \left(\nu_m{}^2 \phi_m u_m + \nu^2_{m+1} \phi_{m+1} u_{m+1} + \ldots\ldots + \nu_n{}^2 \phi_n u_n \right)\ldots (7),$$

as we see by (1); and therefore, since ρ is always positive, we infer that

$$\nu_m{}^2 \phi_m u_m + \nu^2_{m+1} \phi_{m+1} u_{m+1} + \ldots\ldots + \nu_n{}^2 \phi_n u_n \ldots\ldots\ldots (8),$$

has at least μ roots.

Again, since (8) is an expression of the same form as $f(x)$, similar reasoning proves that

$$\nu_m{}^4 \phi_m u_m + \nu^4_{m+1} \phi_{m+1} u_{m+1} + \ldots\ldots + \nu_n{}^4 \phi_n u_n$$

has at least μ internal roots; and the process may be continued to any extent. In this way we obtain a series of functions, all with μ internal roots at least, which differ from the original function $f(x)$ by the continually increasing relative importance of the components of the higher orders. When the process has been carried sufficiently far, we shall arrive at a function, whose form differs as little as we please from that of the normal function of highest order, viz. u_n, and which has therefore $n-1$ internal roots. It follows that, since no roots can be lost in passing down the series of functions, the number of internal roots of $f(x)$ cannot exceed $n-1$.

The other half of the theorem is proved in a similar manner by continuing the series of functions backwards from $f(x)$. In this way we obtain

$$
\begin{aligned}
&\phi_m\, u_m + \qquad\quad \phi_{m+1}\, u_{m+1} + \cdots + \qquad \phi_n\, u_n \\
&\nu_m^{-2}\, \phi_m\, u_m + \nu^{-2}_{m+1}\, \phi_{m+1}\, u_{m+1} + \cdots + \nu_n^{-2}\, \phi_n\, u_n \\
&\nu_m^{-4}\, \phi_m\, u_m + \nu^{-4}_{m+1}\, \phi_{m+1}\, u_{m+1} + \cdots + \nu_n^{-4}\, \phi_n\, u_n \\
&\cdots\cdots\cdots\cdots\cdots\cdots\cdots\cdots\cdots\cdots\cdots\cdots\cdots\cdots ,
\end{aligned}
$$

arriving at last at a function sensibly coincident in form with the normal function of *lowest* order, viz. u_m, and having therefore $m-1$ internal roots. Since no roots can be lost in passing up the series from this function to $f(x)$, it follows that $f(x)$ cannot have fewer internal roots than $m-1$; but it must be understood that any number of the $m-1$ roots may be equal.

We will now prove that $f(x)$ cannot be identically zero, unless all the coefficients ϕ vanish. Suppose that ϕ_r is not zero. Multiply (6) by $\rho\, u_r$, and integrate with respect to x between the limits 0 and l. Then by (5)

$$
\int_0^l \rho\, u_r f(x)\, dx = \phi_r \int_0^l \rho\, u_r^2 dx \quad \cdots\cdots\cdots (9);
$$

from which, since the integral on the right-hand side is finite, we see that $f(x)$ cannot vanish for all values of x included within the range of integration.

Liouville has made use of Sturm's theorem to shew how a series of normal functions may be compounded so as to have an arbitrary sign at all points lying between $x=0$ and $x=l$. His method is somewhat as follows.

The values of x for which the function is to change sign being $a, b, c, \ldots$, quantities which without loss of generality we may suppose to be all different, let us consider the series of determinants,

$$
\begin{vmatrix} u_1(a), & u_1(x) \\ u_2(a), & u_2(x) \end{vmatrix}
\qquad
\begin{vmatrix} u_1(a), & u_1(b), & u_1(x) \\ u_2(a), & u_2(b), & u_2(x) \\ u_3(a), & u_3(b), & u_3(x) \end{vmatrix}, \&c.
$$

The first is a linear function of $u_1(x)$ and $u_2(x)$, and by Sturm's theorem has therefore one internal root at most, which root is evidently a. Moreover the determinant is not identically zero, since the coefficient of $u_2(x)$, viz. $u_1(a)$, does not vanish, whatever be the value of a. We have thus obtained a function, which changes sign at an arbitrary point a, and there only, internally.

The second determinant vanishes when $x = a$, and when $x = b$, and, since it cannot have more than two internal roots, it changes sign, when x passes through these values, and there only. The coefficient of $u_3(x)$ is the value assumed by the first determinant when $x = b$, and is therefore finite. Hence the second determinant is not identically zero.

Similarly the third determinant in the series vanishes and changes sign when $x = a$, when $x = b$, and when $x = c$, and at these internal points only. The coefficient of $u_4(x)$ is finite, being the value of the second determinant when $x = c$.

It is evident that by continuing this process we can form functions compounded of the normal functions, which shall vanish and change sign for any arbitrary values of x, and not elsewhere internally; or, in other words, we can form a function whose sign is arbitrary over the whole range from $x = 0$ to $x = l$.

On this theorem Liouville founds his demonstration of the possibility of representing an arbitrary function between $x = 0$ and $x = l$ by a series of normal functions. If we assume the possibility of the expansion and take

$$f(x) = \phi_1 u_1(x) + \phi_2 u_2(x) + \phi_3 u_3(x) + \dots \dots \dots \dots (10),$$

the necessary values of ϕ_1, ϕ_2, &c. are determined by (9), and we find

$$f(x) = \Sigma \left\{ u_r(x) \int_0^l \rho u_r(x) f(x)\, dx \div \int_0^l \rho u_r^2(x)\, dx \right\} \dots \dots \dots (11).$$

If the series on the right be denoted by $F(x)$, it remains to establish the identity of $f(x)$ and $F(x)$.

If the right-hand member of (11) be multiplied by $\rho u_r(x)$ and integrated with respect to x from $x = 0$ to $x = l$, we see that

$$\int_0^l \rho u_r(x) F(x)\, dx = \int_0^l \rho u_r(x) f(x)\, dx,$$

or, as we may also write it,

$$\int_0^l \{F(x) - f(x)\} \rho u_r(x)\, dx = 0 \dots \dots \dots \dots \dots (12),$$

where $u_r(x)$ is *any* normal function. From (12) it follows that

$$\int_0^l \{F(x) - f(x)\} \{A_1 u_1(x) + A_2 u_2(x) + A_3 u_3(x) + \dots\} \rho\, dx = 0 \dots (13),$$

where the coefficients A_1, A_2, &c. are arbitrary.

R.

12

Now if $F(x) - f(x)$ be not identically zero, it will be possible so to choose the constants A_1, A_2, &c. that $A_1 u_1(x) + A_2 u_2(x) + \dots$ has throughout the same sign as $F(x) - f(x)$, in which case every element of the integral would be positive, and equation (13) could not be true. It follows that $F(x) - f(x)$ cannot differ from zero, or that the series of normal functions forming the right-hand member of (11) is identical with $f(x)$ for all values of x from $x = 0$ to $x = l$.

The arguments and results of this section are of course applicable to the particular case of a uniform string for which the normal functions are circular.

143.   When the vibrations of a string are not confined to one plane, it is usually most convenient to resolve them into two sets executed in perpendicular planes, which may be treated independently. There is, however, one case of this description worth a passing notice, in which the motion is most easily conceived and treated without resolution.

Suppose that

$$y = \sin \frac{s\pi x}{l} \cos \frac{2s\pi t}{\tau} \left.\begin{array}{c} \\ \\ \end{array}\right\} \dots\dots\dots\dots\dots(1).$$
$$z = \sin \frac{s\pi x}{l} \sin \frac{2s\pi t}{\tau}$$

Then

$$r = \sqrt{y^2 + z^2} = \sin \frac{s\pi x}{l} \dots\dots\dots\dots\dots(2),$$

and

$$z : y = \tan \frac{2s\pi t}{\tau} \dots\dots\dots\dots\dots\dots(3),$$

shewing that the whole string is at any moment in one plane, which revolves uniformly, and that each particle describes a circle with radius $\sin \frac{s\pi x}{l}$. In fact, the whole system turns without relative displacement about its position of equilibrium, completing each revolution in the time $\tau \div s$. The mechanics of this case is quite as simple as when the motion is confined to one plane, the resultant of the tensions acting at the extremities of any small portion of the string's length being balanced by the centrifugal force.

144. The general differential equation for a uniform string, viz.

$$\frac{d^2y}{dt^2} = a^2 \frac{d^2y}{dx^2} \dots\dots\dots\dots\dots\dots\dots(1),$$

may be transformed by a change of variables into

$$\frac{d^2y}{du\,dv} = 0 \dots\dots\dots\dots\dots\dots\dots(2),$$

where $u = x - at$, $v = x + at$. The general solution of (2) is

$$y = f(u) + F(v) = f(x - at) + F(x + at)\dots\dots\dots(3),$$

f, F being two arbitrary functions.

Let us consider first the case in which F vanishes. When t has any particular value, the equation

$$y = f(x - at) \dots\dots\dots\dots\dots\dots\dots(4),$$

expressing the relation between x and y, represents the form of the string. A change in the value of t is merely equivalent to an alteration in the origin of x, so that (4) indicates that a certain *form* is propagated along the string with uniform velocity a in the positive direction. Whatever the value of y may be at the point x and at the time t, the same value of y will obtain at the point $x + a\,\Delta t$ at the time $t + \Delta t$.

The form thus perpetuated may be any whatever, so long as it does not violate the restrictions on which (1) depends.

When the motion consists of the propagation of a wave in the positive direction, a certain relation subsists between the inclination and the velocity at any point. Differentiating (4) we find

$$\frac{dy}{dt} = -a\,\frac{dy}{dx} \dots\dots\dots\dots\dots\dots\dots(5).$$

Initially, $\frac{dy}{dt}$ and $\frac{dy}{dx}$ may both be given arbitrarily, but if the above relation be not satisfied, the motion cannot be represented by (4).

In a similar manner the equation

$$y = F(x + at)\dots\dots\dots\dots\dots\dots\dots(6),$$

denotes the propagation of a wave in the *negative* direction, and the relation between $\frac{dy}{dt}$ and $\frac{dy}{dx}$ corresponding to (5) is

$$\frac{dy}{dt} = a\,\frac{dy}{dx} \dots\dots\dots\dots\dots\dots\dots(7).$$

In the general case the motion consists of the simultaneous propagation of two waves with velocity a, the one in the positive, and the other in the negative direction; and these waves are entirely independent of one another. In the first $\frac{dy}{dt} = -a\frac{dy}{dx}$, and in the second $\frac{dy}{dt} = a\frac{dy}{dx}$. The initial values of $\frac{dy}{dt}$ and $\frac{dy}{dx}$ must be conceived to be divided into two parts, which satisfy respectively the relations (5) and (7). The first constitutes the wave which will advance in the positive direction without change of form; the second, the negative wave. Thus, initially,

$$\left.\begin{array}{c} f'(x) + F'(x) = \dfrac{dy}{dx} \\[2mm] f'(x) - F'(x) = -\dfrac{1}{a}\dfrac{dy}{dt} \end{array}\right\},$$

whence

$$\left.\begin{array}{c} f'(x) = \tfrac{1}{2}\left(\dfrac{dy}{dx} - \dfrac{1}{a}\dfrac{dy}{dt}\right) \\[2mm] F'(x) = \tfrac{1}{2}\left(\dfrac{dy}{dx} + \dfrac{1}{a}\dfrac{dy}{dt}\right) \end{array}\right\} \quad\ldots\ldots\ldots\ldots\ldots(8),$$

equations which determine the functions f' and F' for all values of the argument from $x = -\infty$ to $x = \infty$, if the initial values of $\frac{dy}{dx}$ and $\frac{dy}{dt}$ be known.

If the disturbance be originally confined to a finite portion of the string, the positive and negative waves separate after the interval of time required for each to traverse half the disturbed portion.

Fig. 23.

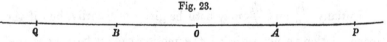

Suppose, for example, that AB is the part initially disturbed. A point P on the positive side remains at rest until the positive wave has travelled from A to P, is disturbed during the passage of the wave, and ever after remains at rest. The negative wave never affects P at all. Similar statements apply, *mutatis mutandis*, to a point Q on the negative side of AB. If the character of the original disturbance be such that $\frac{dy}{dx} - \frac{1}{a}\frac{dy}{dt}$ vanishes initially, there is no positive wave, and the point P is never disturbed at all; and if $\frac{dy}{dx} + \frac{1}{a}\frac{dy}{dt}$ vanish initially, there is no negative wave. If $\frac{dy}{dt}$ vanish initially, the positive and the negative waves are similar and equal, and then neither can vanish. In cases where either wave vanishes, its evanescence may be considered to be due to the mutual destruction of two component waves, one depending on the initial displacements, and the other on the initial velocities. On the one side these two waves conspire, and on the other they destroy one another. This explains the apparent paradox, that P can fail to be affected sooner or later after AB has been disturbed.

The subsequent motion of a string that is initially displaced without velocity, may be readily traced by graphical methods. Since the positive and the negative waves are equal, it is only necessary to divide the original disturbance into two equal parts, to displace these, one to the right, and the other to the left, through a space equal to at, and then to recompound them. We shall presently apply this method to the case of a plucked string of finite length.

145.   Vibrations are called *stationary*, when the motion of each particle of the system is proportional to some function of the time, the same for all the particles.   If we endeavour to satisfy

$$\frac{d^2y}{dt^2} = a^2 \frac{d^2y}{dx^2} \dots\dots\dots\dots\dots\dots\dots\dots(1),$$

by assuming $y = XT$, where X denotes a function of x only, and T a function of t only, we find

$$\frac{1}{T}\frac{d^2T}{d(at)^2} = \frac{1}{X}\frac{d^2X}{dx^2} = m^2 \quad \text{(a constant)},$$

so that

$$\begin{aligned} T &= A\cos mat + B\sin mat \\ X &= C\cos mx + D\sin mx \end{aligned} \Bigg\} \dots\dots\dots\dots(2),$$

proving that the vibrations must be simple harmonic, though of arbitrary period.   The value of y may be written

$$y = P\cos(mat - \epsilon)\ \cos(mx - \alpha)$$
$$= \tfrac{1}{2}P\cos(mat + mx - \epsilon - \alpha) + \tfrac{1}{2}P\cos(mat - mx - \epsilon + \alpha)\dots\dots(3),$$

shewing that the most general kind of stationary vibration may be regarded as due to the superposition of equal progressive vibra-

182   TRANSVERSE VIBRATIONS OF STRINGS.   [145.

tions, whose directions of propagation are opposed. Conversely, two stationary vibrations may combine into a progressive one.

The solution $y = f(x - at) + F(x + at)$ applies in the first instance to an infinite string, but may be interpreted so as to give the solution of the problem for a finite string in certain cases. Let us suppose, for example, that the string terminates at $x = 0$, and is held fast there, while it extends to infinity in the positive direction only. Now so long as the point $x = 0$ actually remains at rest, it is a matter of indifference whether the string be prolonged on the negative side or not. We are thus led to regard the given string as forming part of one doubly infinite, and to seek whether and how the initial displacements and velocities on the negative side can be taken, so that on the whole there shall be no displacement at $x = 0$ throughout the subsequent motion. The initial values of y and $\dot{y}$ on the positive side determine the corresponding parts of the positive and negative waves, into which we know that the whole motion can be resolved. The former has no influence at the point $x = 0$. On the negative side the positive and the negative waves are initially at our disposal, but with the latter we are not concerned. The problem is to determine the positive wave on the negative side, so that in conjunction with the given negative wave on the positive side of the origin, it shall leave that point undisturbed.

Let $OPQRS...$ be the line (of any form) representing the wave in OX, which advances in the negative direction. It is

Fig. 24.

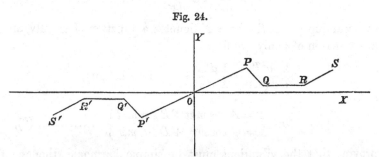

evident that the requirements of the case are met by taking on the other side of O what may be called the *contrary* wave, so that O is the geometrical centre, bisecting every chord (such as PP') which passes through it. Analytically, if $y = f(x)$ is the equation of $OPQRS......$, $-y = f(-x)$ is the equation of $OP'Q'R'S'......$

When after a time t the curves are shifted to the left and to the right respectively through a distance at, the co-ordinates corresponding to $x = 0$ are necessarily equal and opposite, and therefore when compounded give zero resultant displacement.

The effect of the constraint at O may therefore be represented by supposing that the negative wave moves through undisturbed, but that a positive wave at the same time emerges from O. This reflected wave may at any time be found from its parent by the following rule:

Let $APQRS...$ be the position of the parent wave. Then the reflected wave is the position which this would assume, if it were

Fig. 25.

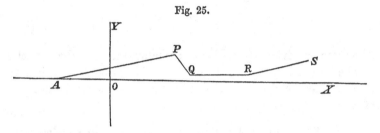

turned through two right angles, first about OX as an axis of rotation, and then through the same angle about OY. In other words, the return wave is the image of $APQRS$ formed by successive optical reflection in OX and OY, regarded as plane mirrors.

The same result may also be obtained by a more analytical process. In the general solution

$$y = f(x - at) + F(x + at),$$

the functions $f(z)$, $F(z)$ are determined by the initial circumstances for all positive values of z. The condition at $x = 0$ requires that

$$f(-at) + (F(at) = 0$$

for all positive values of t, or

$$f(-z) = -F(z)$$

for positive values of z. The functions f and F are thus determined for all positive values of x and t.

There is now no difficulty in tracing the course of events when *two* points of the string A and B are held fast. The initial disturbance in AB divides itself into positive and negative waves, which are reflected backwards and forwards between the fixed

points, changing their character from positive to negative, and
vice versâ, at each reflection. After an even number of reflec-
tions in each case the original form and motion is completely
recovered. The process is most easily followed in imagination
when the initial disturbance is confined to a small part of the
string, more particularly when its character is such as to give rise
to a wave propagated in one direction only. The *pulse* travels with
uniform velocity (a) to and fro along the length of the string, and
after it has returned *a second time* to its starting point the
original condition of things is exactly restored. The period of
the motion is thus the time required for the pulse to traverse
the length of the string twice, or

$$\tau = \frac{2l}{a} \quad \dots\dots\dots\dots\dots\dots\dots\dots\dots\dots(1).$$

The same law evidently holds good whatever may be the character
of the original disturbance, only in the general case it may
happen that the *shortest* period of recurrence is some aliquot part
of τ.

146. The method of the last few sections may be advantageously applied to the case of a plucked string. Since the initial
velocity vanishes, half of the displacement belongs to the positive
and half to the negative wave. The manner in which the wave
must be completed so as to produce the same effect as the constraint, is shewn in the figure, where the upper curve represents

Fig. 26.

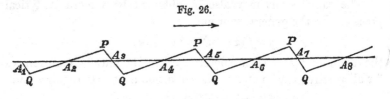

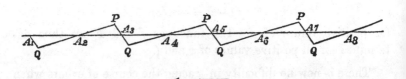

the positive, and the lower the negative wave in their initial
positions. In order to find the configuration of the string at any

future time, the two curves must be superposed, after the upper has been shifted to the right and the lower to the left through a space equal to at.

The resultant curve, like its components, is made up of straight pieces. A succession of six at intervals of a twelfth of the period,

Fig. 27.

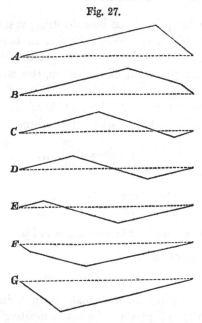

shewing the course of the vibration, is given in the figure (Fig. 27), taken from Helmholtz. From G the string goes back again to A through the same stages[1].

It will be observed that the inclination of the string at the points of support alternates between two constant values.

147. If a small disturbance be made at the time t at the point x of an infinite stretched string, the effect will not be felt at O until after the lapse of the time $x \div a$, and will be in all respects the same as if a like disturbance had been made at the point $x + \Delta x$ at time $t - \Delta x \div a$. Suppose that similar disturbances are communicated to the string at intervals of time τ at points whose distances from O increase each time by $a\,\delta\tau$, then

[1] This method of treating the vibration of a plucked string is due to Young. *Phil. Trans.*, 1800. The student is recommended to make himself familiar with it by actually constructing the forms of Fig. 27.

it is evident that the result at O will be the same as if the disturbances were all made at the same point, provided that the time-intervals be increased from τ to $\tau + \delta\tau$. This remark contains the theory of the alteration of pitch due to motion of the source of disturbance; a subject which will come under our notice again in connection with aerial vibrations.

148.   When one point of an infinite string is subject to a forced vibration, trains of waves proceed from it in both directions according to laws, which are readily investigated.   We shall suppose that the origin is the point of excitation, the string being there subject to the forced motion $y = Ae^{ipt}$; and it will be sufficient to consider the positive side.   If the motion of each element ds be resisted by the frictional force $\kappa \dot{y}\,ds$, the differential equation is

$$\frac{d^2y}{dt^2} + \kappa\frac{dy}{dt} = a^2\frac{d^2y}{dx^2}\dots\dots\dots\dots\dots\dots(1);$$

or since $y \propto e^{ipt}$,

$$\frac{d^2y}{dx^2} = \left(\frac{i\kappa p}{a^2} - \frac{p^2}{a^2}\right)y = \lambda^2 y\dots\dots\dots\dots\dots(2),$$

if for brevity we write λ^2 for the coefficient of y.

The general solution is

$$y = \{Ce^{-\lambda x} + De^{+\lambda x}\}\,e^{ipt}\dots\dots\dots\dots\dots(3).$$

Now since y is supposed to vanish at an infinite distance, D must vanish, if the real part of λ be taken positive.   Let

$$\lambda = \alpha + i\beta,$$

where α is positive.

Then the solution is

$$y = Ae^{-(\alpha + i\beta)x + ipt}\dots\dots\dots\dots\dots\dots(4),$$

or, on throwing away the imaginary part,

$$y = Ae^{-\alpha x}\ \cos(pt - \beta x)\dots\dots\dots\dots\dots(5),$$

corresponding to the forced motion at the origin

$$y = A\cos pt\dots\dots\dots\dots\dots\dots(6).$$

An arbitrary constant may, of course, be added to t.

To determine α and β, we have

$$\alpha^2 - \beta^2 = -\frac{p^2}{a^2};\qquad 2\alpha\beta = \frac{\kappa p}{a^2}\dots\dots\dots\dots\dots(7).$$

If we suppose that κ is small,

$$\beta = \frac{p}{a}, \qquad \alpha = \frac{\kappa}{2a} \text{ nearly,}$$

and
$$y = A e^{-\frac{\kappa}{2a}x} \cos\left(pt - \frac{p}{a}x\right) \dots\dots\dots(8).$$

This solution shews that there is propagated along the string a wave, whose amplitude slowly diminishes on account of the exponential factor. If $\kappa = 0$, this factor disappears, and we have simply

$$y = A \cos\left(pt - \frac{px}{a}\right)\dots\dots\dots(9).$$

This result stands in contradiction to the general law that, when there is no friction, the forced vibrations of a system (due to a single simple harmonic force) must be synchronous in phase throughout. According to (9), on the contrary, the phase varies continuously in passing from one point to another along the string. The fact is, that we are not at liberty to suppose $\kappa = 0$ in (8), inasmuch as that equation was obtained on the assumption that the real part of λ in (3) is positive, and not zero. However long a finite string may be, the coefficient of friction may be taken so small that the vibrations are not damped before reaching the further end. After this point of smallness, reflected waves begin to complicate the result, and when the friction is diminished indefinitely, an infinite series of such must be taken into account, and would give a resultant motion of the same phase throughout.

This problem may be solved for a string whose mass is supposed to be concentrated at equidistant points, by the method of § 120. The co-ordinate ψ_1 may be supposed to be given ($= A e^{ipt}$), and it will be found that the system of equations (5) of § 120 may all be satisfied by taking

$$\psi_r = \theta^{r-1}\psi_1,$$

where θ is a complex constant determined by a quadratic equation. The result for a continuous string may be afterwards deduced.

CHAPTER VII.

LONGITUDINAL AND TORSIONAL VIBRATIONS OF BARS.

149. THE next system to the string in order of simplicity is the bar, by which term is usually understood in Acoustics a mass of matter of uniform substance and elongated cylindrical form. At the ends the cylinder is cut off by planes perpendicular to the generating lines. The centres of inertia of the transverse sections lie on a straight line which is called the *axis*.

The vibrations of a bar are of three kinds—longitudinal, torsional, and lateral. Of these the last are the most important, but at the same time the most difficult in theory. They are considered by themselves in the next chapter, and will only be referred to here so far as is necessary for comparison and contrast with the other two kinds of vibrations.

Longitudinal vibrations are those in which the axis remains unmoved, while the transverse sections vibrate to and fro in the direction perpendicular to their planes. The moving power is the resistance offered by the rod to extension or compression.

One peculiarity of this class of vibrations is at once evident. Since the force necessary to produce a given extension in a bar is proportional to the area of the section, while the mass to be moved is also in the same proportion, it follows that for a bar of given length and material the periodic times and the modes of vibration are independent of the area and of the form of the transverse section. A similar law obtains, as we shall presently see, in the case of torsional vibrations.

It is otherwise when the vibrations are lateral. The periodic times are indeed independent of the thickness of the bar in the direction perpendicular to the plane of flexure, but the motive power in this case, viz. the resistance to bending, increases more rapidly than the thickness in that plane, and therefore an increase in thickness is accompanied by a rise of pitch.

In the case of longitudinal and lateral vibrations, the mechanical constants concerned are the density of the material and the value of Young's modulus. For small extensions (or compressions) Hooke's law, according to which the tension varies as the extension, holds good. If the extension, viz. $\dfrac{\text{actual length} - \text{natural length}}{\text{natural length}}$, be called ϵ, we have $T = q\epsilon$, where q is Young's modulus, and T is the tension per unit area necessary to produce the extension ϵ. Young's modulus may therefore be defined as the force which would have to be applied to a bar of unit section, in order to double its length, if Hooke's law continued to hold good for so great extensions; its dimensions are accordingly those of a force divided by an area.

The torsional vibrations depend also on a second elastic constant μ, whose interpretation will be considered in the proper place.

Although in theory the three classes of vibrations, depending respectively on resistance to extension, to torsion, and to flexure are quite distinct, and independent of one another so long as the squares of the strains may be neglected, yet in actual experiments with bars which are neither uniform in material nor accurately cylindrical in figure it is often found impossible to excite longitudinal or torsional vibrations without the accompaniment of some measure of lateral motion. In bars of ordinary dimensions the gravest lateral motion is far graver than the gravest longitudinal or torsional motion, and consequently it will generally happen that the principal tone of either of the latter kinds agrees more or less perfectly in pitch with some overtone of the former kind. Under such circumstances the regular modes of vibrations become unstable, and a small irregularity may produce a great effect. The difficulty of exciting purely longitudinal vibrations in a bar is similar to that of getting a string to vibrate in one plane.

With this explanation we may proceed to consider the three classes of vibrations independently, commencing with longitudinal vibrations, which will in fact raise no mathematical questions beyond those already disposed of in the previous chapters.

150.   When a rod is stretched by a force parallel to its length, the stretching is in general accompanied by lateral contraction in such a manner that the augmentation of volume is less than if the displacement of every particle were parallel to the axis.   In the case of a short rod and of a particle situated near the cylindrical boundary, this lateral motion would be comparable in magnitude with the longitudinal motion, and could not be overlooked without risk of considerable error.   But where a rod, whose length is great in proportion to the linear dimensions of its section, is subject to a stretching of one sign throughout, the longitudinal motion accumulates, and thus in the case of ordinary rods vibrating longitudinally in the graver modes, the inertia of the lateral motion may be neglected.   Moreover we shall see later how a correction may be introduced, if necessary.

Let x be the distance of the layer of particles composing any section from the equilibrium position of one end, when the rod is unstretched, either by permanent tension or as the result of vibrations, and let ξ be the displacement, so that the actual position is given by $x + \xi$.   The equilibrium and actual position of a neighbouring layer being $x + \delta x$, $x + \delta x + \xi + \frac{d\xi}{dx}\delta x$ respectively, the *elongation* is $\frac{d\xi}{dx}$; and thus, if T be the tension per unit area acting across the section,

$$T = q\,\frac{d\xi}{dx}\dots\dots\dots\dots\dots\dots(1).$$

Consider now the forces acting on the slice bounded by x and $x + \delta x$.   If the area of the section be ω, the tension at x is by (1) $q\omega\,\dfrac{d\xi}{dx}$, acting in the negative direction, and at $x + \delta x$ the tension is

$$q\omega\left(\frac{d\xi}{dx} + \frac{d^2\xi}{dx^2}\delta x\right),$$

acting in the positive direction; and thus the force on the slice due to the action of the adjoining parts is on the whole

$$q\omega\frac{d^2\xi}{dx^2}\delta x.$$

The mass of the element is $\rho\omega\,\delta x$, if ρ be the original density, and therefore if X be the accelerating force acting on it, the equation of equilibrium is

$$X + \frac{q}{\rho}\frac{d^2\xi}{dx^2} = 0 \dots\dots\dots\dots\dots(2).$$

In what follows we shall not require to consider the operation of an impressed force. To find the equation of motion we have only to replace X by the reaction against acceleration $-\ddot{\xi}$, and thus if $q : \rho = a^2$, we have

$$\frac{d^2\xi}{dt^2} = a^2\frac{d^2\xi}{dx^2}\dots\dots\dots\dots\dots(3).$$

This equation is of the same form as that applicable to the transverse displacements of a stretched string, and indicates the undisturbed propagation of waves of any type in the positive and negative directions. The velocity a is relative to the *unstretched* condition of the bar; the apparent velocity with which a disturbance is propagated in space will be greater in the ratio of the stretched and unstretched lengths of any portion of the bar. The distinction is material only in the case of permanent tension.

151. For the actual magnitude of the velocity of propagation, we have

$$a^2 = q : \rho = q\omega : \rho\omega,$$

which is the ratio of the whole tension necessary (according to Hooke's law) to double the length of the bar and the longitudinal density. If the same bar were stretched with total tension T, and were flexible, the velocity of propagation of waves along it would be $\sqrt{(T : \rho\omega)}$. In order then that the velocity might be the same in the two cases, T must be $q\omega$, or, in other words, the tension would have to be that theoretically necessary in order to double the length. The tones of longitudinally vibrating rods are thus very high in comparison with those obtainable from strings of comparable length.

In the case of steel the value of q is about 22×10^8 grammes weight per square centimetre. To express this in absolute units of force on the c. g. s.[1] system, we must multiply by 980. In the same system the density of steel (identical with its specific gravity referred to water) is 7·8. Hence for steel

$$a = \sqrt{\frac{980 \times 22 \times 10^8}{7\cdot8}} = 530,000$$

---

[1] Centimetre, Gramme, Second. This system is recommended by a Committee of the British Association. Brit. Ass. Report, 1873.

approximately, which shews that the velocity of sound in steel is about 530,000 centimetres per second, or about 16 times greater than the velocity of sound in air. In glass the velocity is about the same as in steel.

It ought to be mentioned that in strictness the value of q determined by statical experiments is not that which ought to be used here. As in the case of gases, which will be treated in a subsequent chapter, the rapid alterations of state concerned in the propagation of sound are attended with thermal effects, one result of which is to increase the effective value of q beyond that obtained from observations on extension conducted at a constant temperature. But the data are not precise enough to make this correction of any consequence in the case of solids.

152. The solution of the general equation for the longitudinal vibrations of an unlimited bar, namely

$$\xi = f(x - at) + F(x + at),$$

being the same as that applicable to a string, need not be further considered here.

When both ends of a bar are free, there is of course no permanent tension, and at the ends themselves there is no temporary tension. The condition for a free end is therefore

$$\frac{d\xi}{dx} = 0 \dots\dots\dots\dots(1).$$

To determine the normal modes of vibration, we must assume that ξ varies as a harmonic function of the time—$\cos nat$. Then as a function of x, ξ must satisfy

$$\frac{d^2\xi}{dx^2} + n^2\xi = 0 \dots\dots\dots\dots(2),$$

of which the complete integral is

$$\xi = A \cos nx + B \sin nx \dots\dots\dots(3),$$

where A and B are independent of x.

Now since $\frac{d\xi}{dx}$ vanishes always when $x = 0$, we get $B = 0$; and again since $\frac{d\xi}{dx}$ vanishes when $x = l$—the natural length of the bar, $\sin nl = 0$, which shews that n is of the form

$$n = \frac{i\pi}{l} \dots\dots\dots\dots(4),$$

i being integral.

Accordingly, the normal modes are given by equations of the form

$$\xi = A \cos \frac{i\pi x}{l} \cos \frac{i\pi a t}{l} \dots\dots\dots\dots\dots(5),$$

in which of course an arbitrary constant may be added to t, if desired.

The complete solution for a bar with both ends free is therefore expressed by

$$\xi = \Sigma_{i=0}^{i=\infty} \cos \frac{i\pi x}{l} \left\{ A_i \cos \frac{i\pi a t}{l} + B_i \sin \frac{i\pi a t}{l} \right\} \dots\dots(6),$$

where A_i and B_i are arbitrary constants, which may be determined in the usual manner, when the initial values of ξ and $\dot{\xi}$ are given.

A zero value of i is admissible; it gives a term representing a displacement ξ constant with respect both to space and time, and amounting in fact only to an alteration of the origin.

The period of the gravest component in (6) corresponding to $i = 1$, is $2l \div a$, which is the time occupied by a disturbance in travelling twice the length of the rod. The other tones found by ascribing integral values to i form a complete harmonic scale; so that according to this theory the note given by a rod in longitudinal vibration would be in all cases musical.

In the gravest mode the centre of the rod, where $x = \frac{1}{2}l$, is a place of no motion, or node; but the periodic elongation or compression $\frac{d\xi}{dx}$ is there a maximum.

153. The case of a bar with one end free and the other fixed may be deduced from the general solution for a bar with both ends free, and of twice the length. For whatever may be the initial state of the bar free at $x = 0$ and fixed at $x = l$, such displacements and velocities may always be ascribed to the sections of a bar extending from 0 to $2l$ and free at both ends as shall make the motions of the parts from 0 to l identical in the two cases. It is only necessary to suppose that from l to $2l$ the displacements and velocities are initially equal and opposite to those found in the portion from 0 to l at an equal distance from the centre $x = l$. Under these circumstances the centre must by the symmetry remain at rest throughout the motion, and then the

portion from 0 to l satisfies all the required conditions. We conclude that the vibrations of a bar free at one end and fixed at the other are identical with those of one half of a bar of twice the length of which both ends are free, the latter vibrating only in the uneven modes, obtained by making i in succession all *odd* integers. The tones of the bar still belong to a harmonic scale, but the even tones (octave, &c. of the fundamental) are wanting.

The period of the gravest tone is the time occupied by a pulse in travelling *four* times the length of the bar.

154. When both ends of a bar are fixed, the conditions to be satisfied at the ends are that the value of ξ is to be invariable. At $x = 0$, we may suppose that $\xi = 0$. At $x = l$, ξ is a small constant α, which is zero if there be no permanent tension. Independently of the vibrations we have evidently $\xi = x\alpha \div l$, and we should obtain our result most simply by assuming this term at once. But it may be instructive to proceed by the general method.

Assuming that as a function of the time ξ varies as

$$A \cos nat + B \sin nat,$$

we see that as a function of x it must satisfy

$$\frac{d^2\xi}{dx^2} + n^2\xi = 0,$$

of which the general solution is

$$\xi = C \cos nx + D \sin nx \dots\dots\dots\dots(1).$$

But since ξ vanishes with x for all values of t, $C = 0$, and thus we may write

$$\xi = \Sigma \sin nx \{A \cos nat + B \sin nat\}.$$

The condition at $x = l$ now gives

$$\Sigma \sin nl \{A \cos nat + B \sin nat\} = \alpha,$$

from which it follows that for every finite admissible value of n

$$\sin nl = 0, \text{ or } n = \frac{i\pi}{l}.$$

But for the zero value of n, we get

$$A_{_0} \sin nl = \alpha,$$

and the corresponding term in ξ is

$$\xi = A_0 \sin nx = \alpha \frac{\sin nx}{\sin nl} = \alpha \frac{x}{l}.$$

The complete value of ξ is accordingly

$$\xi = \alpha \frac{x}{l} + \Sigma_{i=1}^{i=\infty} \sin \frac{i\pi x}{l} \left\{ A \cos \frac{i\pi at}{l} + B \sin \frac{i\pi at}{l} \right\} \dots (2).$$

The series of tones form a complete harmonic scale (from which however any of the members may be missing in any actual case of vibration), and the period of the gravest component is the time taken by a pulse to travel twice the length of the rod, the same therefore as if both ends were free. It must be observed that we have here to do with the *unstretched* length of the rod, and that the period for a given natural length is independent of the permanent tension.

The solution of the problem of the doubly fixed bar in the case of no permanent tension might also be derived from that of a doubly free bar by mere differentiation with respect to x. For in the latter problem $\frac{d\xi}{dx}$ satisfies the necessary differential equation, viz.

$$\frac{d^2}{dt^2}\left(\frac{d\xi}{dx}\right) = a^2 \frac{d^2}{dx^2}\left(\frac{d\xi}{dx}\right),$$

inasmuch as ξ satisfies

$$\frac{d^2\xi}{dt^2} = a^2 \frac{d^2\xi}{dx^2};$$

and at both ends $\frac{d\xi}{dx}$ vanishes. Accordingly $\frac{d\xi}{dx}$ in this problem satisfies all the conditions prescribed for ξ in the case when both ends are fixed. The two series of tones are thus identical.

155. The effect of a small load M attached to any point of the rod is readily calculated approximately, as it is sufficient to assume the type of vibration to be unaltered (§ 88). We will take the case of a rod fixed at $x=0$, and free at $x=l$. The kinetic energy is proportional to

$$\tfrac{1}{2}\int_0^l \rho\omega \sin^2 \frac{i\pi x}{2l}\, dx + \tfrac{1}{2}M \sin^2 \frac{i\pi x}{2l},$$

or to

$$\frac{\rho\omega l}{4}\left(1 + \frac{2M}{\rho\omega l}\sin^2\frac{i\pi x}{2l}\right).$$

Since the potential energy is unaltered, we see by the principles of Chapter IV., that the effect of the small load M at a distance x from the fixed end is to increase the period of the component tones in the ratio

$$1 : 1 + \frac{M}{\rho \omega l} \sin^2 \frac{i\pi x}{2l}.$$

The small quantity $M : \rho \omega l$ is the ratio of the load to the whole mass of the rod.

If the load be attached at the free end, $\sin^2 \frac{i\pi x}{2l} = 1$, and the effect is to depress the pitch of every tone by the same small interval. It will be remembered that i is here an *uneven* integer.

If the point of attachment of M be a node of any component, the pitch of that component remains unaltered by the addition.

156. Another problem worth notice occurs when the load at the free end is great in comparison with the mass of the rod. In this case we may assume as the type of vibration, a condition of uniform extension along the length of the rod.

If ξ be the displacement of the load M, the kinetic energy is

$$T = \tfrac{1}{2} M \dot{\xi}^2 + \tfrac{1}{2} \dot{\xi}^2 \int_0^l \rho \omega \frac{x^2}{l^2} \, dx = \tfrac{1}{2} \dot{\xi}^2 \left(M + \tfrac{1}{3} \rho \omega l \right) \ldots\ldots\ldots(1).$$

The tension corresponding to the displacement ξ is $q\omega \frac{\xi}{l}$, and thus the potential energy of the displacement is

$$V = \frac{q\omega \xi^2}{2l} \ \ldots\ldots\ldots\ldots\ldots\ldots\ldots(2).$$

The equation of motion is

$$\left(M + \tfrac{1}{3}\rho \omega l \right) \ddot{\xi} + \frac{q\omega}{l} \xi = 0,$$

and if $\xi \propto \cos pt$

$$p^2 = \frac{q\omega}{l} \div \left(M + \tfrac{1}{3}\rho \omega l \right) \ \ldots\ldots\ldots\ldots\ldots (3).$$

The correction due to the inertia of the rod is thus equivalent to the addition to M of one-third of the mass of the rod.

157. Our mathematical discussion of longitudinal vibrations may close with an estimate of the error involved in neglecting the lateral motion of the parts of the rod not situated on the

axis. If the ratio of lateral contraction to longitudinal extension be denoted by μ, the lateral displacement of a particle distant r from the axis will be $\mu r \epsilon$, in the case of equilibrium, where ϵ is the extension. Although in strictness this relation will be modified by the inertia of the lateral motion, yet for the present purpose it may be supposed to hold good.

The constant μ is a numerical quantity, lying between 0 and $\frac{1}{2}$. If μ were negative, a longitudinal tension would produce a lateral swelling, and if μ were greater than $\frac{1}{2}$, the lateral contraction would be great enough to overbalance the elongation, and cause a diminution of volume on the whole. The latter state of things would be inconsistent with stability, and the former can scarcely be possible in ordinary solids. At one time it was supposed that μ was necessarily equal to $\frac{1}{4}$, so that there was only one independent elastic constant, but experiments have since shewn that μ is variable. For glass and brass Wertheim found experimentally $\mu = \frac{1}{3}$.

If η denote the lateral displacement of the particle distant r from the axis, and if the section be circular, the kinetic energy due to the lateral motion is

$$\delta T = \pi \rho \int_0^l \int_0^r \dot{\eta}^2 dx \, . \, r dr$$

$$= \frac{\rho \omega \mu^2 r^2}{4} \, . \int_0^l \left(\frac{d\xi}{dx}\right)^2 dx.$$

Thus the whole kinetic energy is

$$T + \delta T = \frac{\rho \omega}{2} \int_0^l \xi^2 dx + \frac{\rho \omega \mu^2 r^2}{4} \int_0^l \left(\frac{d\xi}{dx}\right)^2 dx.$$

In the case of a bar free at both ends, we have

$$\xi \propto \cos \frac{i \pi x}{l}, \quad \frac{d\xi}{dx} \propto -\frac{i\pi}{l} \sin \frac{i \pi x}{l},$$

and thus

$$T + \delta T : T = 1 + \frac{i^2 \mu^2 \pi^2}{2} \frac{r^2}{l^2} \, .$$

The effect of the inertia of the lateral motion is therefore to increase the period in the ratio

$$1 : 1 + \frac{i^2 \mu^2 \pi^2}{4} \frac{r^2}{l^2} \, .$$

This correction will be nearly insensible for the graver modes of bars of ordinary proportions of length to thickness.

158. Experiments on longitudinal vibrations may be made with rods of deal or of glass. The vibrations are excited by friction, with a wet cloth in the case of glass; but for metal or wooden rods it is necessary to use leather charged with powdered rosin. "The longitudinal vibrations of a pianoforte string may be excited by gently rubbing it longitudinally with a piece of india rubber, and those of a violin string by placing the bow obliquely across the string, and moving it along the string longitudinally, keeping the same point of the bow upon the string. The note is unpleasantly shrill in both cases."

"If the peg of the violin be turned so as to alter the pitch of the lateral vibrations very considerably, it will be found that the pitch of the longitudinal vibrations has altered very slightly. The reason of this is that in the case of the lateral vibrations the change of velocity of wave-transmission depends chiefly on the change of tension, which is considerable. But in the case of the longitudinal vibrations, the change of velocity of wave-transmission depends upon the change of extension, which is comparatively slight[1]."

In Savart's experiments on longitudinal vibrations, a peculiar sound, called by him a "son rauque," was occasionally observed, whose pitch was an octave below that of the longitudinal vibration. According to Terquem[2] the cause of this sound is a transverse vibration, whose appearance is due to an approximate agreement between its own period and that of the sub-octave of the longitudinal vibration. If this view be correct, the phenomenon would be one of the second order, probably referable to the fact that longitudinal compression of a bar tends to produce curvature.

159. The second class of vibrations, called torsional, which depend on the resistance opposed to twisting, is of very small importance. A solid or hollow cylindrical rod of circular section may be twisted by suitable forces, applied at the ends, in such a manner that each transverse section remains in its own plane. But if the section be not circular, the effect of a twist is of a more complicated character, the twist being necessarily attended by a warping of the layers of matter originally composing the normal sections. Although the effects of the warping might pro-

---

[1] Donkin's *Acoustics*, p. 154.
[2] *Ann. de Chimie*, LVII. 129—190.

bably be determined in any particular case if it were worth while, we shall confine ourselves here to the case of a circular section, when there is no motion parallel to the axis of the rod.

The force with which twisting is resisted depends upon an elastic constant different from q, called the rigidity. If we denote it by n, the relation between q, n, and μ may be written

$$n = \frac{q}{2(\mu + 1)} \dots\dots\dots\dots\dots\dots(1)^1,$$

shewing that n lies between $\frac{1}{2}q$ and $\frac{1}{3}q$. In the case of $\mu = \frac{1}{3}$, $n = \frac{3}{8}q$.

Let us now suppose that we have to do with a rod in the form of a thin tube of radius r and thickness dr, and let θ denote the angular displacement of any section, distant x from the origin. The rate of twist at x is represented by $\frac{d\theta}{dx}$, and the shear of the material composing the pipe by $r\frac{d\theta}{dx}$. The opposing force per unit of area is $nr\frac{d\theta}{dx}$; and since the area is $2\pi r\,dr$, the moment round the axis is

$$2n\pi r^3\,dr\,\frac{d\theta}{dx}.$$

Thus the force of restitution acting on the slice dx has the moment

$$2n\pi r^3\,dr\,dx\,\frac{d^2\theta}{dx^2}.$$

Now the moment of inertia of the slice under consideration is $2\pi r\,dr\,.\,dx\,.\,\rho\,.\,r^2$, and therefore the equation of motion assumes the form

$$\rho\,\frac{d^2\theta}{dt^2} = n\,\frac{d^2\theta}{dx^2} \dots\dots\dots\dots\dots\dots(2).$$

Since this is independent of r, the same equation applies to a cylinder of finite thickness or to one solid throughout.

The velocity of wave propagation is $\sqrt{\dfrac{n}{\rho}}$, and the whole theory is precisely similar to that of longitudinal vibrations, the condition

---

[1] Thomson and Tait, § 683. This, it should be remarked, applies to isotropic material only.

for a free end being $\dfrac{d\theta}{dx} = 0$, and for a fixed end $\theta = 0$, or, if a permanent twist be contemplated, $\theta = $ constant.

The velocity of longitudinal vibrations is to that of torsional vibrations in the ratio $\sqrt{q} : \sqrt{n}$ or $\sqrt{(2+2\mu)} : 1$. The same ratio applies to the frequencies of vibration for bars of equal length vibrating in corresponding modes under corresponding terminal conditions. If $\mu = \tfrac{1}{3}$, the ratio of frequencies would be

$$\sqrt{q} : \sqrt{n} = \sqrt{8} : \sqrt{3} = 1\cdot63,$$

corresponding to an interval rather greater than a fifth.

In any case the ratio of frequencies must lie between

$$\sqrt{2} : 1 = 1\cdot414, \quad \text{and } \sqrt{3} : 1 = 1\cdot732.$$

Longitudinal and torsional vibrations were first investigated by Chladni.

CHAPTER VIII.

LATERAL VIBRATIONS OF BARS.

160. In the present chapter we shall consider the lateral vibrations of thin elastic rods, which in their natural condition are straight. Next to those of strings, this class of vibrations is perhaps the most amenable to theoretical and experimental treatment. There is difficulty sufficient to bring into prominence some important points connected with the general theory, which the familiarity of the reader with circular functions may lead him to pass over too lightly in the application to strings; while at the same time the difficulties of analysis are not such as to engross attention which should be devoted to general mathematical and physical principles.

Daniel Bernoulli[1] seems to have been the first who attacked the problem. Euler, Riccati, Poisson, Cauchy, and more recently Strehlke[2], Lissajous[3], and A. Seebeck[4] are foremost among those who have advanced our knowledge of it.

161. The problem divides itself into two parts, according to the presence, or absence, of a permanent longitudinal tension. The consideration of permanent tension entails additional complication, and is of interest only in its application to stretched strings, whose stiffness, though small, cannot be neglected altogether. Our attention will therefore be given principally to the two extreme cases, (1) when there is no permanent tension, (2) when the tension is the chief agent in the vibration.

---

[1] *Comment. Acad. Petrop.* t. xiii. [2] Pogg. *Ann.* Bd. xxvii.

[3] *Ann. d. Chimie* (3), xxx. 385.

[4] *Abhandlungen d. Math. Phys. Classe d. K. Sächs. Gesellschaft d. Wissenschaften.* Leipzig, 1852.

With respect to the section of the rod, we shall suppose that one principal axis lies in the plane of vibration, so that the bending at every part takes place in a direction of maximum or minimum (or stationary) flexural rigidity. For example, the surface of the rod may be one of revolution, each section being circular, though not necessarily of constant radius. Under these circumstances the potential energy of the bending for each element of length is proportional to the square of the curvature multiplied by a quantity depending on the material of the rod, and on the moment of inertia of the transverse section about an axis through its centre of inertia perpendicular to the plane of bending. If ω be the area of the section, $\kappa^2\omega$ its moment of inertia, q Young's modulus, ds the element of length, and dV the corresponding potential energy for a curvature $1 \div R$ of the axis of the rod,

$$dV = \tfrac{1}{2}\, q\, \kappa^2 \omega\, \frac{ds}{R^2} \quad\dotfill(1).$$

This result is readily obtained by considering the extension of the various filaments of which the bar may be supposed to be made up. Let η be the distance from the axis of the projection on the plane of bending of a filament of section $d\omega$. Then the length of the filament is altered by the bending in the ratio

$$1 : 1 + \frac{\eta}{R},$$

R being the radius of curvature. Thus on the side of the axis for which η is positive, viz. on the *outward* side, a filament is extended, while on the other side of the axis there is compression. The force necessary to produce the extension $\frac{\eta}{R}$ is $q\,\frac{\eta}{R}\,d\omega$ by the definition of Young's modulus; and thus the whole couple by which the bending is resisted amounts to

$$\int q\frac{\eta}{R}\,.\,\eta\,.\,d\omega = \frac{q}{R}\kappa^2\omega,$$

if ω be the area of the section and κ its radius of gyration about a line through the axis, and perpendicular to the plane of bending. The angle of bending corresponding to a length of axis ds is $ds \div R$, and thus the work required to bend ds to curvature $1 \div R$ is

$$\tfrac{1}{2}\, q\, \kappa^2 \omega\, \frac{ds}{R^2},$$

since the *mean* is half the *final* value of the couple.

For a circular section κ is one-half the radius.

That the potential energy of the bending would be proportional, *cœteris paribus*, to the square of the curvature, is evident beforehand. If we call the coefficient B, we may take

$$V = \tfrac{1}{2} \int B \, \frac{ds}{R^2},$$

or, in view of the approximate straightness,

$$V = \tfrac{1}{2} \int B \left(\frac{d^2 y}{dx^2}\right)^2 dx \quad \dots\dots\dots\dots\dots\dots(2),$$

in which y is the lateral displacement of that point on the axis of the rod whose abscissa, measured parallel to the undisturbed position, is x. In the case of a rod whose sections are similar and similarly situated B is a constant, and may be removed from under the integral sign.

The kinetic energy of the moving rod is derived partly from the motion of translation, parallel to y, of the elements composing it, and partly from the rotation of the same elements about axes through their centres of inertia perpendicular to the plane of vibration. The former part is expressed by

$$\tfrac{1}{2} \int \rho\omega \, \dot{y}^2 dx \quad \dots\dots\dots\dots\dots\dots\dots(3),$$

if ρ denote the volume-density. To express the latter part, we have only to observe that the angular displacement of the element dx is $\dfrac{dy}{dx}$, and therefore its angular velocity $\dfrac{d}{dt}\dfrac{dy}{dx}$. The square of this quantity must be multiplied by half the moment of inertia of the element, that is, by $\tfrac{1}{2}\kappa^2\rho\omega \, dx$. We thus obtain

$$T = \tfrac{1}{2}\int \rho\omega \, \dot{y}^2 dx + \tfrac{1}{2}\int \kappa^2 \rho\omega \left(\frac{d}{dt}\frac{dy}{dx}\right)^2 dx \quad \dots\dots\dots(4).$$

162.   In order to form the equation of motion we may avail ourselves of the principle of virtual velocities. If for simplicity we confine ourselves to the case of uniform section, we have

$$\delta V = B \int \frac{d^2 y}{dx^2} \frac{d^2 \delta y}{dx^2} \, dx$$

$$= B \frac{d^2 y}{dx^2} \cdot \frac{d\delta y}{dx} - B \frac{d^3 y}{dx^3} \delta y + B \int \frac{d^4 y}{dx^4} \, \delta y \, dx \dots\dots\dots\dots(1),$$

where the terms free from the integral sign are to be taken between the limits. This expression includes only the internal forces due to the bending. In what follows we shall suppose that there are no forces acting from without, or rather none that do work upon the system. A force of constraint, such as that necessary to hold any point of the bar at rest, need not be regarded, as it does no work and therefore cannot appear in the equation of virtual velocities.

The virtual moment of the accelerations is

$$\int \rho \omega \frac{d^2y}{dt^2} \delta y \, dx + \int \rho \omega \kappa^2 \frac{d^2}{dt^2} \left(\frac{dy}{dx}\right) \delta \left(\frac{dy}{dx}\right) dx$$

$$= \int \rho \omega \left(\frac{d^2y}{dt^2} - \kappa^2 \frac{d^4y}{dx^2 dt^2}\right) \delta y \, dx + \rho \omega \kappa^2 \delta y \frac{d^3y}{dt^2 dx} \ldots\ldots\ldots (2).$$

Thus the variational equation of motion is

$$\int \left\{ B \frac{d^4y}{dx^4} + \rho \omega \left(\frac{d^2y}{dt^2} - \kappa^2 \frac{d^4y}{dx^2 dt^2}\right) \right\} \delta y \, dx$$

$$+ B \frac{d^2y}{dx^2} \delta \left(\frac{dy}{dx}\right) + \left\{ \rho \omega \kappa^2 \frac{d^3y}{dt^2 dx} - B \frac{d^3y}{dx^3} \right\} \delta y = 0 \ldots\ldots\ldots (3),$$

in which the terms free from the integral sign are to be taken between the limits. From this we derive as the equation to be satisfied at all points of the length of the bar

$$B \frac{d^4y}{dx^4} + \rho \omega \left(\frac{d^2y}{dt^2} - \kappa^2 \frac{d^4y}{dx^2 dt^2}\right) = 0,$$

while at each end

$$B \frac{d^2y}{dx^2} \delta \left(\frac{dy}{dx}\right) + \left\{ \rho \omega \kappa^2 \frac{d^3y}{dt^2 dx} - B \frac{d^3y}{dx^3} \right\} \delta y = 0;$$

or, if we introduce the value of B, viz. $q \kappa^2 \omega$, and write $q \div \rho = b^2$,

$$\frac{d^2y}{dt^2} + b^2 \kappa^2 \frac{d^4y}{dx^4} - \kappa^2 \frac{d^4y}{dx^2 dt^2} = 0 \ldots\ldots\ldots\ldots\ldots(4),$$

and for each end

$$b^2 \frac{d^2y}{dx^2} \delta \left(\frac{dy}{dx}\right) + \left\{ \frac{d^3y}{dt^2 dx} - b^2 \frac{d^3y}{dx^3} \right\} \delta y = 0 \ldots\ldots\ldots(5).$$

In these equations b expresses the velocity of transmission of longitudinal waves.

The condition (5) to be satisfied at the ends assumes different forms according to the circumstances of the case. It is possible to conceive a constraint of such a nature that the ratio $\delta \left(\frac{dy}{dx}\right) : \delta y$ has a prescribed finite value. The second boundary condition is then obtained from (5) by introduction of this ratio. But in all the cases that we shall have to consider, there is either no constraint or the constraint is such that either $\delta \left(\frac{dy}{dx}\right)$ or δy vanishes, and then the boundary conditions take the form

$$\frac{d^2y}{dx^2} \delta \left(\frac{dy}{dx}\right) = 0, \qquad \left\{\frac{d^3y}{dt^2dx} - b^2 \frac{d^3y}{dx^3}\right\} \delta y = 0 \ldots\ldots\ldots(6).$$

We must now distinguish the special cases that may arise. If an end be free, δy and $\delta \left(\frac{dy}{dx}\right)$ are both arbitrary, and the conditions become

$$\frac{d^2y}{dx^2} = 0, \qquad \frac{d^3y}{dt^2dx} - b^2 \frac{d^3y}{dx^3} = 0 \ldots\ldots\ldots\ldots(7),$$

the first of which may be regarded as expressing that no couple acts at the free end, and the second that no force acts.

If the direction at the end be free, but the end itself be constrained to remain at rest by the action of an applied force of the necessary magnitude, in which case for want of a better word the rod is said to be *supported*, the conditions are

$$\frac{d^2y}{dx^2} = 0, \qquad \delta y = 0 \ldots\ldots\ldots\ldots\ldots(8),$$

by which (5) is satisfied.

A third case arises when an extremity is constrained to maintain its direction by an applied couple of the necessary magnitude, but is free to take any position. We have then

$$\delta \left(\frac{dy}{dx}\right) = 0, \qquad \frac{d^3y}{dt^2dx} - b^2 \frac{d^3y}{dx^3} = 0 \ldots\ldots\ldots (9).$$

Fourthly, the extremity may be constrained both as to position and direction, in which case the rod is said to be *clamped*. The conditions are plainly

$$\delta \left(\frac{dy}{dx}\right) = 0, \qquad \delta y = 0 \ldots\ldots\ldots\ldots (10).$$

Of these four cases the first and last are the more important; the third we shall omit to consider, as there are no experimental means by which the contemplated constraint could be realized. Even with this simplification a considerable variety of problems remain for discussion, as either end of the bar may be free, clamped or supported, but the complication thence arising is not so great as might have been expected. We shall find that different cases may be treated together, and that the solution for one case may sometimes be derived immediately from that of another.

In experimenting on the vibrations of bars, the condition for a clamped end may be realized with the aid of a vice of massive construction. In the case of a free end there is of course no difficulty so far as the end itself is concerned; but, when both ends are free, a question arises as to how the weight of the bar is to be supported. In order to interfere with the vibration as little as possible, the supports must be confined to the neighbourhood of the nodal points. It is sometimes sufficient merely to lay the bar on bridges, or to pass a loop of string round the bar and draw it tight by screws attached to its ends. For more exact purposes it would perhaps be preferable to carry the weight of the bar on a pin traversing a hole drilled through the middle of the thickness in the plane of vibration.

When an end is to be 'supported,' it may be pressed into contact with a fixed plate whose plane is perpendicular to the length of the bar.

163. Before proceeding further we shall introduce a supposition, which will greatly simplify the analysis, without seriously interfering with the value of the solution. We shall assume that the terms depending on the angular motion of the sections of the bar may be neglected, which amounts to supposing the *inertia* of each section concentrated at its centre. We shall afterwards (§ 186) investigate a correction for the rotatory inertia, and shall prove that under ordinary circumstances it is small. The equation of motion now becomes

$$\frac{d^2y}{dt^2} + \kappa^2 b^2 \frac{d^4y}{dx^4} = 0 \dots\dots\dots\dots\dots(1),$$

and the boundary conditions for a free end

$$\frac{d^2y}{dx^2} = 0, \qquad \frac{d^3y}{dx^3} = 0 \dots\dots\dots\dots\dots(2).$$

The next step in conformity with the general plan will be the assumption of the harmonic form of y. We may conveniently take

$$y = u \cos\left(\frac{\kappa b}{l^2} m^2 t\right) \quad\dots\dots\dots\dots\dots(3),$$

where l is the length of the bar, and m is an abstract number, whose value has to be determined. Substituting in (1), we obtain

$$\frac{d^4 u}{dx^4} = \frac{m^4}{l^4} u \quad\dots\dots\dots\dots\dots\dots(4).$$

If $u = e^{p\frac{mx}{l}}$ be a solution, we see that p is one of the fourth roots of unity, viz. $+1, -1, +i, -i$; so that the complete solution is

$$u = A \cos m \frac{x}{l} + B \sin m \frac{x}{l} + C e^{\frac{mx}{l}} + D e^{-\frac{mx}{l}},$$

containing four arbitrary constants.

We have still to satisfy the four boundary conditions,—two for each end. These determine the ratios $A : B : C : D$, and furnish besides an equation which m must satisfy. Thus a series of particular values of m are alone admissible, and for each m the corresponding u is determined in everything except a constant multiplier. We shall distinguish the different functions u belonging to the same system by suffixes.

The value of y at any time may be expanded in a series of the functions u (§§ 92, 93). If ϕ_1, ϕ_2, &c. be the normal co-ordinates, we have

$$y = \phi_1 u_1 + \phi_2 u_2 + \dots \quad\dots\dots\dots\dots(5),$$

and $T = \frac{1}{2}\rho\omega \int (\dot{\phi}_1 u_1 + \dot{\phi}_2 u_2 + \dots)^2 \, dx$

$$= \frac{1}{2}\rho\omega \left\{ \dot{\phi}_1^2 \int u_1^2 dx + \dot{\phi}_2^2 \int u_2^2 dx + \dots \right\} \quad\dots\dots\dots (6).$$

We are fully justified in asserting at this stage that each integrated product of the functions vanishes, and therefore the process of the following section need not be regarded as more than a *verification*. It is however required in order to determine the value of the integrated squares.

164. Let u_m, $u_{m'}$ denote two of the normal functions corresponding respectively to m and m'. Then

$$\frac{d^4 u_m}{dx^4} = \frac{m^4}{l^4} u_m, \qquad \frac{d^4 u_{m'}}{dx^4} = \frac{m'^4}{l^4} u_{m'} \quad \ldots\ldots\ldots\ldots(1);$$

or, if dashes indicate differentiation with respect to $m\frac{x}{l}$, $m'\frac{x}{l}$,

$$u_m'''' = u_m, \qquad u_{m'}'''' = u_{m'} \quad \ldots\ldots\ldots\ldots\ldots (2).$$

If we subtract equations (2) after multiplying them by $u_{m'}$, u_m respectively, and then integrate over the length of the bar, we have

$$\frac{m'^4 - m^4}{l^4} \int u_m u_{m'}\, dx = \int \left(u_m \frac{d^4 u_{m'}}{dx^4} - u_{m'} \frac{d^4 u_m}{dx^4} \right) dx$$

$$= u_m \frac{d^3 u_{m'}}{dx^3} - u_{m'} \frac{d^3 u_m}{dx^3} + \frac{du_{m'}}{dx} \frac{d^2 u_m}{dx^2} - \frac{du_m}{dx} \frac{d^2 u_{m'}}{dx^2} \quad \ldots\ldots (3),$$

the integrated terms being taken between the limits.

Now whether the end in question be clamped, supported, or free[1], each term vanishes on account of one or other of its factors. We may therefore conclude that, if u_m, $u_{m'}$ refer to two modes of vibration (corresponding of course to the same terminal conditions) of which a rod is capable, then

$$\int u_m u_{m'} dx = 0 \quad \ldots\ldots\ldots\ldots\ldots(4),$$

provided m and m' be different.

The attentive reader will perceive that in the process just followed, we have in fact retraced the steps by which the fundamental differential equation was itself proved in § 162. It is the

---

[1] The reader should observe that the cases here specified are particular, and that the right-hand member of (3) vanishes, provided that

$$u_m : \frac{d^3 u_m}{dx^3} = u_{m'} : \frac{d^3 u_{m'}}{dx^3},$$

and

$$\frac{du_m}{dx} : \frac{d^2 u_m}{dx^2} = \frac{du_{m'}}{dx} : \frac{d^2 u_{m'}}{dx^2}$$

These conditions include, for instance, the case of a rod whose end is urged towards its position of equilibrium by a force proportional to the displacement, as by a spring without inertia.

original *variational* equation that has the most immediate connection with the conjugate property. If we denote y by u and δy by v,

$$\delta V = B \int \frac{d^2u}{dx^2} \frac{d^2v}{dx^2}\, dx,$$

and the equation in question is

$$B \int \frac{d^2u}{dx^2} \frac{d^2v}{dx^2}\, dx + \rho\omega \int \ddot{u}\, v\, dx = 0 \ldots\ldots\ldots\ldots\ldots(5).$$

Suppose now that u relates to a normal component vibration, so that $\ddot{u} + n^2 u = 0$, where n is some constant; then

$$n^2 \rho\omega \int u v\, dx = B \int \frac{d^2u}{dx^2} \frac{d^2v}{dx^2}\, dx\,.$$

By similar reasoning, if v be a normal function, and u represent any displacement possible to the system,

$$n'^2 \rho\omega \int u v\, dx = B \int \frac{d^2u}{dx^2} \frac{d^2v}{dx^2}\, dx\,.$$

We conclude that if u and v be both normal functions, *which have different periods,*

$$\int u v\, dx = 0 \ldots\ldots\ldots\ldots\ldots\ldots\ldots(6);$$

and this proof is evidently as direct and general as could be desired.

The reader may investigate the formula corresponding to (6), when the term representing the rotatory inertia is retained.

By means of (6) we may verify that the admissible values of n^2 are real. For if n^2 were complex, and $u = \alpha + i\beta$ were a normal function, then $\alpha - i\beta$, the conjugate of u, would be a normal function also, corresponding to the conjugate of n^2, and then the product of the two functions, being a sum of squares, would not vanish, when integrated [1].

If in (3) m and m' be the same, the equation becomes identically true, and we cannot at once infer the value of $\int u_m^2 dx$.

---

[1] This method is, I believe, due to Poisson.

We must take m' equal to $m + \delta m$, and trace the limiting form of the equation as δm tends to vanish. In this way we find

$$\frac{4m^3}{l^4}\int u_m{}^2 dx =$$

$$u\,\frac{d}{dm}\frac{d^3u}{dx^3} - \frac{du}{dm}\frac{d^3u}{dx^3} + \frac{d^2u}{dx^2}\frac{d}{dm}\frac{du}{dx} - \frac{du}{dx}\frac{d}{dm}\frac{d^2u}{dx^2},$$

the right-hand side being taken between the limits.

Now $\qquad \dfrac{du}{dx} = \dfrac{m}{l}u'$, &c., $\qquad \dfrac{du}{dm} = \dfrac{x}{l}u'$, &c.,

and thus

$$\frac{4m^3}{l^4}\int u_m{}^2 dx = \frac{3m^2}{l^3}u\,u''' + \frac{m^3x}{l^4}u\,u'''' - \frac{m^3x}{l^4}u'u'''$$

$$+ \frac{m^2}{l^3}u'u'' + \frac{m^3x}{l^4}(u'')^2 - \frac{2m^2}{l^3}u'u'' - \frac{m^3x}{l^4}u'u''',$$

in which $u'''' = u$, so that

$$\frac{4m}{l}\int u_m{}^2 dx = 3u\,u''' + \frac{mx}{l}u^2 - \frac{2mx}{l}u'u''' - u'u'' + \frac{mx}{l}(u'')^2 \ldots (7),$$

between the limits.

Now whether an end be clamped, supported, or free,

$$u\,u''' = 0, \qquad u'u'' = 0,$$

and thus, if we take the origin of x at one end of the rod,

$$\int_0^l u^2 dx = \left\{\frac{x}{4}(u^2 - 2u'u''' + u''^2)\right\}_0$$

$$= \tfrac{1}{4}l\,(u^2 - 2u'u''' + u''^2)_{x=l} \ldots\ldots\ldots\ldots(8).$$

The form of our integral is independent of the terminal condition at $x = 0$. If the end $x = l$ be free, u'' and u''' vanish, and accordingly

$$\int_0^l u^2 dx = \tfrac{1}{4}l\,u^2\,(l) \ldots\ldots\ldots\ldots\ldots\ldots(9),$$

that is to say, for a rod with one end free the mean value of u^2 is one-fourth of the terminal value, and that whether the other end be clamped, supported, or free.

Again, if we suppose that the rod is clamped at $x = l$, u and u' vanish, and (8) gives

$$\int_0^l u^2 dx = \tfrac{1}{4} l \, [u''(l)]^2.$$

Since this must hold good whatever be the terminal condition at the other end, we see that for a rod, one end of which is fixed and the other free,

$$\int_0^l u^2 dx = \tfrac{1}{4} l u^2 \text{ (free end)} = \tfrac{1}{4} l u''^2 \text{ (fixed end)},$$

shewing that in this case u^2 at the free end is the same as u''^2 at the clamped end.

The annexed table gives the values of four times the mean of u^2 in the different cases.

clamped, free.........	u^2 (free end), or u''^2 (clamped end)
free, free	u^2 (free end)
clamped, clamped ...	u''^2 (clamped end)
supported, supported	$-2u'u'''$ (supported end) $= 2u'^2$
supported, free	u^2 (free end), or $-2u'u'''$ (supported end)
supported, clamped	u''^2 (clamped end), or $-2u'u'''$ (supported end)

By the introduction of these values the expression for T assumes a simpler form. In the case, for example, of a clamped-free or a free-free rod,

$$T = \frac{\rho l \omega}{8} \left\{ \dot{\phi}_1^2 u_1^2 (l) + \dot{\phi}_2^2 u_2^2 (l) + \ldots \right\} \ldots \ldots \ldots \ldots (10),$$

where the end $x = l$ is supposed to be free.

165. A similar method may be applied to investigate the values of $\int u'^2 dx$, and $\int u''^2 dx$. In the derivation of equation (7) of the preceding section nothing was assumed beyond the truth of the equation $u'''' = u$, and since this equation is equally true of any of the derived functions, we are at liberty to replace u by u' or u''. Thus

$$\frac{4m}{l} \int_0^l u'^2 dx = 3u'u + \frac{mx}{l} u'^2 - 2 \frac{mx}{l} u''u - u''u''' + \frac{mx}{l} u'''^2$$

$$= 3uu' + \frac{mx}{l} u'^2 - u''u''' + \frac{mx}{l} u'''^2,$$

taken between the limits, since the term $u\,u''$ vanishes in all three cases.

For a free-free rod

$$\frac{4m}{l}\int_0^l u'^2 dx = 3\,(uu')_l - 3(uu')_0 + m\,(u'^2)_l$$

$$= 6\,(uu')_l + m\,(u'^2)_l \dots\dots\dots\dots(1),$$

for, as we shall see, the values of $u\,u'$ must be equal and opposite at the two ends. Whether u be positive or negative at $x=l$, $u\,u'$ is positive.

For a rod which is clamped at $x=0$ and free at $x=l$

$$\frac{4m}{l}\int_0^l u'^2 dx = 3\,(uu')_l + mu_l'^2 + (u''u''')_0.$$

We have already seen that $u_0'' = u_l$, and it will appear (§ 173) that $u_0''' = -u_l'$, so that

$$\frac{4m}{l}\int_0^l u'^2 dx = 2\,(uu')_l + mu_l'^2 \dots\dots\dots\dots(2),$$

a result that we shall have occasion to use later.

By applying the same equation to the evaluation of $\int u''^2 dx$, we find

$$\frac{4m}{l}\int u''^2 dx = 3u''u' + \frac{mx}{l}u''^2 - 2\frac{mx}{l}u'''u' - u'''u + \frac{mx}{l}u^2$$

$$= m\,(u''^2 - 2u'u''' + u^2)_l,$$

since $u'u''$ and $u\,u'''$ vanish.

Comparing this with (8) § 164, we see that

$$\int u''^2 dx = \int u^2 dx \dots\dots\dots\dots(3),$$

whatever the terminal conditions may be.

The same result may be arrived at more directly by integrating by parts the equation

$$\frac{m^4}{l^4} u^2 = u\frac{d^4u}{dx^4}.$$

166. We may now form the expression for V in terms of the normal co-ordinates.

$$V = \frac{b^2 \kappa^2 \rho \omega}{2} \int \left\{ \phi_1 \frac{d^2 u_1}{dx^2} + \phi_2 \frac{d^2 u_2}{dx^2} + \ldots \right\}^2 dx$$

$$= \frac{b^2 \kappa^2 \rho \omega}{2} \left\{ \phi_1^2 \int \left(\frac{d^2 u_1}{dx^2} \right)^2 dx + \phi_2^2 \int \left(\frac{d^2 u_2}{dx^2} \right)^2 dx + \ldots \right\}$$

$$= \frac{b^2 \kappa^2 \rho \omega}{2 l^4} \left\{ m_1^4 \phi_1^2 \int u_1^2 dx + m_2^4 \phi_2^2 \int u_2^2 dx + \ldots \right\} \quad \ldots\ldots\ldots (1).$$

If the functions u be those proper to a rod free at $x = l$, this expression reduces to

$$V = \frac{b^2 \kappa^2 \rho \omega}{8 l^3} \left\{ m_1^4 [u_1(l)]^2 \phi_1^2 + m_2^4 [u_2(l)]^2 \phi_2^2 + \ldots \right\} \ldots\ldots\ldots (2).$$

In any case the equations of motion are of the form

$$\rho \omega \int u_1^2 dx \; \ddot{\phi}_1 + \frac{b^2 \kappa^2 \rho \omega}{l^4} m_1^4 \int u_1^2 dx \; \phi_1 = \Phi_1 \ldots\ldots\ldots (3),$$

and, since $\Phi_1 \delta \phi_1$ is by definition the work done by the impressed forces during the displacement $\delta \phi_1$,

$$\Phi_1 = \int Y u_1 \rho \omega \, dx \ldots\ldots\ldots\ldots\ldots (4),$$

if $Y \rho \omega \, dx$ be the lateral force acting on the element of mass $\rho \omega \, dx$. If there be no impressed forces, the equation reduces to

$$\ddot{\phi}_1 + \frac{b^2 \kappa^2 m_1^4}{l^4} \phi_1 = 0 \ldots\ldots\ldots\ldots\ldots\ldots (5),$$

as we know it ought to do.

167. The significance of the reduction of the integrals $\int u^2 dx$ to dependence on the terminal values of the function and its derivatives may be placed in a clearer light by the following line of argument. To fix the ideas, consider the case of a rod clamped at $x = 0$, and free at $x = l$, vibrating in the normal mode expressed by u. If a small addition Δl be made to the rod at the free end, the form of u (considered as a function of x) is changed, but, in accordance with the general principle established in Chapter IV. (§ 88), we may calculate the period

under the altered circumstances without allowance for the change
of type, if we are content to neglect the square of the change.
In consequence of the straightness of the rod at the place where
the addition is made, there is no alteration in the potential
energy, and therefore the alteration of period depends entirely
on the variation of T. This quantity is increased in the ratio

$$\int_0^l u^2 dx \ : \ \int_0^{l+\Delta l} u^2 dx,$$

or $\quad 1 : 1 + \dfrac{u_i^2 \Delta l}{\int_0^l u^2 dx}$,

which is also the ratio in which the square of the period is
augmented.   Now, as we shall see presently, the actual period
varies as l^2, and therefore the change in the square of the period
is in the ratio

$$1 : 1 + \frac{4\Delta l}{l}.$$

A comparison of the two ratios shews that

$$u_i^2 : \int u^2 dx = 4 : l.$$

The above reasoning is not insisted upon as a demonstration,
but it serves at least to explain the reduction of which the integral is susceptible.   Other cases in which such integrals occur
may be treated in a similar manner, but it would often require
care to predict with certainty what amount of discontinuity in the
varied type might be admitted without passing out of the range
of the principle on which the argument depends.   The reader
may, if he pleases, examine the case of a string in the middle
of which a small piece is interpolated.

168.   In treating problems relating to vibrations the usual
course has been to determine in the first place the forms of the
normal functions, viz. the functions representing the normal
types, and afterwards to investigate the integral formulæ by
means of which the particular solutions may be combined to
suit arbitrary initial circumstances.   I have preferred to follow
a different order, the better to bring out the generality of the
method, *which does not depend upon a knowledge of the normal
functions*.   In pursuance of the same plan, I shall now investigate

the connection of the arbitrary constants with the initial circumstances, and solve one or two problems analogous to those treated under the head of Strings.

The general value of y may be written

$$y = \left(A_1 \cos \frac{\kappa b}{l^2} m_1^2 t + B_1 \sin \frac{\kappa b}{l^2} m_1^2 t \right) u_1$$

$$+ \left(A_2 \cos \frac{\kappa b}{l^2} m_2^2 t + B_2 \sin \frac{\kappa b}{l^2} m_2^2 t \right) u_2$$

$$+ \ldots\ldots\ldots\ldots\ldots\ldots\ldots\ldots\ldots\ldots\ldots\ldots\ldots\ldots\ldots \quad (1),$$

so that initially

$$y_0 = A_1 u_1 + A_2 u_2 + \ldots \quad\ldots\ldots\ldots\ldots\ldots\ldots\ldots (2),$$

$$\dot{y}_0 = \frac{\kappa b}{l^2} \{ m_1^2 B_1 u_1 + m_2^2 B_2 u_2 + \ldots \} \quad\ldots\ldots\ldots\ldots (3).$$

If we multiply (2) by u_r and integrate over the length of the rod, we get

$$\int y_0 u_r \, dx = A_r \int u_r^2 \, dx \quad\ldots\ldots\ldots\ldots\ldots (4),$$

and similarly from (3)

$$\frac{l^2}{\kappa b} \int \dot{y}_0 u_r \, dx = m_r^2 B_r \int u_r^2 \, dx \quad\ldots\ldots\ldots\ldots (5),$$

formulæ which determine the arbitrary constants A_r, B_r.

It must be observed that we do not need to prove analytically the possibility of the expansion expressed by (1). If *all* the particular solutions are included, (1) necessarily represents the most general vibration possible, and may therefore be adapted to represent any admissible initial state.

Let us now suppose that the rod is originally at rest, in its position of equilibrium, and is set in motion by a blow which imparts velocity to a small portion of it. Initially, that is, at the moment when the rod becomes free, $y_0 = 0$, and $\dot{y}_0$ differs from zero only in the neighbourhood of one point ($x = c$).

From (4) it appears that the coefficients A vanish, and from (5) that

$$m_r^2 B_r \int u_r^2 \, dx = \frac{l^2}{\kappa b} u_r(c) \int \dot{y}_0 \, dx.$$

Calling $\int \dot{y}_0 \rho \omega \, dx$, the whole momentum of the blow, Y, we have

$$B_r = \frac{l^2 Y}{\kappa b \rho \omega} \; \frac{u_r(c)}{m_r^2 \int u_r^2 dx} \; \dots\dots\dots\dots\dots (6),$$

and for the final solution

$$y = \frac{l^2 Y}{\kappa b \rho \omega} \left\{ \frac{u_1(c) \, u_1(x)}{m_1^2 \int u_1^2 dx} \sin\left(\frac{\kappa b}{l^2} m_1^2 t\right) + \dots \right.$$

$$\left. + \frac{u_r(c) \, u_r(x)}{m_r^2 \int u_r^2 dx} \sin\left(\frac{\kappa b}{l^2} m_r^2 t\right) + \dots\dots \right\} \dots\dots (7).$$

In adapting this result to the case of a rod free at $x = l$, we may replace

$$\int u_r^2 dx \quad \text{by} \quad \tfrac{1}{4} l \, [u_r(l)]^2.$$

If the blow be applied at a node of one of the normal components, that component is missing in the resulting motion. The present calculation is but a particular case of the investigation of § 101.

169. As another example we may take the case of a bar, which is initially at rest but deflected from its natural position by a lateral force acting at $x = c$. Under these circumstances the coefficients B vanish, and the others are given by (4), § 168.

Now

$$\int_0^l y_0 u_r dx = \frac{l^4}{m_r^4} \int_0^l y_0 \frac{d^4 u_r}{dx^4} \, dx,$$

and on integrating by parts

$$\int_0^l y_0 \frac{d^4 u_r}{dx^4} \, dx = y_0 \frac{d^3 u_r}{dx^3} - \frac{dy_0}{dx} \frac{d^2 u_r}{dx^2}$$

$$+ \frac{d^2 y_0}{dx^2} \frac{du_r}{dx} - \frac{d^3 y_0}{dx^3} u_r + \int_0^l \frac{d^4 y_0}{dx^4} u_r dx,$$

in which the terms free from the integral sign are to be taken between the limits; by the nature of the case y_0 satisfies the same terminal conditions as does u_r, and thus all these terms vanish at both limits. If the external force initially applied to the element dx be $Y dx$, the equation of equilibrium of the bar gives

$$\rho \omega \, \kappa^2 b^2 \frac{d^4 y_0}{dx^4} = Y \quad \ldots\ldots\ldots\ldots\ldots\ldots (1),$$

and accordingly

$$\int_0^l y_0 u_r \, dx = \frac{l^4}{\rho \omega \kappa^2 b^2 m_r^4} \int_0^l Y u_r(x) \, dx.$$

If we now suppose that the initial displacement is due to a force applied in the immediate neighbourhood of the point $x = c$, we have

$$\int_0^l y_0 u_r \, dx = \frac{l^4 u_r(c)}{\rho \omega \kappa^2 b^2 m_r^4} \int Y dx,$$

and for the complete value of y at time t,

$$y = \Sigma \left\{ \frac{l^4 u_r(c) u_r(x)}{m_r^4 \kappa^2 b^2 \int \rho \omega u_r^2 \, dx} \cos \frac{\kappa b}{l^2} m_r^2 t \right\} \int Y dx \quad \ldots\ldots\ldots (2).$$

In deriving the above expression we have not hitherto made any special assumptions as to the conditions at the ends, but if we now confine ourselves to the case of a bar which is clamped at $x = 0$ and free at $x = l$, we may replace

$$\int u_r^2 dx \quad \text{by} \quad \tfrac{1}{4} l \, [u_r(l)]^2.$$

If we suppose further that the force to which the initial deflection is due acts at the end, so that $c = l$, we get

$$y = 4 \Sigma \left\{ \frac{l^3 u_r(x)}{m_r^4 \kappa^2 b^2 \rho \omega u_r(l)} \cos \frac{\kappa b}{l^2} m_r^2 t \right\} \int Y dx \quad \ldots\ldots (3).$$

When $t = 0$, this equation must represent the initial displacement. In cases of this kind a difficulty may present itself as to how it is possible for the series, every term of which satisfies the condition $y''' = 0$, to represent an initial displacement in which this condition is violated. The fact is, that after triple differentiation with respect to x, the series no longer converges for $x = l$, and accordingly the value of y''' is not to be arrived at by making the differentiations first and summing the terms afterwards. The truth of this statement will be apparent if we consider a point distant dl from the end, and replace

$$u'''(l - dl) \text{ by } u'''(l) - u^{\text{iv}}(l)\, dl,$$

in which $u^{\text{iv}}(l)$ is equal to

$$\frac{m^4}{l^4}\, u\,(l).$$

For the solution of the present problem by normal co-ordinates the reader is referred to § 101.

170. The forms of the normal functions in the various particular cases are to be obtained by determining the ratios of the four constants in the general solution of

$$\frac{d^4u}{dx^4} = \frac{m^4}{l^4}\, u\,.$$

If for the sake of brevity x' be written for $\frac{mx}{l}$, the solution may be put into the form

$$u = A\,(\cos x' + \cosh x') + B\,(\cos x' - \cosh x')$$
$$+ C\,(\sin x' + \sinh x') + D\,(\sin x' - \sinh x') \,\ldots\ldots(1).$$

$\cosh x$ and $\sinh x$ are the hyperbolic cosine and sine of x, defined by the equations

$$\cosh x = \tfrac{1}{2}(e^x + e^{-x}), \quad \sinh x = \tfrac{1}{2}(e^x - e^{-x})\ldots\ldots\ldots\ldots(2).$$

I have followed the usual notation, though the introduction of a special symbol might very well be dispensed with, since

$$\cosh x = \cos ix, \quad \sinh x = -i \sin ix \,\ldots\ldots\ldots\ldots(3),$$

where $i = \sqrt{-1}$; and then the connection between the formulæ of circular and hyperbolic trigonometry would be more apparent. The rules for differentiation are expressed in the equations

$$\frac{d}{dx}\cosh x = \sinh x, \quad \frac{d}{dx}\sinh x = \cosh x$$

$$\frac{d^2}{dx^2}\cosh x = \cosh x, \quad \frac{d^2}{dx^2}\sinh x = \sinh x.$$

In differentiating (1) any number of times, the same four compound functions as there occur are continually reproduced. The only one of them which does not vanish with x' is $\cos x' + \cosh x'$, whose value is then 2.

Let us take first the case in which both ends are free. Since $\frac{d^2u}{dx^2}$ and $\frac{d^3u}{dx^3}$ vanish with x, it follows that $B = 0$, $D = 0$, so that

$$u = A\,(\cos x' + \cosh x') + C\,(\sin x' + \sinh x')\ldots\ldots (4).$$

We have still to satisfy the necessary conditions when $x = l$, or $x' = m$. These give

$$\left.\begin{aligned}A\,(-\cos m + \cosh m) + C\,(-\sin m + \sinh m) = 0 \\ A\,(\ \ \sin m + \sinh m) + C\,(-\cos m + \cosh m) = 0\end{aligned}\right\}\ \ldots\ldots(5),$$

equations whose compatibility requires that

$$(\cosh m - \cos m)^2 = \sinh^2 m - \sin^2 m,$$

or in virtue of the relation

$$\cosh^2 m - \sinh^2 m = 1\ldots\ldots\ldots\ldots\ldots\ldots(6),$$

$$\cos m\,\cosh m = 1\ldots\ldots\ldots\ldots\ldots\ldots(7).$$

This is the equation whose roots are the admissible values of m. If (7) be satisfied, the two ratios of $A : C$ given in (5) are equal, and either of them may be substituted in (4). The constant multiplier being omitted, we have for the normal function

$$u = (\sin m - \sinh m)\left\{\cos\frac{mx}{l} + \cosh\frac{mx}{l}\right\}$$

$$- (\cos m - \cosh m)\left\{\sin\frac{mx}{l} + \sinh\frac{mx}{l}\right\}\ldots\ldots\ldots(8),$$

or, if we prefer it,

$$u = (\cos m - \cosh m)\left\{\cos\frac{mx}{l} + \cosh\frac{mx}{l}\right\}$$

$$+ (\sin m + \sinh m)\left\{\sin\frac{mx}{l} + \sinh\frac{mx}{l}\right\}\ldots\ldots\ldots(9);$$

and the simple harmonic component of this type is expressed by

$$y = Pu\cos\left(\frac{\kappa b}{l^2}m^2t + \epsilon\right)\ldots\ldots\ldots\ldots\ldots(10).$$

171. The frequency of the vibration is $\frac{\kappa b}{2\pi l^2}m^2$, in which b is a velocity depending only on the material of which the bar is formed, and m is an abstract number. Hence for a given material and mode of vibration the frequency varies directly as κ—the radius of gyration of the section about an axis perpendicular to the

plane of bending—and inversely as the *square* of the length. These results might have been anticipated by the argument from dimensions, if it were considered that the frequency is necessarily determined by the value of l, together with that of κb—the only quantity depending on space, time and mass, which occurs in the differential equation. If everything concerning a bar be given, except its absolute magnitude, the frequency varies inversely as the linear dimension.

These laws find an important application in the case of tuning forks, whose prongs vibrate as rods, fixed at the ends where they join the stalk, and free at the other ends. Thus the period of vibration of forks of the same material and shape varies as the linear dimension. The period will be approximately independent of the thickness perpendicular to the plane of bending, but will vary inversely with the thickness in the plane of bending. When the thickness is given, the period is as the square of the length.

In order to lower the pitch of a fork we may, for temporary purposes, load the ends of the prongs with soft wax, or file away the metal near the base, thereby weakening the spring. To raise the pitch, the ends of the prongs, which act by inertia, may be filed.

The value of b attains its maximum in the case of steel, for which it amounts to about 5237 metres per second. For brass the velocity would be less in about the ratio $1\cdot5 : 1$, so that a tuning fork made of brass would be about a fifth lower in pitch than if the material were steel.

172. The solution for the case when both ends are clamped may be immediately derived from the preceding by a double differentiation. Since y satisfies at both ends the terminal conditions

$$\frac{d^2y}{dx^2}=0, \qquad \frac{d^3y}{dx^3}=0,$$

it is clear that y'' satisfies

$$y''=0, \qquad \frac{dy''}{dx}=0,$$

which are the conditions for a clamped end. Moreover the general differential equation is also satisfied by y''. Thus we may take, omitting a constant multiplier, as before,

$$u = (\sin m - \sinh m)\{\cos x' - \cosh x'\}$$
$$- (\cos m - \cosh m)\{\sin x' - \sinh x'\} \dots\dots\dots(1),$$

while m is given by the same equation as before, namely,

$$\cos m \cosh m = 1 \ldots\ldots\ldots\ldots\ldots\ldots(2).$$

We conclude that the component tones have the same pitch in the two cases.

In each case there are four systems of points determined by the evanescence of y and its derivatives. When y vanishes, there is a node; where y' vanishes, a loop, or place of maximum displacement; where y'' vanishes, a point of inflection; and where y''' vanishes, a place of maximum curvature. Where there are in the first case (free-free) points of inflection and of maximum curvature, there are in the second (clamped-clamped) nodes and loops respectively; and *vice versâ*, points of inflection and of maximum curvature for a doubly-clamped rod correspond to nodes and loops of a rod whose ends are free.

173.   We will now consider the vibrations of a rod clamped at $x = 0$, and free at $x = l$. Reverting to the general integral (1) § 170 we see that A and C vanish in virtue of the conditions at $x = 0$, so that

$$u = B\,(\cos x' - \cosh x') + D\,(\sin x - \sinh x') \ldots\ldots\ldots\ldots(1).$$

The remaining conditions at $x = l$ give

$$\left.\begin{array}{l} B\,(\ \ \cos m + \cosh m) + D\,(\sin m + \sinh m) = 0 \\ B\,(-\sin m + \sinh m) + D\,(\cos m + \cosh m) = 0 \end{array}\right\},$$

whence, omitting the constant multiplier,

$$u = (\sin m + \sinh m)\left\{\cos \frac{mx}{l} - \cosh \frac{mx}{l}\right\}$$
$$- (\cos m + \cosh m)\left\{\sin \frac{mx}{l} - \sinh \frac{mx}{l}\right\} \ldots\ldots\ldots(2),$$

or

$$u = (\cos m + \cosh m)\left\{\cos \frac{mx}{l} - \cosh \frac{mx}{l}\right\}$$
$$+ (\sin m - \sinh m)\left\{\sin \frac{mx}{l} - \sinh \frac{mx}{l}\right\} \ldots\ldots\ldots(3),$$

where m must be a root of

$$\cos m \cosh m + 1 = 0 \ldots\ldots\ldots\ldots\ldots\ldots(4).$$

The periods of the component tones in the present problem are thus different from, though, as we shall see presently, nearly related to, those of a rod both whose ends are clamped, or free.

If the value of u in (2) or (3) be differentiated twice, the result (u'') satisfies of course the fundamental differential equation. At $x = 0$, $\frac{d^2}{dx^2} u''$, $\frac{d^3}{dx^3} u''$ vanish, but at $x = l$ u'' and $\frac{d}{dx} u''$ vanish. The function u'' is therefore applicable to a rod clamped at l and free at 0, proving that the points of inflection and of maximum curvature in the original curve are at the same distances from the clamped end, as the nodes and loops respectively are from the free end.

174. In default of tables of the hyperbolic cosine or its logarithm, the admissible values of m may be calculated as follows. Taking first the equation

$$\cos m \, \cosh m = 1 \quad\ldots\ldots\ldots\ldots\ldots\ldots(1),$$

we see that m, when large, must approximate in value to $\frac{1}{2}(2i + 1)\pi$, i being an integer. If we assume

$$m = \tfrac{1}{2}(2i + 1)\pi - (-1)^i\beta \quad\ldots\ldots\ldots\ldots\ldots(2),$$

β will be positive and comparatively small in magnitude.

Substituting in (1), we find

$$\cot \tfrac{1}{2}\beta = e^m = e^{\frac{1}{2}(2i+1)\pi} e^{-(-1)^i\beta};$$

or, if $e^{\frac{1}{2}(2i+1)\pi}$ be called a,

$$a \tan \tfrac{1}{2}\beta = e^{(-1)^i\beta} \quad\ldots\ldots\ldots\ldots\ldots\ldots(3),$$

an equation which may be solved by successive approximation after expanding $\tan \tfrac{1}{2}\beta$ and $e^{(-1)^i\beta}$ in ascending powers of the small quantity β. The result is

$$\beta = \frac{2}{a} + (-1)^i\frac{4}{a^2} + \frac{34}{3a^3} + (-1)^i\frac{112}{3a^4} + \quad\ldots\ldots\ldots\ldots(4)^1,$$

which is sufficiently accurate, even when $i = 1$.

By calculation

$$\beta_1 = \cdot0179666 - \cdot0003228 + \cdot0000082 - \cdot0000002 = \cdot0176518.$$

β_2, β_3, β_4, β_5 are found still more easily. After β_5 the first term of the series gives β correctly as far as six significant figures. The

---

[1] This process is somewhat similar to that adopted by Strehlke.

table contains the value of β, the angle whose circular measure is β, and the value of $\sin \frac{1}{2}\beta$, which will be required further on.

Free-Free Bar.

	β.	β expressed in degrees, minutes, and seconds.	$\sin \dfrac{\beta}{2}$.
1	$10^{-1} \times \cdot176518$	$1° \; 0' \; 40''\cdot94$	$10^{-2} \times \cdot88258$
2	$10^{-3} \times \cdot777010$	$2' \; 40''\cdot2699$	$10^{-3} \times \cdot38850$
3	$10^{-4} \times \cdot335505$	$6''\cdot92029$	$10^{-4} \times \cdot16775$
4	$10^{-5} \times \cdot144989$	$\cdot299062$	$10^{-6} \times \cdot72494$
5	$10^{-7} \times \cdot626556$	$\cdot0129237$	$10^{-7} \times \cdot31328$

The values of m which satisfy (1) are

$$m_1 = 4\cdot7123890 + \beta_1 = 4\cdot7300408$$
$$m_2 = 7\cdot8539816 - \beta_2 = 7\cdot8532046$$
$$m_3 = 10\cdot9955743 + \beta_3 = 10\cdot9956078$$
$$m_4 = 14\cdot1371669 - \beta_4 = 14\cdot1371655$$
$$m_5 = 17\cdot2787596 + \beta_5 = 17\cdot2787596$$

after which $m = \frac{1}{2}(2i + 1)\pi$ to seven decimal places.

We will now consider the roots of the equation

$$\cos m \, \cosh m = -1 \quad\dotfill\quad (5).$$

Assuming

$$m = \tfrac{1}{2}(2i - 1)\pi - (-1)^i \alpha \dotfill (6)$$

we obtain the same result as before

$$e^m = \cot \tfrac{1}{2}\alpha = a' e^{-(-1)^i \alpha},$$

where however $\qquad a' = e^{\frac{1}{2}(2i-1)\pi}$

From this it appears that the series of values of α is the same as that of β, though the corresponding suffixes are not the same. In fact

$$\alpha_2 = \beta_1, \quad \alpha_3 = \beta_2, \;\dots\dots\; \alpha_{i+1} = \beta_i^{1},$$

so that we have nothing further to calculate than α_1, for which however the series (4) is not sufficiently convergent. The value

---

[1] This connexion between α and β does not appear to have been hitherto noticed.

of α_1 may be obtained by trial and error from the equation

$$\log_{10} \cot \tfrac{1}{2}\alpha_1 - \cdot 6821882 - \cdot 43429448\, \alpha_1 = 0,$$

and will be found to be

$$\alpha_1 = \cdot 3043077.$$

Another method by which m_1 may be obtained directly will be given presently.

The values of m, which satisfy (5), are

$$m_1 = 1\cdot5707963 + \alpha_1 = 1\cdot875104$$
$$m_2 = 4\cdot7123890 - \alpha_2 = 4\cdot694737$$
$$m_3 = 7\cdot8539816 + \alpha_3 = 7\cdot854758$$
$$m_4 = 10\cdot9955743 - \alpha_4 = 10\cdot995541$$
$$m_5 = 14\cdot1371669 + \alpha_5 = 14\cdot137168$$
$$m_6 = 17\cdot2787596 - \alpha_6 = 17\cdot278759,$$

after which $m = \tfrac{1}{2}(2i - 1)\pi$ sensibly. The frequencies are proportional to m^2, and are therefore for the higher tones nearly in the ratio of the squares of the odd numbers. However, in the case of overtones of very high order, the pitch may be slightly disturbed by the rotatory inertia, whose effect is here neglected.

175. Since the component vibrations of a system, not subject to dissipation, are necessarily of the harmonic type, all the values of m^2, which satisfy

$$\cos m \cosh m = \pm\, 1 \dots\dots\dots\dots\dots\dots(1),$$

must be real. We see further that, if m be a root, so are also $-m$, $m\sqrt{-1}$, $-m\sqrt{-1}$. Hence, taking first the lower sign, we have

$$\tfrac{1}{2}(\cos m \cosh m + 1) = 1 - \frac{m^4}{12} + \frac{m^8}{12^2.35} - \dots\dots$$

$$= \left(1 - \frac{m^4}{m_1^{\,4}}\right)\left(1 - \frac{m^4}{m_2^{\,4}}\right) \&\text{c.} \dots\dots\dots\dots\dots\dots(2).$$

If we take the logarithms of both sides, expand, and equate coefficients, we get

$$\Sigma\frac{1}{m^4} = \frac{1}{12}; \quad \Sigma\frac{1}{m^8} = \frac{1}{12^2}\cdot\frac{33}{35}; \quad \&\text{c.}\dots\dots\dots\dots(3).$$

This is for a clamped-free rod.

From the known value of Σm^{-8}, the value of m_1 may be derived with the aid of approximate values of m_2, m_3,...... We find

$$\Sigma m^{-8} = \cdot006547621,$$

and

$$m_2^{-8} = \cdot000004237$$
$$m_3^{-8} = \cdot000000069$$
$$m_4^{-8} = \cdot000000005,$$

whence

$$m_1^{-8} = \cdot006543310$$

giving

$$m_1 = \cdot1875105, \quad \text{as before.}$$

In like manner, if both ends of the bar be clamped or free,

$$1 - \frac{m^4}{12.35} + \dots = \left(1 - \frac{m^4}{m_1^4}\right)\left(1 - \frac{m^4}{m_2^4}\right) \&c. \dots\dots\dots (4),$$

whence $\Sigma \dfrac{1}{m^4} = \dfrac{1}{12.35}$ &c, where of course the summation is exclusive of the zero value of m.

176. The frequencies of the series of tones are proportional to m^2. The interval between any tone and the gravest of the series may conveniently be expressed in octaves and fractions of an octave. This is effected by dividing the difference of the logarithms of m^2 by the logarithm of 2. The results are as follows:

1·4629	2·6478
2·4358	4·1332
3·1590	5·1036
3·7382, &c.	5·8288, &c.

where the first column relates to the tones of a rod both whose ends are clamped, or free; and the second column to the case of a rod clamped at one end but free at the other. Thus from the second column we find that the first overtone is 2·6478 octaves higher than the gravest tone. The fractional part may be reduced to mean semitones by multiplication by 12. The interval is then two octaves + 7·7736 mean semitones. It will be seen that the rise of pitch is much more rapid than in the case of strings.

If a rod be clamped at one end and free at the other, the pitch of the gravest tone is 2 (log 4·7300 − log 1·8751) ÷ log 2 or 2·6698 octaves lower than if both ends were clamped, or both free.

177.   In order to examine more closely the curve in which the rod vibrates, we will transform the expression for u into a form more convenient for numerical calculation, taking first the case when both ends are free.   Since $m = \frac{1}{2}(2i+1)\pi - (-1)^i\beta$, $\cos m = \sin\beta$, $\sin m = \cos i\pi \times \cos\beta$; and therefore, m being a root of $\cos m \cosh m = 1$, $\cosh m = \operatorname{cosec}\beta$.

Also
$$\sinh^2 m = \cosh^2 m - 1 = \tan^2 m = \cot^2 \beta,$$

or, since $\cot\beta$ is positive,
$$\sinh m = \cot\beta.$$

Thus

$$\frac{\sin m - \sinh m}{\cos m - \cosh m} = \frac{1 - \cos i\pi \sin\beta}{\cos\beta}$$

$$= \frac{(\cos\frac{1}{2}\beta - \cos i\pi \sin\frac{1}{2}\beta)^2}{(\cos\frac{1}{2}\beta - \cos i\pi \sin\frac{1}{2}\beta)(\cos\frac{1}{2}\beta + \cos i\pi \sin\frac{1}{2}\beta)}$$

$$= \frac{\cos\frac{1}{2}\beta \cos i\pi - \sin\frac{1}{2}\beta}{\cos\frac{1}{2}\beta \cos i\pi + \sin\frac{1}{2}\beta}.$$

We may therefore take, omitting the constant multiplier,

$$u = (\cos\tfrac{1}{2}\beta \cos i\pi + \sin\tfrac{1}{2}\beta)\left\{\sin\frac{mx}{l} + \sinh\frac{mx}{l}\right\}$$

$$- (\cos\tfrac{1}{2}\beta \cos i\pi - \sin\tfrac{1}{2}\beta)\left\{\cos\frac{mx}{l} + \cosh\frac{mx}{l}\right\}$$

$$= \sqrt{2} \cos i\pi \sin\left\{\frac{mx}{l} - \frac{\pi}{4} + (-1)^i\frac{\beta}{2}\right\}$$

$$+ \sin\tfrac{1}{2}\beta\, e^{\frac{mx}{l}} - \cos i\pi \cos\tfrac{1}{2}\beta\, e^{-\frac{mx}{l}} \quad\ldots\ldots\ldots\ldots\ldots(1).$$

If we further throw out the factor $\sqrt{2}$, and put $l=1$, we may take
$$u = F_1 + F_2 + F_3,$$
where

$$\left.\begin{aligned}
F_1 &= \cos i\pi \sin\left\{mx - \tfrac{1}{4}\pi + \tfrac{1}{2}(-1)^i\beta\right\} \\
\log F_2 &= mx \log e + \log\sin\tfrac{1}{2}\beta - \log\sqrt{2} \\
\log \pm F_3 &= - mx \log e + \log\cos\tfrac{1}{2}\beta - \log\sqrt{2}
\end{aligned}\right\}\ldots\ldots\ldots(2),$$

from which u may be calculated for different values of i and x.

At the centre of the bar, $x = \frac{1}{2}$, and F_2, F_3 are numerically equal in virtue of $e^m = \cot \frac{1}{2} \beta$. When i is *even*, these terms cancel. For F_1, we have $F_1 = (-1)^i \sin \frac{1}{2} i\pi$, which is equal to zero when i is even, and to ± 1 when i is odd. When i is even, therefore, the sum of the three terms vanishes, and there is accordingly a node in the middle.

When $x = 0$, u reduces to $-2 (-1)^i \sin \{\frac{1}{4} \pi - \frac{1}{2} (-1)^i \beta\}$, which (since β is always small) shews that for no value of i is there a node at the end. If a long bar of steel (held, for example, at the centre) be gently tapped with a hammer while varying points of its length are damped with the fingers, an unusual deadness in the sound will be noticed, as the end is closely approached.

178. We will now take some particular cases.

Vibration with two nodes. $i = 1$.

If $i = 1$, the vibration is the gravest of which the rod is capable. Our formulæ become

$$F_1 = - \sin \{x \, (270^\circ + 1^\circ 0' 40'' \cdot 94) - 45^\circ - 30' 20''\cdot 47\}$$

$$\log F_2 = \quad 2\,054231 \, x + \bar{3}\cdot 7952391$$

$$\log F_3 = - 2\cdot 054231 \, x + \bar{1}\cdot 8494681,$$

from which is calculated the following table, giving the values of u for x equal to $\cdot 00$, $\cdot 05$, $\cdot 10$, &c.

The values of $u : u(\cdot 5)$ for the intermediate values of x (in the last column) were found by interpolation formulæ. If o, p, q, r, s, t be six consecutive terms, that intermediate between q and r is

$$\frac{q + r}{2} + \frac{q + r - (p + s)}{4^2} + \frac{3}{4^4} \left\{ 2 \left[q + r - (p + s)\right] - (p + s) + o + t \right\}.$$

x	F_1	F_2	F_3	u	$u : u(\cdot 5)$
·000	+ ·7133200	+ ·0062408	+ ·7070793	+ 1·4266401	+ 1·645219
·025	...	...	...	...	1·454176
·050	·5292548	·0079059	·5581572	1·0953179	1·263134
·075	...	...	...	...	1·072162
·100	·3157243	·0100153	·4406005	·7663401	·8837528
·125	...	...	...	...	·6969004
·150	+ ·0846166	·0126874	·3478031	·4451071	·5133028
·175	...	...	...	...	·3341625
·200	− ·1512020	·0160726	·2745503	+ ·1394209	+ ·1607819
·225	...	...	...	...	− ·0054711
·250	·3786027	·0203609	·2167256	− ·1415162	·1631982
·275	...	...	...	...	·3109982
·300	·5849255	·0257934	·1710798	·3880523	·4475066
·325	...	...	...	..	·5714137
·350	·7586838	·0326753	·1350477	·5909608	·6815032
·375	...	...	...	...	.7766629
·400	·8902038	·0413934	·1066045	·7422059	·8559210
·425	...	...	...	...	·9184491
·450	·9721635	·0524376	·0841519	·8355740	·9635940
·475	...	`·...`	...	...	·9908730
·500	− 1·000000	+ ·0664285	·0664282	− ·8671433	− 1·0000000

Since the vibration curve is symmetrical with respect to the middle of the rod, it is unnecessary to continue the table beyond $x = \cdot 5$. The curve itself is shewn in fig. 28.

Fig. 28.

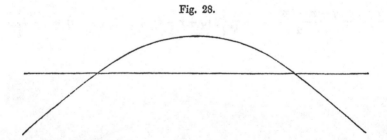

To find the position of the node, we have by interpolation

$$x = \cdot 200 + \frac{\cdot 1607819}{\cdot 1662530} \times \cdot 025 = \cdot 22418,$$

which is the fraction of the whole length by which the node is distant from the nearer end.

Vibration with three nodes. $i = 2.$

$$F_1 = \sin\{(450^\circ - 2'\,40''\cdot27)\,x - 45^\circ + 1'\,20''\cdot135\}$$

$$\log F_2 = 3\cdot410604\,x + \bar{4}\cdot4388816$$

$$\log(-F_3) = -3\cdot410604\,x + \bar{1}\cdot8494850.$$

x	$u : -u(0)$	$\dot{x}$	$u : -u(0)$
·000	− 1·0000	·250	+ ·5847
·025	·8040	·275	·6374
·050	·6079	·300	·6620
·075	·4147	·325	·6569
·100	·2274	·350	·6245
·125	− ·0487	·375	·5652
·150	+ ·1175	·400	·4830
·175	·2672	·425	·3805
·200	·3972	·450	·2627
·225	·5037	·475	·1340
		·500	·0000

In this table, as in the preceding, the values of u were calculated directly for $x = \cdot000,\ \cdot050,\ \cdot100$ &c., and interpolated for the intermediate values. For the position of the node the table gives by ordinary interpolation $x = \cdot132$. Calculating from the above formulæ, we find

$$u\,(\cdot1321) = -\cdot000076,$$

$$u\,(\cdot1322) = +\cdot000881,$$

whence $x = \cdot132108$, agreeing with the result obtained by Strehlke. The place of maximum excursion may be found from the derived function. We get

$$u'\,(\cdot3083) = +\cdot0006077, \quad u'\,(308\tfrac{1}{4}) = -\cdot0002227,$$

whence $\qquad u'\,(\cdot308373) = 0.$

Hence u is a maximum, when $x = \cdot308373$; it then attains the value $\cdot6636$, which, it should be observed, is much less than the excursion at the end.

The curve is shewn in fig. 29.

Fig. 29.

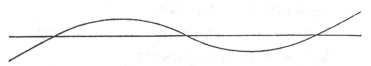

Vibration with four nodes. $i = 3$.

$$F_1 = - \sin \{ (630^\circ + 6''\!\cdot\!92)\, x - 45^\circ - 3''\!\cdot\!46 \},$$
$$\log F_2 = 4\!\cdot\!775332\, x + \bar{5}\!\cdot\!0741527,$$
$$\log F_3 = - 4\!\cdot\!775332\, x + \bar{1}\!\cdot\!8494850.$$

From this $u\,(0) = 1\!\cdot\!41424$, $u\,(\tfrac{1}{2}) = 1\!\cdot\!00579$. The positions of the nodes are readily found by trial and error. Thus

$$u\,(\cdot 3558) = - \cdot 000037 \qquad u\,(\cdot 3559) = + \cdot 001047,$$

whence $u\,(\cdot 355803) = 0$. The value of x for the node near the end is $\cdot 0944$, (Seebeck).

The position of the loop is best found from the derived function. It appears that $u' = 0$, when $x = \cdot 2200$, and then $u = - \cdot 9349$. There is also a loop at the centre, where however the excursion is not so great as at the two others.

Fig. 30.

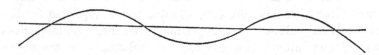

We saw that at the centre of the bar F_2 and F_3 are numerically equal. In the neighbourhood of the middle, F_3 is evidently very small, if i be moderately great, and thus the equation for the nodes reduces approximately to

$$\frac{mx}{l} - \frac{\pi}{4} + (-1)^i \frac{\beta}{2} = \pm n\pi,$$

n being an integer. If we transform the origin to the centre of the rod, and replace m by its approximate value $\tfrac{1}{2}\,(2\,i + 1)\,\pi$, we find

$$\frac{x}{l} = \frac{\pm 2n - i}{2i + 1},$$

shewing that near the middle of the bar the nodes are uniformly spaced, the interval between consecutive nodes being $2l \div (2i+1)$. This theoretical result has been verified by the measurements of Strehlke and Lissajous.

For methods of approximation applicable to the nodes near the ends, when i is greater than 3, the reader is referred to the memoir by Seebeck already mentioned § 160, and to Donkin's *Acoustics* (p. 194).

179. The calculations are very similar for the case of a bar clamped at one end and free at the other If $u \propto F$, and $F = F_1 + F_2 + F_3$, we have in general

$$F_1 = \cos \left\{ mx + \tfrac{1}{4}\pi + \tfrac{1}{2}(-1)^i a \right\},$$

$$F_2 = \frac{(-1)^i}{\sqrt{2}} \sin \tfrac{1}{2}a \, e^{mx} ; \quad F_3 = -\frac{1}{\sqrt{2}} \cos \tfrac{1}{2}a \, e^{-mx}$$

If $i = 1$, we obtain for the calculation of the gravest vibration-curve

$$F_1 = \cos \left\{ \frac{180}{\pi} mx^0 + 45^0 - 8^0 \, 43'\cdot 0665 \right\},$$

$$\log(-F_2) = \quad mx \log e + \bar{1}\cdot 0300909.$$

$$\log(-F_3) = - mx \log e + \bar{1}\cdot 8444383.$$

These give on calculation

$$F(0) = \cdot 000000, \qquad F(\cdot 6) = \cdot 743452,$$

$$F(\cdot 2) = \cdot 102974, \qquad F(\cdot 8) = 1\cdot 169632,$$

$$F(\cdot 4) = \cdot 370625, \qquad F(1\cdot 0) = 1\cdot 612224,$$

from which fig. 31 was constructed.

Fig. 31.

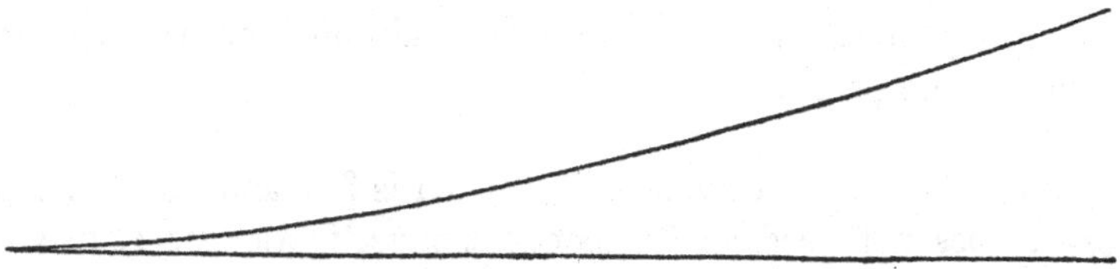

The distances of the nodes from the free end in the case of a rod clamped at the other end are given by Seebeck and by Donkin.

2nd tone ·2261.

3rd tone ·1321, ·4999.

4th tone ·0944, ·3558, ·6439.

i^{th} tone $\dfrac{1\cdot3222}{4i-2}$, $\dfrac{4\cdot9820}{4i-2}$, $\dfrac{9\cdot0007}{4i-2}$, $\dfrac{4j-3}{4i-2}$, $\dfrac{4i-10\cdot9993}{4i-2}$, $\dfrac{4i-7\cdot0175}{4i-2}$

"The last row in this table must be understood as meaning that $\dfrac{4j-3}{4i-2}$ may be taken as the distance of the jth node from the free end, except for the first three and the last two nodes."

When both ends are free, the distances of the nodes from the nearer end are

1st tone ·2242.

2nd tone ·1321   ·5.

3rd tone ·0944   ·3558.

i^{th} tone $\dfrac{1\cdot3222}{4i+2}$ $\dfrac{4\cdot9820}{4i+2}$ $\dfrac{9\cdot0007}{4i+2}$ $\dfrac{4j-3}{4i+2}$.

The points of inflection for a free-free rod (corresponding to the nodes of a clamped-clamped rod) are also given by Seebeck;—

	1st point.	2nd point.	κ^{th} point.
1st tone	No inflection point.		
2nd tone......	·5000		
3rd tone	·3593		
i^{th} tone	$\dfrac{5\cdot0175}{4i+2}$	$\dfrac{8\cdot9993}{4i+2}$	$\dfrac{4\kappa+1}{4i+2}$

Except in the case of the extreme nodes (which have no corresponding inflection-point), the nodes and inflection-points always occur in close proximity.

180.   The case where one end of a rod is free and the other *supported* does not need an independent investigation, as it may be

referred to that of a rod with both ends free *vibrating in an even mode*, that is, with a node in the middle. For at the central node y and y'' vanish, which are precisely the conditions for a supported end. In like manner the vibrations of a clamped-supported rod are the same as those of one-half of a rod both whose ends are clamped, vibrating with a central node.

181. The last of the six combinations of terminal conditions occurs when both ends are supported. Referring to (1) § 170, we see that the conditions at $x = 0$, give $A = 0$, $B = 0$; so that

$$u = (C + D) \sin x' + (C - D) \sinh x'.$$

Since u and u'' vanish when $x' = m$, $C - D = 0$, and $\sin m = 0$.

Hence the solution is

$$y = \sin \frac{i\pi x}{l} \; \cos \frac{i^2 \pi^2 \kappa b}{l^2} \, t \; \dots\dots\dots\dots\dots \; (1),$$

where i is an integer. An arbitrary constant multiplier may of course be prefixed, and a constant may be added to t.

It appears that the normal curves are the same as in the case of a string stretched between two fixed points, but the sequence of tones is altogether different, the frequency varying as the *square* of i. The nodes and inflection-points coincide, and the loops (which are also the points of maximum curvature) bisect the distances between the nodes.

182. The theory of a vibrating rod may be applied to illustrate the general principle that the natural periods of a system fulfil the maximum-minimum condition, and that the greatest of the natural periods exceeds any that can be obtained by a variation of type. Suppose that the vibration curve of a clamped-free rod is that in which the rod would dispose itself if deflected by a force applied at its free extremity. The equation of the curve may be taken to be

$$y = -3lx^2 + x^3,$$

which satisfies $\dfrac{d^4 y}{dx^4} = 0$ throughout, and makes y and y' vanish at 0, and y'' at l. Thus, if the configuration of the rod at time t be

$$y = (-3lx^2 + x^3) \cos pt \; \dots\dots\dots\dots\dots \; (1),$$

the potential energy is by (1) § 161, $6 \, q\kappa^2 \, \omega l^3 \cos^2 pt$, while the kinetic energy is $\frac{33}{70} \rho \omega \, l^7 p^2 \sin^2 pt$; and thus $p^2 = \frac{140}{11} \frac{\kappa^2 b^2}{l^4}$.

Now p_1 (the true value of p for the gravest tone) is equal to

$$\frac{\kappa b}{l^2} \times (1\cdot8751)^2;$$

so that

$$p_1 : p = (1\cdot8751)^2 \sqrt{\frac{11}{140}} = \cdot98556,$$

shewing that the real pitch of the gravest tone is rather (but comparatively little) lower than that calculated from the hypothetical type. It is to be observed that the hypothetical type in question violates the terminal condition $y''' = 0$. This circumstance, however, does not interfere with the application of the principle, for the assumed type may be any which would be admissible as an initial configuration; but it tends to prevent a very close agreement of periods.

We may expect a better approximation, if we found our calculation on the curve in which the rod would be deflected by a force acting at some little distance from the free end, between which and the point of action of the force $(x = c)$ the rod would be straight, and therefore without potential energy. Thus

$$\text{potential energy} = 6 \, q\kappa^2\omega c^3 \cos^2 pt.$$

The kinetic energy can be readily found by integration from the value of y.

From 0 to c $y = -3cx^2 + x^3$;

and from c to l $y = c^2(c - 3x)$,

as may be seen from the consideration that y and y' must not suddenly change at $x = c$. The result is

$$\text{kinetic energy} = \rho \omega \, p^2 \sin^2 pt \left[\frac{33}{70} c^7 + \tfrac{1}{2} c^4 (l - c)(c^2 + 3l^2) \right]$$

whence

$$\frac{1}{p^2} = \frac{1}{6\kappa^2 b^2} \left[\frac{33}{70} c^4 + \frac{c}{2}(l - c)(c^2 + 3l^2) \right] \dots\dots\dots\dots (2).$$

The maximum value of $1 \div p^2$ will occur when the point of application of the force is in the neighbourhood of the node of the second normal component vibration. If we take $c = \frac{3}{4} l$, we obtain a result which is too high in the musical scale by the interval

expressed by the ratio 1 : ·9977, and is accordingly extremely near
the truth. This example may give an idea how nearly the period
of a vibrating system may be calculated by simple means without
the solution of differential or transcendental equations.

The type of vibration just considered would be that actually
assumed by a bar which is itself devoid of inertia, but carries a
load M at its free end, provided that the rotatory inertia of M could
be neglected. We should have, in fact,

$$V = 6q\kappa^2\omega l^3 \cos^2 pt, \qquad T = 2Ml^6 p^2 \sin^2 pt,$$

so that
$$p^2 = \frac{3q\kappa^2\omega}{Ml^3} \dots\dots\dots\dots\dots\dots\dots (3).$$

Even if the inertia of the bar be not altogether negligible in
comparison with M, we may still take the same type as the basis of
an approximate calculation :

$$V = 6q\kappa^2\omega l^3 \cos^2 pt,$$

$$T = \left(2Ml^6 + \frac{33}{70}\rho\omega l^7\right)p^2 \sin^2 pt,$$

whence

$$\frac{1}{p^2} = \frac{l^3}{3q\kappa^2\omega}\left(M + \frac{33}{140}\rho\omega l\right)\dots\dots\dots\dots\dots(4),$$

that is, M is to be increased by about one quarter of the mass of
the rod. Since this result is accurate when M is infinite, and does
not differ much from the truth, even when $M = 0$, it may be regarded as generally applicable as an approximation. The error
will always be on the side of estimating the pitch too high.

183.   But the neglect of the rotatory inertia of M could not
be justified under the ordinary conditions of experiment. It is as
easy to imagine, though not to construct, a case in which the inertia
of translation should be negligible in comparison with the inertia of
rotation, as the opposite extreme which has just been considered.
If both kinds of inertia in the mass M be included, even though
that of the bar be neglected altogether, the system possesses two
distinct and independent periods of vibration.

Let z and θ denote the values of y and $\dfrac{dy}{dx}$ at $x = l$. Then the
equation of the curve of the bar is

$$y = \frac{3z - l\theta}{l^2}x^2 + \frac{l\theta - 2z}{l^3}x^3,$$

and

$$V = \frac{2q\kappa^2\omega}{l^3}\{3z^2 - 3zl\theta + l^2\theta^2\} \dots\dots\dots(1);$$

while for the kinetic energy

$$T = \tfrac{1}{2}M\dot{z}^2 + \tfrac{1}{2}M\kappa'^2\dot{\theta}^2 \dots\dots\dots\dots(2),$$

if κ' be the radius of gyration of M about an axis perpendicular to the plane of vibration.

The equations of motion are therefore

$$\left.\begin{aligned} M\ddot{z} \quad + \frac{2q\kappa^2\omega}{l^3}\ (6z - 3l\theta) = 0 \\ M\kappa'^2\ddot{\theta} + \frac{2q\kappa^2\omega}{l^3}\ (-3lz + 2l^2\theta) = 0 \end{aligned}\right\}\dots\dots\dots(3),$$

whence, if z and θ vary as $\cos pt$, we find

$$p^2 = \frac{2q\kappa^2\omega}{Ml\kappa'^2}\left\{1 + \frac{3\kappa'^2}{l^2} \pm \sqrt{1 + \frac{3\kappa'^2}{l^2} + \frac{9\kappa'^4}{l^4}}\right\}\dots\dots\dots(4),$$

corresponding to the two periods, which are always different.

If we neglect the rotatory inertia by putting $\kappa' = 0$, we fall back on our previous result

$$p^2 = \frac{3q\kappa^2\omega}{Ml^3}.$$

The other value of p^2 is then infinite.

If $\kappa' : l$ be merely small, so that its higher powers may be neglected,

$$\left.\begin{aligned} p^2 &= \frac{4q\kappa^2\omega}{Ml\kappa'^2}\left(1 + \frac{9}{4}\frac{\kappa'^2}{l^2}\right) \\ p^2 &= \frac{3q\kappa^2\omega}{Ml^3}\left(1 - \frac{9}{4}\frac{\kappa'^2}{l^2}\right) \end{aligned}\right\}\dots\dots\dots\dots(5).$$

If on the other hand κ'^2 be very great, so that rotation is prevented,

$$p^2 = \frac{12\,q\kappa^2\omega}{Ml^3} \quad \text{or} \quad \frac{q\kappa^2\omega}{Ml\kappa'^2}\dots\dots\dots\dots(6),$$

the latter of which is very small. It appears that when rotation is prevented, the pitch is an octave higher than if there were no rotatory inertia at all. These conclusions might also be derived directly from the differential equations; for if $\kappa' = \infty$, $\theta = 0$, and then

$$M\ddot{z} + \frac{12q\kappa^2\omega}{l^3}\, z = 0;$$

but if $\kappa' = 0$, $\theta = \frac{3}{2l}\, z$, by the second of equations (3), and in that case

$$M\ddot{z} + \frac{3q\kappa^2\omega}{l^3}\, z = 0.$$

184. If any addition to a bar be made at the end, the period of vibration is prolonged. If the end in question be free, suppose first that the piece added is without inertia. Since there would be no alteration in either the potential or kinetic energies, the pitch would be unchanged; but in proportion as the additional part acquires inertia, the pitch falls (§ 88).

In the same way a small continuation of a bar beyond a clamped end would be without effect, as it would acquire no motion. No change will ensue if the new end be also clamped; but as the first clamping is relaxed, the pitch falls, in consequence of the diminution in the potential energy of a given deformation.

The case of a 'supported' end is not quite so simple. Let the original end of the rod be A, and let the added piece which is at first supposed to have no inertia, be AB. Initially the end A is fixed, or held, if we like so to regard it, by a spring of infinite stiffness. Suppose that this spring, which has no inertia, is gradually relaxed. During this process the motion of the new end B diminishes, and at a certain point of relaxation, B comes to rest. During this process the pitch falls. B, being now at rest, may be supposed to become fixed, and the abolition of the spring at A entails another fall of pitch, to be further increased as AB acquires inertia.

185. The case of a rod which is not quite uniform may be treated by the general method of § 90. We have in the notation there adopted

$$c_r = \int B_0 \left(\frac{d^2 u_r}{dx^2}\right)^2 dx, \qquad \delta c_r = \int \delta B \left(\frac{d^2 u_r}{dx^2}\right)^2 dx$$

$$a_r = \int \overline{\rho\omega_0} u_r{}^2 dx, \qquad \delta a_r = \int \delta\overline{\rho\omega}\, u_r{}^2 dx,$$

whence, P_r being the uncorrected value of p_r,

$$p_r^2 = P_r^2 \left\{ 1 + \frac{\int \delta B \left(\frac{d^2 u_r}{dx^2}\right)^2 dx}{\int B_0 \left(\frac{d^2 u_r}{dx^2}\right)^2 dx} - \frac{\int \delta \overline{\rho \omega}\, u_r^2 dx}{\int \overline{\rho \omega}_0 u_r^2 dx} \right\}$$

$$= P_r^2 \left\{ 1 + \frac{\int \delta B u_r''^2 dx}{B_0 \int u_r^2 dx} - \frac{\int \delta \overline{\rho \omega}\, u_r^2 dx}{\overline{\rho \omega}_0 \int u_r^2 dx} \right\} \ldots\ldots\ldots\ldots (1).$$

For example, if the rod be clamped at 0 and free at l,

$$p_r^2 = \frac{B_0 m^4}{\rho \omega_0 l^4} \left\{ 1 + \frac{4}{l u_l^2} \int_0^l \frac{\delta B}{B_0} u''^2 dx - \frac{4}{l u_l^2} \int_0^l \frac{\delta \overline{\rho \omega}}{\rho \omega_0} u^2 dx \right\}$$

The same formula applies to a doubly free bar.

The effect of a small load dM is thus given by

$$p^2 = \frac{B_0 m^4}{\rho \omega_0 l^4} \left\{ 1 - 4 \frac{u^2\, dM}{u_l^2\, M'} \right\} \ldots\ldots\ldots\ldots (2),$$

where M' denotes the mass of the whole bar. If the load be at the end, its effect is the same as a lengthening of the bar in the ratio $M' : M' + dM$. (Compare § 167.)

186. The same principle may be applied to estimate the correction due to the rotatory inertia of a uniform rod. We have only to find what addition to make to the kinetic energy, supposing that the bar vibrates according to the same law as would obtain, were there no rotatory inertia.

Let us take, for example, the case of a bar clamped at 0 and free at l, and assume that the vibration is of the type,

$$y = u \cos pt,$$

where u is one of the functions investigated in § 179. The kinetic energy of the rotation is

$$\tfrac{1}{2} \int \rho \omega \kappa^2 \left(\frac{d^2 y}{dx\, dt}\right)^2 dx = \frac{\rho \omega \kappa^2 m^2 p^2}{2 l^2} \sin^2 pt \int_0^l u'^2 dx$$

$$= \frac{\rho \omega \kappa^2 m p^2}{8 l} \sin^2 pt \, (2uu' + mu'^2)_l,$$

by (2) § 165.

To this must be added

$$\frac{\rho\omega}{2}\,p^2\sin^2 pt\int_0^l u^2\,dx, \quad\text{or}\quad \frac{\rho\omega l}{8}p^2\sin^2 pt\,u_l^2;$$

so that the kinetic energy is increased in the ratio

$$1\ :\ 1+\frac{m\kappa^2}{l^2}\left(2\frac{u'}{u}+m\frac{u'^2}{u^2}\right)_l.$$

The altered frequency bears to that calculated without allowance for rotatory inertia a ratio which is the square root of the reciprocal of the preceding. Thus

$$p:P=1-\tfrac{1}{2}\,\frac{m\kappa^2}{l^2}\left(2\frac{u'}{u}+m\frac{u'^2}{u^2}\right)_l\dots\dots\dots\dots(1).$$

By use of the relations $\cosh m = -\sec m$, $\sinh m = \cos i\pi\,.\,\tan m$, we may express $u':u$ when $x=l$ in the form

$$\frac{u'}{u}=\frac{-\sin m}{\cos i\pi+\cos m}=\frac{\cos\alpha}{1-\cos i\pi\sin\alpha}$$

if we substitute for m from

$$m=\tfrac{1}{2}\,(2\,i-1)\,\pi-(-1)^i\,\alpha.$$

In the case of the gravest tone, $\alpha = \cdot 3043$, or, in degrees and minutes, $\alpha = 17^\circ\ 26'$, whence

$$\frac{u'}{u}=\cdot 73413,\qquad 2\frac{u'}{u}+m\frac{u'^2}{u^2}=2\cdot 4789.$$

Thus

$$p:P=1-2\cdot 3241\frac{\kappa^2}{l^2}\dots\dots\dots\dots\dots(2),$$

which gives the correction for rotatory inertia in the case of the gravest tone.

When the order of the tone is moderate, α is very small, and then

$$u':u=1\quad\text{sensibly,}$$

and

$$p:P=1-\left(1+\frac{m}{2}\right)\frac{m\kappa^2}{l^2}\dots\dots\dots\dots(3),$$

shewing that the correction increases in importance with the order of the component.

In all ordinary bars $\kappa:l$ is very small, and the term depending on its square may be neglected without sensible error.

187. When the rigidity and density of a bar are variable from point to point along it, the normal functions cannot in general be expressed analytically, but their nature may be investigated by the methods of Sturm and Liouville explained in § 142.

If, as in § 162, B denote the variable flexural rigidity at any point of the bar, and $\rho\omega\,dx$ the mass of the element, whose length is dx, we find as the general differential equation

$$\frac{d^2}{dx^2}\left(B\frac{d^2y}{dx^2}\right) + \rho\omega\frac{d^2y}{dt^2} = 0 \ldots\ldots\ldots\ldots\ldots\ldots(1),$$

the effects of rotatory inertia being omitted. If we assume that $y \propto \cos\nu t$, we obtain as the equation to determine the form of the normal functions

$$\frac{d^2}{dx^2}\left(B\frac{d^2y}{dx^2}\right) = \nu^2\rho\omega\, y \ldots\ldots\ldots\ldots\ldots\ldots(2),$$

in which ν^2 is limited by the terminal conditions to be one of an infinite series of definite quantities $\nu_1^2, \nu_2^2, \nu_3^2 \ldots\ldots$

Let us suppose, for example, that the bar is clamped at both ends, so that the terminal values of y and $\dfrac{dy}{dx}$ vanish. The first normal function, for which ν^2 has its lowest value ν_1^2, has no internal root, so that the vibration-curve lies entirely on one side of the equilibrium-position. The second normal function has one internal root, the third function has two internal roots, and, generally, the rth function has $r-1$ internal roots.

Any two different normal functions are conjugate, that is to say, their product will vanish when multiplied by $\rho\omega\,dx$, and integrated over the length of the bar.

Let us examine the number of roots of a function $f(x)$ of the form

$$f(x) = \phi_m u_m(x) + \phi_{m+1} u_{m+1}(x) + \ldots + \phi_n u_n(x)\ldots\ldots\ldots(3),$$

compounded of a finite number of normal functions, of which the function of lowest order is $u_m(x)$ and that of highest order is $u_n(x)$. If the number of internal roots of $f(x)$ be μ, so that there are $\mu + 4$ roots in all, the derived function $f'(x)$ cannot have less than $\mu + 1$ internal roots besides two roots at the extremities, and the second derived function cannot have less than $\mu + 2$ roots

No roots can be lost when the latter function is multiplied by B, and another double differentiation with respect to x will leave at least μ internal roots. Hence by (2) and (3) we conclude that

$$\nu_m{}^2 \phi_m u_m (x) + \nu_{m+1}{}^2 \phi_{m+1} u_{m+1} (x) + \ldots + \nu_n{}^2 \phi_n u_n (x) \ldots (4)$$

has at least as many roots as $f(x)$. Since (4) is a function of the same form as $f(x)$, the same argument may be repeated, and a series of functions obtained, every member of which has at least as many roots as $f(x)$ has. When the operation by which (4) was derived from (3) has been repeated sufficiently often, a function is arrived at whose form differs as little as we please from that of the component normal function of highest order $u_n(x)$; and we conclude that $f(x)$ cannot have more than $n-1$ internal roots. In like manner we may prove that $f(x)$ cannot have less than $m-1$ internal roots.

The application of this theorem to demonstrate the possibility of expanding an arbitrary function in an infinite series of normal functions would proceed exactly as in § 142.

188. When the bar, whose lateral vibrations are to be considered, is subject to longitudinal tension, the potential energy of any configuration is composed of two parts, the first depending on the stiffness by which the bending is directly opposed, and the second on the reaction against the extension, which is a necessary accompaniment of the bending, when the ends are nodes. The second part is similar to the potential energy of a deflected string; the first is of the same nature as that with which we have been occupied hitherto in this Chapter, though it is not entirely independent of the permanent tension.

Consider the extension of a filament of the bar of section $d\omega$, whose distance from the axis projected on the plane of vibration is η. Since the sections, which were normal to the axis originally, remain normal during the bending, the length of the filament bears to the corresponding element of the axis the ratio $R + \eta : R$, R being the radius of curvature. Now the axis itself is extended in the ratio $q : q + T$, reckoning from the unstretched state, if $T\omega$ denote the whole tension to which the bar is subjected. Hence the actual tension on the filament is $\left\{ T + \dfrac{\eta}{R} (T + q) \right\} d\omega$.

from which we find for the moment of the couple acting across the section

$$\int \left\{ T + \frac{\eta}{R} \left(T + q \right) \right\} \eta \, d\omega = \frac{q+T}{R} \kappa^2 \omega,$$

and for the whole potential energy due to stiffness

$$\tfrac{1}{2} \left(q + T \right) \kappa^2 \omega \int \left(\frac{d^2 y}{dx^2} \right)^2 dx \dots\dots\dots\dots\dots\dots (1),$$

an expression differing from that previously used (§ 162) by the substitution of $q + T$ for q.

Since q is the tension required to stretch a bar of unit area to twice its natural length, it is evident that in most practical cases T would be negligible in comparison with q.

The expression (1) denotes the work that would be gained during the straightening of the bar, if the length of each element of the axis were preserved constant during the process. But when a stretched bar or string is allowed to pass from a displaced to the natural position, the length of the axis is decreased. The amount of the decrease is $\tfrac{1}{2} \int \left(\frac{dy}{dx} \right)^2 dx$, and the corresponding gain of work is

$$\tfrac{1}{2} T \omega \int \left(\frac{dy}{dx} \right)^2 dx.$$

Thus

$$V = \tfrac{1}{2} \left(q + T \right) \kappa^2 \omega \int \left(\frac{d^2 y}{dx^2} \right)^2 dx + \tfrac{1}{2} T \omega \int \left(\frac{dy}{dx} \right)^2 dx \dots (2).$$

The variation of the first part due to a hypothetical displacement is given in § 162. For the second part, we have

$$\tfrac{1}{2} \delta \int \left(\frac{dy}{dx} \right)^2 dx = \int \frac{dy}{dx} \frac{d\delta y}{dx} \, dx = \left\{ \frac{dy}{dx} \delta y \right\} - \int \frac{d^2 y}{dx^2} \delta y \, dx \dots\dots (3).$$

In all the cases that we have to consider, δy vanishes at the limits. The general differential equation is accordingly

$$\kappa^2 \left(q + T \right) \frac{d^4 y}{dx^4} - T \frac{d^2 y}{dx^2} + \rho \frac{d^2 y}{dt^2} - \kappa^2 \rho \frac{d^4 y}{dx^2 \, dt^2} = 0,$$

or, if we put $q + T = b^2 \rho, \quad T = a^2 \rho,$

$$\kappa^2 \left(b^2 \frac{d^4 y}{dx^4} - \frac{d^4 y}{dx^2 \, dt^2} \right) - a^2 \frac{d^2 y}{dx^2} + \frac{d^2 y}{dt^2} = 0 \dots\dots\dots (4).$$

For a more detailed investigation of this equation the reader is referred to the writings of Clebsch[1] and Donkin.

[1] *Theorie der Elasticität fester Körper.* Leipzig, 1862.

189.　If the ends of the rod, or wire, be clamped, $\frac{dy}{dx} = 0$, and the terminal conditions are satisfied. If the nature of the support be such that, while the extremity is constrained to be a node, there is no couple acting on the bar, $\frac{d^2y}{dx^2}$ must vanish, that is to say, the end must be straight. This supposition is usually taken to represent the case of a string stretched over bridges, as in many musical instruments; but it is evident that the part beyond the bridge must partake of the vibration, and that therefore its length cannot be altogether a matter of indifference.

If in the general differential equation we take y proportional to $\cos nt$, we get

$$\kappa^2 \left(b^2 \frac{d^4y}{dx^4} + n^2 \frac{d^2y}{dx^2} \right) - a^2 \frac{d^2y}{dx^2} - n^2 y = 0 \dots\dots\dots\dots (1),$$

which is evidently satisfied by

$$y = \sin i \frac{\pi x}{l} \ \cos nt \dots\dots\dots\dots\dots (2),$$

if n be suitably determined. The same solution also makes y and y'' vanish at the extremities. By substitution we obtain for n,

$$n^2 = \frac{i^2 \pi^2}{l^2} \ \frac{a^2 l^2 + i^2 \pi^2 \kappa^2 b^2}{l^2 + i^2 \pi^2 \kappa^2} \dots\dots\dots\dots\dots (3),$$

which determines the frequency.

If we suppose the wire infinitely thin, $n^2 = i^2 \pi^2 a^2 \div l^2$, the same as was found in Chapter VI., by starting from the supposition of perfect flexibility. If we treat $\kappa : l$ as a very small quantity, the approximate value of n is

$$n = \frac{i\pi a}{l} \left\{ 1 + i^2 \frac{\pi^2 \kappa^2}{2l^2} \left(\frac{b^2}{a^2} - 1 \right) \right\}.$$

For a wire of circular section of radius r, $\kappa^2 = \frac{1}{4} r^2$, and if we replace b and a by their values in terms of q, T, and ρ,

$$n = \frac{i\pi a}{l} \left\{ 1 + \frac{i^2 \pi^2}{8} \ \frac{r^2}{l^2} \ \frac{q}{T} \right\} \dots\dots\dots\dots (4),$$

which gives the correction for rigidity[1]. Since the expression within brackets involves i, it appears that the harmonic relation of the component tones is disturbed by the stiffness.

---

[1] Donkin's *Acoustics*, Art. 184.

190. The investigation of the correction for stiffness when the ends of the wire are clamped is not so simple, in consequence of the change of type which occurs near the ends. In order to pass from the case of the preceding section to that now under consideration an additional constraint must be introduced, with the effect of still further raising the pitch. The following is, in the main, the investigation of Seebeck and Donkin.

If the rotatory inertia be neglected, the differential equation becomes

$$\left(D^4 - \frac{a^2}{\kappa^2 b^2} D^2 - \frac{n^2}{b^2 \kappa^2}\right) y = 0 \dots\dots\dots\dots(1),$$

where D stands for $\frac{d}{dx}$ In the equation

$$D^4 - \frac{a^2}{\kappa^2 b^2} D^2 - \frac{n^2}{b^2 \kappa^2} = 0,$$

one of the values of D^2 must be positive, and the other negative. We may therefore take

$$D^4 - \frac{a^2}{\kappa^2 b^2} D^2 - \frac{n^2}{b^2 \kappa^2} = (D^2 - \alpha^2)(D^2 + \beta^2) \dots\dots\dots(2),$$

and for the complete integral of (1)

$$y = A \cosh \alpha x + B \sinh \alpha x$$
$$+ C \cos \beta x + D \sin \beta x \dots\dots\dots\dots(3),$$

where α and β are functions of n determined by (2).

The solution must now be made to satisfy the four boundary conditions, which, as there are only three disposable ratios, lead to an equation connecting α, β, l. This may be put into the form

$$\frac{\sinh \alpha l \ \sin \beta l}{1 - \cosh \alpha l \ \cos \beta l} + \frac{2\alpha\beta}{\alpha^2 - \beta^2} = 0 \dots\dots\dots\dots(4).$$

The value of $\frac{2\alpha\beta}{\alpha^2 - \beta^2}$, determined by (2), is $\frac{2nb\kappa}{a^2}$, so that

$$\frac{\sinh \alpha l \ \sin \beta l}{1 - \cosh \alpha l \ \cos \beta l} + \frac{2nb\kappa}{a^2} = 0 \dots\dots\dots\dots(5).$$

From (2) we find also that

$$\alpha^2 = \frac{a^2}{2b^2\kappa^2}\left\{\sqrt{1 + 4\frac{n^2 b^2 \kappa^2}{a^4}} + 1\right\}$$
$$\beta^2 = \frac{a^2}{2b^2\kappa^2}\left\{\sqrt{1 + 4\frac{n^2 b^2 \kappa^2}{a^4}} - 1\right\}$$
$$\dots\dots\dots\dots(6).$$

Thus far our equations are rigorous, or rather as rigorous as the differential equation on which they are founded; but we shall now introduce the supposition that the vibration considered is but slightly affected by the existence of rigidity. This being the case, the approximate expression for y is

$$y = \sin\frac{i\pi x}{l}\,\cos\left(\frac{i\pi}{l}\,at\right),$$

and therefore

$$\beta = \frac{i\pi}{l}, \qquad n = \frac{i\pi a}{l}, \quad \dots\dots\dots\dots\dots(7).$$

nearly.

The introduction of these values into the second of equations (6) proves that $n^2\dfrac{b^2\kappa^2}{a^4}$ or $\dfrac{b^2}{a^2}\dfrac{\kappa^2}{l^2}$ is a small quantity under the circumstances contemplated, and therefore that $a^2 l^2$ is a large quantity. Since $\cosh al$, $\sinh al$ are both large. equation (5) reduces to

$$\tan\beta l = \frac{2nb\kappa}{a^2},$$

or, on substitution of the approximate value for β derived from (6),

$$\tan\frac{nl}{a} = 2\frac{nb\kappa}{a^2}.$$

The approximate value of $\dfrac{nl}{a}$ is $i\pi$. If we take $\dfrac{nl}{a} = i\pi + \theta$, we get

$$\tan(i\pi + \theta) = \tan\theta = \theta = 2\frac{nb\kappa}{a^2} = 2i\pi\frac{b}{a}\frac{\kappa}{l},$$

so that

$$n = i\frac{\pi a}{l}\left(1 + 2\frac{b}{a}\frac{\kappa}{l}\right)\dots\dots\dots\dots\dots(8).$$

According to this equation the component tones are all raised in pitch by the same small interval, and therefore the harmonic relation is not disturbed by the rigidity. It would probably be otherwise if terms involving $\kappa^2 : l^2$ were retained; it does not therefore follow that the harmonic relation is better preserved in spite of rigidity when the ends are clamped than when they are free, but only that there is no additional disturbance in the former case, though the absolute alteration of pitch is much greater. It should be remarked that $b : a$ or $\sqrt{(q + T)} : \sqrt{T}$, is a large quantity, and that, if our result is to be correct, $\kappa : l$ must be small enough to bear multiplication by $b : a$ and yet remain small.

The theoretical result embodied in (8) has been compared with experiment by Seebeck, who found a satisfactory agreement. The constant of stiffness was deduced from observations of the rapidity of the vibrations of a small piece of the wire, when one end was clamped in a vice.

191. It has been shewn in this chapter that the theory of bars, even when simplified to the utmost by the omission of unimportant quantities, is decidedly more complicated than that of perfectly flexible strings. The reason of the extreme simplicity of the vibrations of strings is to be found in the fact that waves of the harmonic type are propagated with a velocity independent of the wave length, so that an arbitrary wave is allowed to travel without decomposition. But when we pass from strings to bars, the constant in the differential equation, viz. $\dfrac{d^2y}{dt^2} + \kappa^2 b^2 \dfrac{d^4y}{dx^4} = 0$, is no longer expressible as a velocity, and therefore the velocity of transmission of a train of harmonic waves cannot depend on the differential equation alone, but must vary with the wave length. Indeed, if it be admitted that the train of harmonic waves can be propagated at all, this consideration is sufficient by itself to prove that the velocity must vary inversely as the wave length. The same thing may be seen from the solution applicable to waves propagated in one direction, viz. $y = \cos \dfrac{2\pi}{\lambda} (Vt - x)$, which satisfies the differential equation if

$$V = \frac{2\pi \kappa b}{\lambda} \quad \dotfill (1).$$

Let us suppose that there are two trains of waves of equal amplitudes, but of different wave lengths, travelling in the same direction. Thus

$$y = \cos 2\pi \left(\frac{t}{\tau} - \frac{x}{\lambda} \right) + \cos 2\pi \left(\frac{t}{\tau'} - \frac{x}{\lambda'} \right)$$

$$= 2 \cos \pi \left\{ t\left(\frac{1}{\tau} - \frac{1}{\tau'}\right) - x\left(\frac{1}{\lambda} - \frac{1}{\lambda'}\right) \right\} \cos \pi \left\{ t\left(\frac{1}{\tau} + \frac{1}{\tau'}\right) - x\left(\frac{1}{\lambda} + \frac{1}{\lambda'}\right) \right\} \dots (2).$$

If $\tau' - \tau$, $\lambda' - \lambda$ be small, we have a train of waves, whose amplitude slowly varies from one point to another between the values 0 and 2, forming a series of groups separated from one another by regions comparatively free from disturbance. In the case of a string or of a column of air, λ varies as τ, and then the groups move forward with the same velocity as the component trains, and there is no change of type. It is otherwise when, as in the case of a bar vibrating transversely, the velocity of propagation is a function of the wave length. The position at time t of the middle of the group which was initially at the origin is given by

$$t\left(\frac{1}{\tau} - \frac{1}{\tau'}\right) - x\left(\frac{1}{\lambda} - \frac{1}{\lambda'}\right) = 0,$$

which shews that the velocity of the group is

$$\left(\frac{1}{\tau} - \frac{1}{\tau'}\right) \div \left(\frac{1}{\lambda} - \frac{1}{\lambda'}\right) = \delta\left(\frac{1}{\tau}\right) \div \delta\left(\frac{1}{\lambda}\right).$$

If we suppose that the velocity V of a train of waves varies as λ^n, we find

$$\frac{d\left(\frac{1}{\tau}\right)}{d\left(\frac{1}{\lambda}\right)} = \frac{d\left(\frac{V}{\lambda}\right)}{d\left(\frac{1}{\lambda}\right)} = -(n-1)V \quad\text{................. (3).}$$

In the present case $n = -1$, and accordingly the velocity of the groups is *twice* that of the component waves[1].

192. On account of the dependence of the velocity of propagation on the wave length, the condition of an infinite bar at any time subsequent to an initial disturbance confined to a limited portion, will have none of the simplicity which characterises the corresponding problem for a string; but nevertheless Fourier's investigation of this question may properly find a place here.

It is required to determine a function of x and t, so as to satisfy

$$\frac{d^2y}{dt^2} + \frac{d^4y}{dx^4} = 0 \quad\text{.......................... (1),}$$

and make initially $y = \phi(x)$, $\dot{y} = \psi(x)$.

A solution of (1) is

$$y = \cos q^2 t \, \cos q(x - \alpha) \quad\text{.....................(2),}$$

where q and α are constants, from which we conclude that

$$y = \int_{-\infty}^{+\infty} d\alpha \, F(\alpha) \int_{-\infty}^{+\infty} dq \cos q^2 t \, \cos q(x - \alpha)$$

---

[1] In the corresponding problem for waves on the surface of deep water, the velocity of propagation varies directly as the square root of the wave length, so that $n = \frac{1}{2}$. The velocity of a group of such waves is therefore *one half* of that of the component trains.

is also a solution, where $F'(\alpha)$ is an arbitrary function of α. If now we put $t = 0$,

$$y_0 = \int_{-\infty}^{+\infty} d\alpha\, F(\alpha) \int_{-\infty}^{+\infty} dq \cos q\,(x - \alpha),$$

which shews that $F(\alpha)$ must be taken to be $\dfrac{1}{2\pi}\,\phi\,(\alpha)$, for then by Fourier's double integral theorem $y_0 = \phi\,(x)$. Moreover, $\dot{y} = 0$; hence

$$y = \frac{1}{2\pi}\int_{-\infty}^{+\infty} d\alpha\, \phi(\alpha) \int_{-\infty}^{+\infty} dq \cos q^2 t\, \cos q\,(x - \alpha) \quad\ldots\ldots\ldots(3)$$

satisfies the differential equation and makes initially,

$$y = \phi\,(x), \quad \dot{y} = 0.$$

By Stokes' theorem (§ 95), or independently, we may now supply the remaining part of the solution, which has to satisfy the differential equation while it makes initially $y = 0$, $\dot{y} = \psi\,(x)$; it is

$$y = \frac{1}{2\pi}\int_{-\infty}^{+\infty} d\alpha\, \psi\,(\alpha) \int_{-\infty}^{+\infty} dq\, \frac{1}{q^2} \sin q^2 t\, \cos q(x - \alpha) \ldots\ldots\ldots(4).$$

The final result is obtained by adding the right-hand members of (3) and (4).

In (3) the integration with respect to q may be effected by means of the formula

$$\int_{-\infty}^{+\infty} dq \cos q^2 t\, \cos qz = \sqrt{\frac{\pi}{t}}\, \sin\left(\frac{\pi}{4} + \frac{z^2}{4t}\right) \ldots\ldots\ldots\ldots(5),$$

which may be proved as follows. If in the well-known integral formula

$$\int_{-\infty}^{+\infty} e^{-a^2 x^2} dx = \frac{\sqrt{\pi}}{a},$$

we put $x + b$ for x, we get

$$\int_{-\infty}^{+\infty} e^{-a^2\,(x^2 + 2bx)}\, dx = \frac{\sqrt{\pi}}{a} e^{a^2 b^2}$$

Now suppose that $a^2 = i = e^{\frac{1}{2}i\pi}$, where $i = \sqrt{-1}$, and retain only the real part of the equation. Thus

$$\int_{-\infty}^{+\infty} \cos (x^2 + 2bx)\, dx = \sqrt{\pi} \sin\left(b^2 + \frac{\pi}{4}\right),$$

whence

$$\int_{-\infty}^{+\infty} \cos x^2 \cos 2bx \, dx = \sqrt{\pi} \sin\left(b^2 + \frac{\pi}{4}\right),$$

from which (5) follows by a simple change of variable. Thus equation (3) may be written

$$y = \frac{1}{2\sqrt{\pi t}} \int_{-\infty}^{+\infty} d\alpha \, \phi(\alpha) \sin\left\{\frac{\pi}{4} + \frac{(x-\alpha)^2}{4t}\right\},$$

or, if $\dfrac{\alpha - x}{2\sqrt{t}} = \mu$,

$$y = \frac{1}{\sqrt{2\pi}} \int_{-\infty}^{+\infty} d\mu \, (\cos \mu^2 + \sin \mu^2) \, \phi(x + 2\mu\sqrt{t}) \dots\dots (6).$$

CHAPTER IX.

VIBRATIONS OF MEMBRANES.

193. THE theoretical membrane is a perfectly flexible and infinitely thin lamina of solid matter, of uniform material and thickness, which is stretched in all directions by a tension so great as to
remain sensibly unaltered during the vibrations and displacements
contemplated. If an imaginary line be drawn across the membrane in any direction, the mutual action between the two portions
separated by an element of the line is proportional to the length of
the element and perpendicular to its direction. If the force in
question be $T_1\, ds$, T_1 may be called the *tension of the membrane;*
it is a quantity of one dimension in mass and -2 in time.

The principal problem in connection with this subject is the
investigation of the transverse vibrations of membranes of different
shapes, whose boundaries are fixed. Other questions indeed may
be proposed, but they are of comparatively little interest; and,
moreover, the methods proper for solving them will be sufficiently illustrated in other parts of this work. We may therefore
proceed at once to the consideration of a membrane stretched over
the area included within a fixed, closed, plane boundary.

194. Taking the plane of the boundary as that of xy, let w
denote the small displacement therefrom of any point P of the
membrane. Round P take a small area S, and consider the forces
acting upon it parallel to z. The resolved part of the tension is
expressed by

$$T_1 \int \frac{dw}{dn}\, ds,$$

where ds denotes an element of the boundary of S, and dn an
element of the normal to the curve drawn outwards. This is
balanced by the reaction against acceleration measured by $\rho S \ddot{y}$,

ρ being a symbol of one dimension in mass and -2 in length denoting the superficial density. Now by Green's theorem, if

$$\nabla^2 = \frac{d^2}{dx^2} + \frac{d^2}{dy^2},$$

$$\int \frac{dw}{dn}\, ds = \iint \nabla^2 w\, dS = \nabla^2 w \,.\, S \quad \text{ultimately,}$$

and thus the equation of motion is

$$\frac{d^2w}{dt^2} = \frac{T_1}{\rho}\left(\frac{d^2w}{dx^2} + \frac{d^2w}{dy^2}\right) \quad \text{.................(1).}$$

The condition to be satisfied at the boundary is of course $w = 0$.

The differential equation may also be investigated from the expression for the potential energy, which is found by multiplying the tension by the superficial stretching. The altered area is

$$\iint \sqrt{1 + \left(\frac{dw}{dx}\right)^2 + \left(\frac{dw}{dy}\right)^2}\, dx\, dy \,;$$

and thus

$$V = \tfrac{1}{2} T_1 \iint \left\{ \left(\frac{dw}{dx}\right)^2 + \left(\frac{dw}{dy}\right)^2 \right\} dx\, dy \quad \text{...........(2),}$$

from which δV is easily found by an integration by parts.

If we write $T_1 \div \rho = c^2$, then c is of the nature of a velocity, and the differential equation is

$$\frac{d^2w}{dt^2} = c^2 \left(\frac{d^2w}{dx^2} + \frac{d^2w}{dy^2}\right) \quad \text{.....................(3).}$$

195. We shall now suppose that the boundary of the membrane is the rectangle formed by the coordinate axes and the lines $x = a$, $y = b$. For every point within the area (3) § 194 is satisfied, and for every point on the boundary $w = 0$.

A particular integral is evidently

$$w = \sin \frac{m\pi x}{a} \sin \frac{n\pi y}{b} \cos pt \quad \text{...............(1),}$$

where

$$p^2 = c^2 \pi^2 \left(\frac{m^2}{a^2} + \frac{n^2}{b^2}\right) \quad \text{.........................(2),}$$

and m and n are integers; and from this the general solution may be derived. Thus

$$w = \sum_{m=1}^{m=\infty} \sum_{n=1}^{n=\infty} \sin \frac{m\pi x}{a} \sin \frac{n\pi y}{b} \{A_{mn} \cos pt + B_{mn} \sin pt\} \quad \text{.........(3).}$$

That this result is really general may be proved *a posteriori*, by shewing that it may be adapted to express arbitrary initial circumstances.

Whatever function of the co-ordinates w may be, it can be expressed for all values of x between the limits 0 and a by the series

$$Y_1 \sin \frac{\pi x}{a} + Y_2 \sin \frac{2\pi x}{a} + \ldots\ldots,$$

where the coefficients Y_1, Y_2, &c. are independent of x. Again whatever function of y any one of the coefficients Y may be, it can be expanded between 0 and b in the series

$$C_1 \sin \frac{\pi y}{b} + C_2 \sin \frac{2\pi y}{b} + \ldots\ldots,$$

where C_1 &c. are constants. From this we conclude that any function of x and y can be expressed within the limits of the rectangle by the double series

$$\sum_{m=1}^{m=\infty} \sum_{n=1}^{n=\infty} A_{mn} \sin \frac{m\pi x}{a} \sin \frac{n\pi y}{b};$$

and therefore that the expression for w in (3) can be adapted to arbitrary initial values of w and $\dot{w}$. In fact

$$\left. \begin{aligned} A_{mn} &= \frac{4}{ab} \int_0^a \int_0^b w_0 \sin \frac{m\pi x}{a} \sin \frac{n\pi y}{b}\, dx\, dy, \\ B_{mn} &= \frac{4}{abp} \int_0^a \int_0^b w_0 \sin \frac{m\pi x}{a} \sin \frac{n\pi y}{b}\, dx\, dy, \end{aligned} \right\} \ldots\ldots(4).$$

The character of the normal functions of a given rectangle,

$$\sin \frac{m\pi x}{a} \sin \frac{n\pi y}{b},$$

as depending on m and n, is easily understood. If m and n be both unity, w retains the same sign over the whole of the rectangle, vanishing at the edge only; but in any other case there are nodal lines running parallel to the axes of coordinates. The number of the nodal lines parallel to x is $n-1$, their equations being

$$y = \frac{b}{n}, \ \frac{2b}{n}, \ \ldots\ldots \frac{(n-1)b}{n}.$$

In the same way the equations of the nodal lines parallel to y are

$$x = \frac{a}{m}, \; \frac{2a}{m}, \; \ldots\ldots \frac{(m-1)\,a}{m},$$

being $m-1$ in number. The nodal system divides the rectangle into mn equal parts, in each of which the numerical value of w is repeated.

196. The expression for w in terms of the normal functions is

$$w = \Sigma\Sigma \; \phi_{mn} \sin \frac{m\pi x}{a} \sin \frac{n\pi y}{b} \; \ldots\ldots\ldots\ldots (1),$$

where ϕ_{mn} &c. are the normal coordinates. We proceed to form the expression for V in terms of ϕ_{mn}. We have

$$\left(\frac{dw}{dx}\right)^2 = \pi^2 \left\{ \Sigma\Sigma\phi_{mn} \frac{m}{a} \cos \frac{m\pi x}{a} \sin \frac{n\pi y}{b} \right\}^2,$$

$$\left(\frac{dw}{dy}\right)^2 = \pi^2 \left\{ \Sigma\Sigma\phi_{mn} \frac{n}{b} \sin \frac{m\pi x}{a} \cos \frac{n\pi y}{b} \right\}^2.$$

In integrating these expressions over the area of the rectangle the products of the normal coordinates disappear, and we find

$$V = \frac{T_1}{2} \iint \left\{ \left(\frac{dw}{dx}\right)^2 + \left(\frac{dw}{dy}\right)^2 \right\} dx\,dy$$

$$= \frac{T_1}{2} \frac{ab\pi^2}{4} \Sigma\Sigma \left(\frac{m^2}{a^2} + \frac{n^2}{b^2}\right) \phi_{mn}{}^2 \; \ldots\ldots\ldots\ldots\ldots (2),$$

the summation being extended to all integral values of m and n.

The expression for the kinetic energy is proved in the same way to be

$$T = \frac{\rho}{2} \frac{ab}{4} \Sigma\Sigma \dot{\phi}_{mn}{}^2 \; \ldots\ldots\ldots\ldots\ldots (3),$$

from which we deduce as the normal equation of motion

$$\ddot{\phi}_{mn} + c^2\pi^2 \left(\frac{m^2}{a^2} + \frac{n^2}{b^2}\right)\phi_{mn} = \frac{4}{ab\rho} \Phi_{mn} \; \ldots\ldots\ldots\ldots (4).$$

In this equation

$$\Phi_{mn} = \int_0^a\!\!\int_0^b Z \sin \frac{m\pi x}{a} \sin \frac{n\pi y}{b} \, dx\,dy \ldots\ldots\ldots\ldots (5),$$

if $Z\,dx\,dy$ denote the transverse force acting on the element $dx\,dy$.

Let us suppose that the initial condition is one of rest under the operation of a constant force Z, such as may be supposed to arise from gaseous pressure. At the time $t = 0$, the impressed force is removed, and the membrane left to itself. Initially the equation of equilibrium is

$$c^2\pi^2\left(\frac{m^2}{a^2} + \frac{n^2}{b^2}\right)(\phi_{mn})_0 = \frac{4}{ab\rho}\,\Phi_{mn} \quad\text{...............} (6),$$

whence $(\phi_{mn})_0$ is to be found. The position of the system at time t is then given by

$$\phi_{mn} = (\phi_{mn})_0\cos\left(\sqrt{\frac{m^2}{a^2} + \frac{n^2}{b^2}}\,.\,c\pi t\right) \quad\text{...........} (7),$$

in conjunction with (1).

In order to express Φ_{mn}, we have merely to substitute for Z its value in (5), or in this case simply to remove Z from under the integral sign. Thus

$$\Phi_{mn} = Z\int_0^a\int_0^b \sin\frac{m\pi x}{a}\sin\frac{n\pi y}{b}\,dx\,dy,$$

$$= Z\frac{ab}{mn\pi^2}(1 - \cos m\pi)\,(1 - \cos n\pi).$$

We conclude that Φ_{mn} vanishes, unless m and n are *both* odd, and that then

$$\Phi_{mn} = \frac{4ab}{mn\pi^2}\,Z.$$

Accordingly, m and n being both odd,

$$\phi_{mn} = \frac{16\,Z}{\pi^2\rho}\,\frac{\cos pt}{mnp^2} \quad\text{.....................} (8),$$

where $$p^2 = c^2\pi^2\left(\frac{m^2}{a^2} + \frac{n^2}{b^2}\right) \quad\text{.....................} (9).$$

This is an example of (8), § 101.

If the membrane, previously at rest in its position of equilibrium, be set in motion by a blow applied at the point $\alpha\,\beta$, the solution is

$$\phi_{mn} = \frac{4}{ab p}\sin\frac{m\pi\alpha}{a}\sin\frac{n\pi\beta}{b}\iint w_0\,dx\,dy\,.\,\sin pt \quad\text{....} (10).$$

197. The frequency of the natural vibrations is found by ascribing different integral values to m and n in the expression

$$\frac{p}{2\pi} = \frac{c}{2}\sqrt{\frac{m^2}{a^2} + \frac{n^2}{b^2}} \qu................ (1).$$

For a given mode of vibration the pitch falls when either side of the rectangle is increased. In the case of the gravest mode, when $m=1$, $n=1$, additions to the shorter side are the more effective; and when the form is very elongated, additions to the longer side are almost without effect.

When a^2 and b^2 are incommensurable, no two pairs of values of m and n can give the same frequency, and each fundamental mode of vibration has its own characteristic period. But when a^2 and b^2 are commensurable, two or more fundamental modes may have the same periodic time, and may then coexist in any proportions, while the motion still retains its simple harmonic character. In such cases the specification of the period does not completely determine the type. The full consideration of the problem now presenting itself requires the aid of the theory of numbers; but it will be sufficient for the purposes of this work to consider a few of the simpler cases, which arise when the membrane is square. The reader will find fuller information in Riemann's lectures on partial differential equations.

If $a = b$,

$$\frac{p}{2\pi} = \frac{c}{2a}\sqrt{m^2 + n^2} \qu................(2).$$

The lowest tone is found by putting m and n equal to unity, which gives only one fundamental mode :—

$$w = \sin\frac{\pi x}{a}\sin\frac{\pi y}{a}\cos pt \qu................(3).$$

Next suppose that one of the numbers m, n is equal to 2, and the other to unity. In this way two distinct types of vibration are obtained, whose periods are the same. If the two vibrations be synchronous in phase, the whole motion is expressed by

$$w = \left\{ C\sin\frac{2\pi x}{a}\sin\frac{\pi y}{a} + D\sin\frac{\pi x}{a}\sin\frac{2\pi y}{a} \right\} \cos pt ...(4) ;$$

so that, although every part vibrates synchronously with a harmonic motion, the type of vibration is to some extent arbitrary.

Four particular cases may be especially noted.   First, if $D = 0$,

$$w = C \sin \frac{2\pi x}{a} \sin \frac{\pi y}{a} \cos pt \ \dots\dots\dots\dots\dots ..(5),$$

which indicates a vibration with one node along the line $x = \frac{1}{2}a$. Similarly if $C = 0$, we have a node parallel to the other pair of edges.   Next, however, suppose that C and D are finite and equal.   Then w is proportional to

$$\sin \frac{2\pi x}{a} \sin \frac{\pi y}{a} + \sin \frac{\pi x}{a} \sin \frac{2\pi y}{b},$$

which may be put into the form

$$2 \sin \frac{\pi x}{a} \sin \frac{\pi y}{a} \left(\cos \frac{\pi x}{a} + \cos \frac{\pi y}{a} \right).$$

This expression vanishes, when

$$\sin \frac{\pi x}{a} = 0, \ \text{ or } \ \sin \frac{\pi y}{a} = 0$$

or again, when

$$\cos \frac{\pi x}{a} + \cos \frac{\pi y}{a} = 0.$$

The first two equations give the edges, which were originally assumed to be nodal; while the third gives $y + x = a$, representing one diagonal of the square.

In the fourth case, when $C = -D$, we obtain for the nodal lines, the edges of the square together with the diagonal $y = x$. The figures represent the four cases.

Fig. 32.

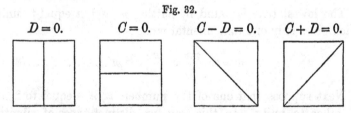

$D = 0.$ $C = 0.$ $C - D = 0.$ $C + D = 0.$

For other relative values of C and D the interior nodal line is curved, but is always analytically expressed by

$$C \cos \frac{\pi x}{a} + D \cos \frac{\pi y}{a} = 0 \ \dots\dots\dots\dots\dots(6),$$

and may be easily constructed with the help of a table of logarithmic cosines.

The next case in order of pitch occurs when $m = 2$, $n = 2$. The values of m and n being equal, no alteration is caused by their interchange, while no other pair of values gives the same frequency of vibration. The only type to be considered is accordingly

$$w = \sin \frac{2\pi x}{a} \sin \frac{2\pi y}{a} \cos pt,$$

whose nodes, determined by the equation

$$\sin \frac{\pi x}{a} \sin \frac{\pi y}{a} \cos \frac{\pi x}{a} \cos \frac{\pi y}{a} = 0,$$

are (in addition to the edges) the straight lines

$$x = \tfrac{1}{2}a \qquad y = \tfrac{1}{2}a.$$

Fig. 33.

The next case which we shall consider is obtained by ascribing to m, n the values 3, 1, and 1, 3 successively. We have

$$w = \left\{ C \sin \frac{3\pi x}{a} \sin \frac{\pi y}{a} + D \sin \frac{\pi x}{a} \sin \frac{3\pi y}{a} \right\} \cos pt.$$

The nodes are given by

$$\sin \frac{\pi x}{a} \sin \frac{\pi y}{a} \left\{ C \left(4 \cos^2 \frac{\pi x}{a} - 1 \right) + D \left(4 \cos^2 \frac{\pi y}{a} - 1 \right) \right\} = 0,$$

or, if we reject the first two factors, which correspond to the edges,

$$C \left(4 \cos^2 \frac{\pi x}{a} - 1 \right) + D \left(4 \cos^2 \frac{\pi y}{a} - 1 \right) = 0 \ldots\ldots\ldots(7).$$

If $C = 0$, we have $\qquad y = \tfrac{1}{3}a, \ y = \tfrac{2}{3}a.$

If $D = 0$, $\qquad\qquad x = \tfrac{1}{3}a, \ x = \tfrac{2}{3}a.$

If $C = -D$, $\qquad\qquad \cos \frac{\pi x}{a} = \pm \cos \frac{\pi y}{a},$

whence, $\qquad\qquad y = x, \ y = a - x,$

which represent the two diagonals.

R.　　　　　　　　　　　　　　　　　　　17

Lastly, if $C = D$, the equation of the node is

$$\cos^2 \frac{\pi x}{a} + \cos^2 \frac{\pi y}{a} = \tfrac{1}{2},$$

or

$$1 + \cos \frac{2\pi x}{a} + \cos \frac{2\pi y}{a} = 0 \dots\dots\dots\dots\dots(8),$$

Fig. 34.

$C = 0.$　　　$D = 0.$　　　$C + D = 0.$　　　$C - D = 0.$

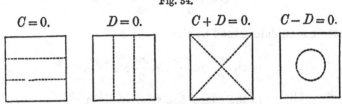

In case (4) when $x = \tfrac{1}{2}a$, $y = \tfrac{1}{4}a$, or $\tfrac{3}{4}a$; and similarly when $y = \tfrac{1}{2}a$, $x = \tfrac{1}{4}a$, or $\tfrac{3}{4}a$. Thus one half of each of the lines joining the middle points of opposite edges is intercepted by the curve.

It should be noticed that in whatever ratio to one another C and D may be taken, the nodal curve always passes through the four points of intersection of the nodal lines of the first two cases, $C = 0$, $D = 0$. If the vibrations of these cases be compounded with corresponding phases, it is evident that in the shaded compartments of Fig. (35) the directions of displacement

Fig. 35.

are the same, and that therefore no part of the nodal curve is to be found there; whatever the ratio of amplitudes, the curve must be drawn through the unshaded portions. When on the other hand the phases are opposed, the nodal curve will pass exclusively through the shaded portions.

When $m = 3$, $n = 3$, the nodes are the straight lines parallel to the edges shewn in Fig. (36).

Fig. 36.

The last case which we shall consider is obtained by putting

$$m = 3, \ n = 2, \quad \text{or} \quad m = 2, \ n = 3.$$

The nodal system is

$$C \sin \frac{3\pi x}{a} \sin \frac{2\pi y}{a} + D \sin \frac{2\pi x}{a} \sin \frac{3\pi y}{a} = 0,$$

or, if the factors corresponding to the edges be rejected,

$$C \left(4 \cos^2 \frac{\pi x}{a} - 1 \right) \cos \frac{\pi y}{a} + D \cos \frac{\pi x}{a} \left(4 \cos^2 \frac{\pi y}{a} - 1 \right) = 0 \ldots \ldots (9).$$

If C or D vanish, we fall back on the nodal systems of the component vibrations, consisting of straight lines parallel to the edges. If $C = D$, our equation may be written

$$\left(\cos \frac{\pi x}{a} + \cos \frac{\pi y}{a} \right) \left(4 \cos \frac{\pi x}{a} \cos \frac{\pi y}{a} - 1 \right) = 0 \ldots \ldots (10),$$

of which the first factor represents the diagonal $y + x = a$, and the second a hyperbolic curve.

If $C = - D$, we obtain the same figure relatively to the other diagonal[1].

198. The pitch of the natural modes of a square membrane, which is nearly, but not quite uniform, may be investigated by the general method of § 90.

We will suppose in the first place that m and n are equal. In this case, when the pitch of a uniform membrane is given, the mode of its vibration is completely determined. If we now conceive a variation of density to ensue, the natural type of vibration is in general modified, but the period may be calculated approximately without allowance for the change of type.

We have

$$T = \tfrac{1}{2} \iint (\rho_0 + \delta\rho) \, \dot\phi_{mm}{}^2 \sin^2 \frac{m\pi x}{a} \sin^2 \frac{m\pi y}{a} \, dx\,dy$$

$$= \tfrac{1}{2} \dot\phi_{mm}{}^2 \left\{ \rho_0 \frac{a^2}{4} + \iint \delta\rho \sin^2 \frac{m\pi x}{a} \sin^2 \frac{m\pi y}{a} \, dx\,dy \right\},$$

of which the second term is the increment of T due to $\delta\rho$. Hence if $w \propto \cos pt$, and P denote the value of p previously to variation, we have

$$p_{mm}{}^2 : P_{mm}{}^2 = 1 - \frac{4}{a^2} \int_0^a \int_0^a \frac{\delta\rho}{\rho_0} \sin^2 \frac{m\pi x}{a} \sin^2 \frac{m\pi y}{a} \, dx\,dy \ldots \ldots (1),$$

[1] Lamé, *Leçons sur l'élasticité*, p. 129.

where $$P_{mm}{}^2 = \frac{2c^2\pi^2 m^2}{a^2}, \quad \text{and} \quad c^2 = T_1 \div \rho_0.$$

For example, if there be a small load M attached to the middle of the square,

$$p_{mm}{}^2 : P_{mm}{}^2 = 1 - \frac{4M}{a^2\rho_0}\sin^4 m\frac{\pi}{2} \quad\dots\dots\dots\dots(2),$$

in which $\sin^4 \frac{1}{2}m\pi$ vanishes, if m be even, and is equal to unity, if m be odd. In the former case the centre is on the nodal line of the unloaded membrane, and thus the addition of the load produces no result.

When, however, m and n are unequal, the problem, though remaining subject to the same general principles, presents a peculiarity different from anything we have hitherto met with. The natural type for the unloaded membrane corresponding to a specified period is now to some extent arbitrary; but the introduction of the load will in general remove the indeterminate element. In attempting to calculate the period on the assumption of the undisturbed type, the question will arise how the selection of the undisturbed type is to be made, seeing that there are an indefinite number, which in the uniform condition of the membrane give identical periods. The answer is that those types must be chosen which differ infinitely little from the actual types assumed under the operation of the load, and such a type will be known by the criterion of its making the period calculated from it a maximum or minimum.

As a simple example, let us suppose that a small load M is attached to the membrane at a point lying on the line $x = \frac{1}{2}a$, and that we wish to know what periods are to be substituted for the two equal periods of the unloaded membrane, found by making

$$m = 1, \; n = 2, \quad \text{or} \quad m = 2, \; n = 1.$$

It is clear that the normal types to be chosen, are those whose nodes are represented in the first two cases of Fig. (32). In the first case the increase in the period due to the load is zero, which is the least that it can be; and in the second case the increase is the greatest possible. If β be the ordinate of M, the kinetic energy is altered in the ratio

$$\frac{\rho}{2}\frac{a^2}{4} : \frac{\rho}{2}\frac{a^2}{4} + \frac{M}{2}\sin^2\frac{2\pi\beta}{a};$$

and thus

$$p_{12}{}^2 : P_{12}{}^2 = 1 + \frac{4M}{a^2\rho}\sin^2\frac{2\pi\beta}{a} \quad\dots\dots\dots\dots(3)$$

while $$p_{21}{}^2 = P_{21}{}^2 = P_{12}{}^2.$$

The ratio characteristic of the interval between the two natural tones of the loaded membrane is thus approximately

$$1 + \frac{2M}{a^2\rho} \sin^2 \frac{2\pi\beta}{a} \quad\ldots\ldots\ldots\ldots\ldots\ldots(4).$$

If $\beta = \frac{1}{2}a$, neither period is affected by the load.

As another example, the case, where the values of m and n are 3 and 1, considered in § 197, may be referred to. With a load in the middle, the two normal types to be selected are those corresponding to the last two cases of Fig. (34), in the former of which the load has no effect on the period.

The problem of determining the vibration of a square membrane which carries a relatively heavy load is more difficult, and we shall not attempt its solution. But it may be worth while to recall to memory the fact that the actual period is greater than any that can be calculated from a hypothetical type, which differs from the actual one.

199. The preceding theory of square membranes includes a good deal more than was at first intended. Whenever in a vibrating system certain parts remain at rest, they may be supposed to be absolutely fixed, and we thus obtain solutions of other questions than those originally proposed. For example, in the present case, wherever a diagonal of the square is nodal, we obtain a solution applicable to a membrane whose fixed boundary is an isosceles right-angled triangle. Moreover, any mode of vibration possible to the triangle corresponds to some natural mode of the square, as may be seen by supposing two triangles put together, the vibrations being equal and opposite at points which are the images of each other in the common hypothenuse. Under these circumstances it is evident that the hypothenuse would remain at rest without constraint, and therefore the vibration in question is included among those of which a complete square is capable.

The frequency of the gravest tone of the triangle is found by putting $m = 1$, $n = 2$ in the formula

$$\frac{p}{2\pi} = \frac{c}{2a} \sqrt{m^2 + n^2} \quad\ldots\ldots\ldots\ldots\ldots\ldots(1),$$

and is therefore equal to $\dfrac{c\sqrt{5}}{2a}$.

The next tone occurs, when $m = 3$, $n = 1$. In this case

$$\frac{p}{2\pi} = \frac{c\sqrt{10}}{2a} \quad\ldots\ldots\ldots\ldots\ldots\ldots\ldots\ldots (2),$$

Fig. 87.

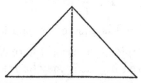

as might also be seen by noticing that the triangle divides itself into two, Fig. (37), whose sides are less than those of the whole triangle in the ratio $\sqrt{2} : 1$.

For the theory of the vibrations of a membrane whose boundary is in the form of an equilateral triangle, the reader is referred to Lamé's 'Leçons sur l'élasticité.' It is proved that the frequency of the gravest tone is $c \div h$, where h is the height of the triangle, which is the same as the frequency of the gravest tone of a square whose diagonal is h.

200. When the fixed boundary of the membrane is circular, the first step towards a solution of the problem is the expression of the general differential equation in polar co-ordinates. This may be effected analytically; but it is simpler to form the polar equation *de novo* by considering the forces which act on the polar element of area $r\,d\theta\,dr$. As in § 194 the force of restitution acting on a small area of the membrane is

$$-T_1 \int \frac{dw}{dn}\,ds = -T_1 \left\{ \frac{d}{dr}\left(\frac{dw}{dr}\,r\,d\theta\right)dr + \frac{d}{d\theta}\left(\frac{dw}{r\,d\theta}\,dr\right)d\theta \right\}$$

$$= -T_1 \cdot r\,d\theta\,dr\left\{\frac{d^2w}{dr^2} + \frac{1}{r}\frac{dw}{dr} + \frac{1}{r^2}\frac{d^2w}{d\theta^2}\right\};$$

and thus, if $T_1 \div \rho = c^2$ as before, the equation of motion is

$$\frac{d^2w}{dt^2} = c^2\left\{\frac{d^2w}{dr^2} + \frac{1}{r}\frac{dw}{dr} + \frac{1}{r^2}\frac{d^2w}{d\theta^2}\right\}\ldots\ldots\ldots\ldots(1).$$

The subsidiary condition to be satisfied at the boundary is that $w = 0$, when $r = a$.

In order to investigate the normal component vibrations we have now to assume that w is a harmonic function of the time. Thus, if $w \propto \cos(pt - \epsilon)$, and for the sake of brevity we write $p \div c = \kappa$, the differential equation appears in the form

$$\frac{d^2w}{dr^2} + \frac{1}{r}\frac{dw}{dr} + \frac{1}{r^2}\frac{d^2w}{d\theta^2} + \kappa^2 w = 0 \dots\dots\dots(2),$$

in which κ is the reciprocal of a linear quantity.

Now whatever may be the nature of w as a function of r and θ, it can be expanded in Fourier's series

$$w = w_0 + w_1 \cos(\theta + \alpha_1) + w_2 \cos 2(\theta + \alpha_2) + \dots\dots(3),$$

in which w_0, w_1, &c. are functions of r, but not of θ. The result of substituting from (3) in (2) may be written

$$\Sigma \left\{ \frac{d^2w_n}{dr^2} + \frac{1}{r}\frac{dw_n}{dr} + \left(\kappa^2 - \frac{n^2}{r^2}\right)w_n \right\} \cos n(\theta + \alpha_n) = 0,$$

the summation extending to all integral values of n. If we multiply this equation by $\cos n(\theta + \alpha_n)$, and integrate with respect to θ between the limits 0 and 2π, we see that each term must vanish separately, and we thus obtain to determine w_n as a function of r

$$\frac{d^2w_n}{dr^2} + \frac{1}{r}\frac{dw_n}{dr} + \left(\kappa^2 - \frac{n^2}{r^2}\right)w_n = 0 \dots\dots\dots(4),$$

in which it is a matter of indifference whether the factor $\cos n(\theta + \alpha_n)$ be supposed to be included in w_n or not.

The solution of (4) involves two distinct functions of r, each multiplied by an arbitrary constant. But one of these functions becomes infinite when r vanishes, and the corresponding particular solution must be excluded as not satisfying the prescribed conditions at the origin of co-ordinates. This point may be illustrated by a reference to the simpler equation derived from (4) by making κ and n vanish, when the solution in question reduces to $w = \log r$, which, however, does not at the origin satisfy $\nabla^2 w = 0$, as may be seen from the value of $\int \frac{dw}{dn}ds$, integrated round a small circle with the origin for centre. In like manner the complete integral of (4) is too general for our present purpose, since it covers the case in which the centre of the membrane is subjected to an external force.

The other function of r, which satisfies (4), is the Bessel's function of the nth order, denoted by $J_n(\kappa r)$, and may be expressed in several ways. The ascending series (obtained immediately from the differential equation) is

$$J_n(z) = \frac{z^n}{2^n \Gamma(n+1)} \left\{ 1 - \frac{z^2}{2 \cdot 2n+2} + \frac{z^4}{2 \cdot 4 \cdot 2n+2 \cdot 2n+4} \right.$$
$$\left. - \frac{z^6}{2 \cdot 4 \cdot 6 \cdot 2n+2 \cdot 2n+4 \cdot 2n+6} + \dots \right\} \dots\dots(5),$$

from which the following relations between functions of consecutive orders may readily be deduced:

$$J_0'(z) = -J_1(z) \dots\dots\dots\dots\dots\dots(6),$$
$$2 J_n'(z) = J_{n-1}(z) - J_{n+1}(z) \dots\dots\dots(7),$$
$$\frac{2n}{z} J_n(z) = J_{n-1}(z) + J_{n+1}(z) \dots\dots\dots(8).$$

When n is an integer, $J_n(z)$ may be expressed by the definite integral

$$J_n(z) = \frac{1}{\pi} \int_0^\pi \cos(z \sin \omega - n\omega) \, d\omega \dots\dots\dots(9),$$

which is Bessel's original form. From this expression it is evident that J_n and its differential coefficients with respect to z are always less than unity.

The ascending series (5), though infinite, is convergent for all values of n and z; but, when z is great, the convergence does not begin for a long time, and then the series becomes useless as a basis for numerical calculation. In such cases another series proceeding by descending powers of z may be substituted with advantage. This series is

$$J_n(z) = \sqrt{\frac{2}{\pi z}} \left\{ 1 - \frac{(1^2 - 4n^2)(3^2 - 4n^2)}{1 \cdot 2 \cdot (8z)^2} + \dots \right\} \cos\left(z - \frac{\pi}{4} - n\frac{\pi}{2}\right)$$
$$+ \sqrt{\frac{2}{\pi z}} \left\{ \frac{1^2 - 4n^2}{1 \cdot 8z} - \frac{(1^2 - 4n^2)(3^2 - 4n^2)(5^2 - 4n^2)}{1 \cdot 2 \cdot 3 \cdot (8z)^3} + \dots \right\}$$
$$\times \sin\left(z - \frac{\pi}{4} - n\frac{\pi}{2}\right) \dots\dots\dots\dots\dots(10);$$

it terminates, if $2n$ be equal to an odd integer, but otherwise, it runs on to infinity, and becomes ultimately divergent. Nevertheless when z is great, the convergent part may be employed in calculation; for it can be proved that the sum of any number of terms differs from the true value of the function by less than the last term included. We shall have occasion later, in connection with another problem, to consider the derivation of this descending series.

As Bessel's functions are of considerable importance in theoretical acoustics, I have thought it advisable to give a table for the functions J_0 and J_1, extracted from Lommel's[1] work, and due

[1] Lommel, *Studien über die Bessel'schen Functionen*. Leipzig, 1868.

originally to Hansen. The functions J_0 and J_1 are connected by the relation $J_0' = -J_1$.

z	$J_0(z)$	$J_1(z)$	z	$J_0(z)$	$J_1(z)$	z	$J_0(z)$	$J_1(z)$
0·0	1·0000	0·0000	4·5	·3205	·2311	9·0	·0903	·2453
0·1	·9975	·0499	4·6	·2961	·2566	9·1	·1142	·2324
0·2	·9900	·0995	4·7	·2693	·2791	9·2	·1367	·2174
0·3	·9776	·1483	4·8	·2404	·2985	9·3	·1577	·2004
0·4	·9604	·1960	4·9	·2097	·3147	9·4	·1768	·1816
0·5	·9385	·2423	5·0	·1776	·3276	9·5	·1939	·1613
0·6	·9120	·2867	5·1	·1443	·3371	9·6	·2090	·1395
0·7	·8812	·3290	5·2	·1103	·3432	9·7	·2218	·1166
0·8	·8463	·3688	5·3	·0758	·3460	9·8	·2323	·0928
0·9	·8075	·4060	5·4	·0412	·3453	9·9	·2403	·0684
1·0	·7652	·4401	5·5	− ·0068	·3414	10·0	·2459	·0435
1·1	·7196	·4709	5·6	+ ·0270	·3343	10·1	·2490	+ ·0184
1·2	·6711	·4983	5·7	·0599	·3241	10·2	·2496	− ·0066
1·3	·6201	·5220	5·8	·0917	·3110	10·3	·2477	·0313
1·4	·5669	·5419	5·9	·1220	·2951	10·4	·2434	·0555
1·5	·5118	·5579	6·0	·1506	·2767	10·5	·2366	·0789
1·6	·4554	·5699	6·1	·1773	·2559	10·6	·2276	·1012
1·7	·3980	·5778	6·2	·2017	·2329	10·7	·2164	·1224
1·8	·3400	·5815	6·3	·2238	·2081	10·8	·2032	·1422
1·9	·2818	·5812	6·4	·2433	·1816	10·9	·1881	·1604
2·0	·2239	·5767	6·5	·2601	·1538	11·0	·1712	·1768
2·1	·1666	·5683	6·6	·2740	·1250	11·1	·1528	·1913
2·2	·1104	·5560	6·7	·2851	·0953	11·2	·1330	·2039
2·3	·0555	·5399	6·8	·2931	·0652	11·3	·1121	·2143
2·4	+ ·0025	·5202	6·9	·2981	·0349	11·4	·0902	·2225
2·5	− ·0484	·4971	7·0	·3001	− ·0047	11·5	·0677	·2284
2·6	·0968	·4708	7·1	·2991	+ ·0252	11·6	·0446	·2320
2·7	·1424	·4416	7·2	·2951	·0543	11·7	− ·0213	·2333
2·8	·1850	·4097	7·3	·2882	·0826	11·8	+ ·0020	·2323
2·9	·2243	·3754	7·4	·2786	·1096	11·9	·0250	·2290
3·0	·2601	·3391	7·5	·2663	·1352	12·0	·0477	·2234
3·1	·2921	·3009	7·6	·2516	·1592	12·1	·0697	·2157
3·2	·3202	·2613	7·7	·2346	·1813	12·2	·0908	·2060
3·3	·3443	·2207	7·8	·2154	·2014	12·3	·1108	·1943
3·4	·3643	·1792	7·9	·1944	·2192	12·4	·1296	·1807
3·5	·3801	·1374	8·0	·1717	·2346	12·5	·1469	·1655
3·6	·3918	·0955	8·1	·1475	·2476	12·6	·1626	·1487
3·7	·3992	·0538	8·2	·1222	·2580	12·7	·1766	·1307
3·8	·4026	+ ·0128	8·3	·0960	·2657	12·8	·1887	·1114
3·9	·4018	− ·0272	8·4	·0692	·2708	12·9	·1988	·0912
4·0	·3972	·0660	8·5	·0419	·2731	13·0	·2069	·0703
4·1	·3887	·1033	8·6	+ ·0146	·2728	13·1	·2129	·0489
4·2	·3766	·1386	8·7	− ·0125	·2697	13·2	·2167	·0271
4·3	·3610	·1719	8·8	·0392	·2641	13·3	·2183	− ·0052
4·4	·3423	·2028	8·9	·0653	·2559	13·4	·2177	+ ·0166

201. In accordance with the notation for Bessel's functions the expression for a normal component vibration may therefore be written

$$w = P J_n (\kappa r) \; \cos n \, (\theta + \alpha) \; \cos (pt + \epsilon) \ldots\ldots\ldots(1);$$

and the boundary condition requires that

$$J_n (\kappa a) = 0 \ldots\ldots\ldots\ldots\ldots\ldots\ldots\ldots(2),$$

an equation whose roots give the admissible values of κ, and therefore of p.

The complete expression for w is obtained by combining the particular solutions embodied in (1) with all admissible values of κ and n, and is necessarily general enough to cover any initial circumstances that may be imagined. We conclude that any function of r and θ may be expanded within the limits of the circle $r = a$ in the series

$$w = \Sigma\Sigma J_n (\kappa r) \{\phi \cos n\theta + \psi \sin n\theta\}\ldots\ldots\ldots\ldots(3).$$

For every integral value of n there are a series of values of κ, given by (2); and for each of these the constants ϕ and ψ are arbitrary.

The determination of the constants is effected in the usual way. Since the energy of the motion is equal to

$$\tfrac{1}{2} \rho \int_0^a \int_0^{2\pi} \dot{w}^2 r \, d\theta \, dr \ldots\ldots\ldots\ldots\ldots\ldots(4),$$

and when expressed by means of the normal co-ordinates can only involve their squares, it follows that the product of any two of the terms in (3) vanishes, when integrated over the area of the circle. Thus, if we multiply (3) by $J_n (\kappa r) \cos n\theta$, and integrate, we find

$$\int_0^a \int_0^{2\pi} w \, J_n (\kappa r) \cos n\theta \, r dr \, d\theta$$

$$= \phi \int\int [J_n (\kappa r)]^2 \cos^2 n\theta \, r dr \, d\theta$$

$$= \phi \, . \, \pi \int_0^a [J_n (\kappa r)]^2 \, r dr \ldots\ldots\ldots\ldots\ldots(5),$$

by which ϕ is determined. The corresponding formula for ψ is obtained by writing $\sin n\theta$ for $\cos n\theta$. A method of evaluating the integral on the right will be given presently. Since ϕ and ψ each contain two terms, one varying as $\cos pt$ and the other as $\sin pt$, it is now evident how the solution may be adapted so as to agree with arbitrary initial values of w and $\dot{w}$.

202. Let us now examine more particularly the character of the fundamental vibrations. If $n = 0$, w is a function of r only, that is to say, the motion is symmetrical with respect to the centre of the membrane. The nodes, if any, are the concentric circles, whose equation is

$$J_0(\kappa r) = 0 \dots\dots\dots\dots\dots\dots(1).$$

When n has an integral value different from zero, w is a function of θ as well as of r, and the equation of the nodal system takes the form

$$J_n(\kappa r)\ \cos n\,(\theta - \alpha) = 0 \dots\dots\dots\dots\dots(2).$$

The nodal system is thus divisible into two parts, the first consisting of the concentric circles represented by

$$J_n(\kappa r) = 0 \dots\dots\dots\dots\dots\dots(3),$$

and the second of the diameters

$$\theta = \alpha + (2m + 1)\,\frac{\pi}{2n} \dots\dots\dots\dots\dots(4),$$

where m is an integer. These diameters are n in number, and are ranged uniformly round the centre; in other respects their position is arbitrary. The radii of the circular nodes will be investigated further on.

203. The important integral formula

$$\int_0^a J_n(\kappa r)\, J_n(\kappa' r)\, r\, dr = 0 \dots\dots\dots\ \dots\dots(1),$$

where κ and κ' are different roots of

$$J_n(\kappa a) = 0 \dots\dots\dots\dots\dots\dots(2),$$

may be verified analytically by means of the differential equations satisfied by $J_n(\kappa r)$, $J_n(\kappa' r)$; but it is both simpler and more instructive to begin with the more general problem, where the boundary of the membrane is not restricted to be circular.

The variational equation of motion is

$$\delta V + \rho \iint \ddot{w}\, \delta w\, dx\, dy = 0 \dots\dots\dots\dots(3),$$

where

$$V = \tfrac{1}{2}\, T_1 \iint \left\{ \left(\frac{dw}{dx}\right)^2 + \left(\frac{dw}{dy}\right)^2 \right\} dx\, dy \dots\dots\dots(4),$$

and therefore

$$\delta V = T_1 \iint \left\{ \frac{dw}{dx}\frac{d\delta w}{dx} + \frac{dw}{dy}\frac{d\delta w}{dy} \right\} dx\,dy \dots\dots\dots(5).$$

In these equations w refers to the actual motion, and δw to a hypothetical displacement consistent with the conditions to which the system is subjected. Let us now suppose that the system is executing one of its normal component vibrations, so that $w = u$, and

$$\ddot{u} + p^2 u = 0 \dots\dots\dots\dots(6),$$

while δw is proportional to another normal function v.

Since $\kappa = p \div c$, we get from (3)

$$\kappa^2 \iint uv\,dx dy = \iint \left\{ \frac{du}{dx}\frac{dv}{dx} + \frac{du}{dy}\frac{dv}{dy} \right\} dx\,dy \dots\dots(7).$$

The integral on the right is symmetrical with respect to u and v, and thus

$$(\kappa'^2 - \kappa^2) \iint u\,v\,dx\,dy = 0 \dots\dots\dots(8),$$

where κ'^2 bears the same relation to v that κ^2 bears to u.

Accordingly, if the normal vibrations represented by u and v have different periods,

$$\iint u\,v\,dx\,dy = 0 \dots\dots\dots(9).$$

In obtaining this result, we have made no assumption as to the boundary conditions beyond what is implied in the absence of reactions against acceleration, which, if they existed, would appear in the fundamental equation (3).

If in (8) we suppose $\kappa' = \kappa$, the equation is satisfied identically, and we cannot infer the value of $\iint u^2 dx dy$. In order to evaluate this integral we must follow a rather different course.

If u and v be functions satisfying within a certain contour the equations $\nabla^2 u + \kappa^2 u = 0$, $\nabla^2 v + \kappa''^2 v = 0$, we have

$$(\kappa'^2 - \kappa^2) \iint u\,v\,dx\,dy = \iint (v\,\nabla^2 u - u\,\nabla^2 v)\,dx\,dy$$
$$= \int \left(v\frac{du}{dn} - u\frac{dv}{dn} \right) ds \dots\dots(10),$$

by Green's theorem. Let us now suppose that v is derived from u by slightly varying κ, so that

$$v = u + \frac{du}{d\kappa}\delta\kappa, \quad \kappa' = \kappa + \delta\kappa;$$

substituting in (10), we find

$$2\kappa \iint u^2 dx\, dy = \int \left(\frac{du}{d\kappa}\frac{du}{dn} - u\frac{d^2u}{dn\,d\kappa}\right) ds \dots\dots\dots(11);$$

or, if u vanish on the boundary,

$$2\kappa \iint u^2 dx\, dy = \int \frac{du}{d\kappa}\frac{du}{dn}\, ds \dots\dots(12);$$

For the application to a circular area of radius r, we have

$$\left.\begin{array}{l} u = \cos n\theta\, J_n(\kappa r)\\ v = \cos n\theta\, J_n(\kappa' r) \end{array}\right\} \dots\dots(13),$$

and thus from (10) on substitution of polar co-ordinates and integration with respect to θ,

$$(\kappa'^2 - \kappa^2)\int_0^r J_n(\kappa r)\, J_n(\kappa' r)\, r dr$$

$$= rJ_n(\kappa r)\frac{d}{dr}J_n(\kappa' r) - rJ_n(\kappa' r)\frac{d}{dr}J_n(\kappa r) \dots\dots(14).$$

Accordingly, if

$$\frac{d}{dr}J_n(\kappa' r) : J_n(\kappa' r) = \frac{d}{dr}J_n(\kappa r) : J_n(\kappa r),$$

and κ and κ' be different,

$$\int_0^r J_n(\kappa r)\, J_n(\kappa' r)\, r dr = 0 \dots\dots(15),$$

an equation first proved by Fourier for the case when

$$J_n(\kappa r) = J_n(\kappa' r) = 0.$$

Again from (12)

$$2\kappa\int_0^r J_n^2(\kappa r)\, r dr = r\frac{dJ}{d\kappa}\frac{dJ}{dr} - rJ\frac{d^2J}{dr\,d\kappa}$$

$$= \kappa r^2 J'^2 - \kappa r^2 J\left(J'' + \frac{1}{\kappa r}J'\right),$$

dashes denoting differentiation with respect to κr. Now

$$J'' + \frac{1}{\kappa r}J' + \left(1 - \frac{n^2}{\kappa^2 r^2}\right)J = 0,$$

and thus

$$2 \int_0^r J_n^2 (\kappa r) \, r dr = r^2 J_n'^2 (\kappa r) + r^2 \left(1 - \frac{n^2}{\kappa^2 r^2}\right) J_n^2 (\kappa r) \quad \ldots\ldots(16).$$

This result is general; but if, as in the application to membranes with fixed boundaries,

$$J_n (\kappa r) = 0,$$

then

$$2 \int_0^r J_n^2 (\kappa r) \, r dr = r^2 J_n'^2 (\kappa r) \quad \ldots\ldots\ldots\ldots(17).$$

204. We may use the result just arrived at to simplify the expressions for T and V. From

$$w = \Sigma\Sigma \{\phi_{mn} J_n (\kappa_{mn} r) \cos n\theta + \psi_{mn} J_n (\kappa_{mn} r) \sin n\theta\} \ldots\ldots\ldots(1),$$

we find

$$T = \tfrac{1}{4} \rho \pi a^2 \Sigma\Sigma J_n'^2 (\kappa_{mn} a) \{\dot{\phi}_{mn}^2 + \dot{\psi}_{mn}^2\} \ldots\ldots\ldots\ldots(2),$$

$$V = \tfrac{1}{4} \rho \pi a^2 \Sigma\Sigma p_{mn}^2 J_n'^2 (\kappa_{mn} a) \{\phi_{mn}^2 + \psi_{mn}^2\} \ldots\ldots(3);$$

whence is derived the normal equation of motion

$$\ddot{\phi}_{mn} + p_{mn}^2 \phi_{mn} = \frac{4 \, \Phi_{mn}}{\rho \pi a^2 J_n'^2 (\kappa_{mn} a)} \ldots\ldots\ldots\ldots(4),$$

and a similar equation for ψ_{mn}. The value of Φ_{mn} is to be found from the consideration that $\Phi_{mn} \delta\phi_{mn}$ denotes the work done by the impressed forces during a hypothetical displacement $\delta\phi_{mn}$; so that if Z be the impressed force, reckoned per unit of area,

$$\Phi_{mn} = \iint Z J_n (\kappa_{mn} r) \cos n\theta \, r dr \, d\theta \ldots\ldots\ldots(5).$$

These expressions and equations do not apply to the case $n = 0$, when ϕ and ψ are amalgamated. We then have

$$\left.\begin{array}{l} T = \tfrac{1}{2} \rho \pi a^2 J_0'^2 (\kappa_{m0} a) \, \dot{\phi}_{m0}^2 \\ V = \tfrac{1}{2} \rho \pi a^2 p_{m0}^2 J_0'^2 (\kappa_{m0} a) \, \phi_{m0}^2 \end{array}\right\} \ldots\ldots\ldots\ldots(6),$$

$$\ddot{\phi}_{m0} + p_{m0}^2 \phi_{m0} = \frac{2 \, \Phi_{m0}}{\rho \pi a^2 J_0'^2 (\kappa_{m0} a)} \ldots\ldots\ldots\ldots(7).$$

As an example, let us suppose that the initial velocities are zero, and the initial configuration that assumed under the influence of a constant pressure Z; thus

$$\Phi_{m0} = Z \, . \, 2\pi \int_0^a J_0 (\kappa_{m0} r) \, r dr.$$

Now by the differential equation,

$$rJ_0(\kappa r) = -\{rJ_0''(\kappa r) + \frac{1}{\kappa}J_0'(\kappa r)\},$$

and thus

$$\int_0^a J_0(\kappa r)\, r\, dr = -\frac{a}{\kappa}J_0'(\kappa a)................(8);$$

so that

$$\Phi_{m0} = -\frac{2\pi a}{\kappa_{m0}} Z J_0'(\kappa_{m0}a).$$

Substituting this in (7), we see that the initial value of ϕ_{m0} is

$$(\phi_{m0})_{t=0} = \frac{-4Z}{\kappa_{m0}p_{m0}^2\rho a\, J_0'(\kappa_{m0}a)} ....(9).$$

For values of n other than zero, Φ and the initial value of ϕ_{mn} vanish. The state of the system at time t is expressed by

$$\phi_{m0} = (\phi_{m0})_{t=0} \cdot \cos p_{m0}t ....(10),$$

$$w = \Sigma \phi_{m0} J_0(\kappa_{m0}r)................(11),$$

the summation extending to all the admissible values of κ_{m0}.

As an example of *forced* vibrations, we may suppose that Z, still constant with respect to space, varies as a harmonic function of the time. This may be taken to represent roughly the circumstances of a small membrane set in vibration by a train of aerial waves. If $Z = \cos qt$, we find, nearly as before,

$$w = \frac{4}{\rho a}\cos qt \Sigma \frac{J_0(\kappa_0 r)}{\kappa_{m0}(q^2 - p_{m0}^2)J_0'(\kappa_{m0}a)} ....(12).$$

The forced vibration is of course independent of θ. It will be seen that, while none of the symmetrical normal components are missing, their relative importance may vary greatly, especially if there be a near approach in value between q and one of the series of quantities p_{m0}. If the approach be very close, the effect of dissipative forces must be included.

205. The pitches of the various simple tones and the radii of the nodal circles depend on the roots of the equation

$$J_n(\kappa a) = J_n(z) = 0.$$

If these (exclusive of zero) taken in order of magnitude be called $z_n^{(1)}, z_n^{(2)}, z_n^{(3)} \ldots\ldots z_n^{(s)}\ldots\ldots$, then the admissible values of p

are to be found by multiplying the quantities $z_n^{(s)}$ by $c \div a$. The particular solution may then be written

$$w = J_n\left(z_n^{(s)}\frac{r}{a}\right)\{A_n^{(s)}\cos n\theta + B_n^{(s)}\sin n\theta\}\cos\left\{\frac{c}{a}z_n^{(s)}t - \epsilon_n^{(s)}\right\} \ldots\ldots (1).$$

The lowest tone of the group n corresponds to $z_n^{(1)}$; and since in this case $J_n\left(z_n^{(1)}\frac{r}{a}\right)$ does not vanish for any value of r less than a, there is no interior nodal circle. If we put $s = 2$, J_n will vanish, when

$$z_n^{(2)}\frac{r}{a} = z_n^{(1)},$$

that is, when

$$r = a\frac{z_n^{(1)}}{z_n^{(2)}},$$

which is the radius of the one interior nodal circle. Similarly if we take the root $z_n^{(s)}$, we obtain a vibration with $s - 1$ nodal circles (exclusive of the boundary) whose radii are

$$a\frac{z_n^{(1)}}{z_n^{(s)}}, \quad a\frac{z_n^{(2)}}{z_n^{(s)}}, \ldots\ldots a\frac{z_n^{(s-1)}}{z_n^{(s)}}.$$

All the roots of the equation $J_n(\kappa a) = 0$ are *real*. For, if possible, let $\kappa a = \lambda + i\mu$ be a root; then $\kappa' a = \lambda - i\mu$ is also a root, and thus by (14) § 203,

$$4i\lambda\mu\int_0^a J_n(\kappa r) J_n(\kappa' r)\, r dr = 0.$$

Now $J_n(\kappa r)$, $J_n(\kappa' r)$ are conjugate complex quantities, whose product is necessarily positive; so that the above equation requires that either λ or μ vanish. That λ cannot vanish appears from the consideration that if κa were a pure imaginary, each term of the ascending series for J_n would be positive, and therefore the sum of the series incapable of vanishing. We conclude that $\mu = 0$, or that κ is real[1]. The same result might be arrived at from the consideration that only circular functions of the time can enter into the analytical expression for a normal component vibration.

The equation $J_n(z) = 0$ has no equal roots (except zero). From equations (7) and (8) § 200 we get

$$J_n' = \frac{n}{z}J_n - J_{n+1},$$

[1] Riemann, p. 260.

whence we see that if J_n, J_n' vanished for the same value of z, J_{n+1} would also vanish for that value. But in virtue of (8) § 200 this would require that *all* the functions J_n vanish for the value of z in question[1].

206. The actual values of z_n may be found by interpolation from Hansen's tables so far as these extend; or formulæ may be calculated from the descending series by the method of successive approximation, expressing the roots directly. For the important case of the symmetrical vibrations $(n = 0)$, the values of z_0 may be found from the following, given by Stokes[2]:

$$\frac{z_0^{(s)}}{\pi} = s - \cdot 25 + \frac{\cdot 050661}{4s - 1} - \frac{\cdot 053041}{(4s - 1)^3} + \frac{\cdot 262051}{(4s - 1)^5} \ldots\ldots (1).$$

For $n = 1$, the formula is

$$\frac{z_1^{(s)}}{\pi} = s + \cdot 25 - \frac{\cdot 151982}{4s + 1} + \frac{\cdot 015399}{(4s + 1)^3} - \frac{\cdot 245835}{(4s + 1)^5} \ldots\ldots (2).$$

The latter series is convergent enough, even for the first root, corresponding to $s = 1$. The series (1) will suffice for values of s greater than unity; but the first root must be calculated independently. The accompanying table (A) is taken from Stokes' paper, with a slight difference of notation.

It will be seen either from the formulæ, or the table, that the difference of successive roots of high order is approximately π. This is true for all values of n, as is evident from the descending series (10) § 200.

M. Bourget has given in his memoir very elaborate tables of the frequencies of the different simple tones and of the radii of the nodal circles. Table B includes the values of z, which satisfy $J_n(z)$, for $n = 0, 1, \ldots 5$, $s = 1, 2, \ldots 9$.

---

[1] Bourget, "Mémoire sur le mouvement vibratoire des membranes circulaires," *Ann. de l'école normale*, t. III., 1866. In one passage M. Bourget implies that he has proved that no two Bessel's functions of integral order can have the same root, but I cannot find that he has done so. The theorem, however, is probably true; in the case of functions, whose orders differ by 1 or 2, it may be easily proved from the formulæ of § 200.

[2] *Camb. Phil. Trans.* Vol. IX. "On the numerical calculation of a class of definite integrals and infinite series."

TABLE A.

s	$\dfrac{z}{\pi}$ for $J_0(z) = 0$.	Diff.	$\dfrac{z}{\pi}$ for $J_1(z) = 0$.	Diff.
1	·7655		1·2197	
2	1·7571	·9916	2·2330	1·0133
3	2·7546	·9975	3·2383	1·0053
4	3·7534	·9988	4·2411	1·0028
5	4·7527	·9993	5·2428	1·0017
6	5·7522	·9995	6·2439	1·0011
7	6·7519	·9997	7·2448	1·0009
8	7·7516	·9997	8·2454	1·0006
9	8·7514	·9998	9·2459	1·0005
10	9·7513	·9999	10·2463	1·0004
11	10·7512	·9999	11·2466	1·0003
12	11·7511	·9999	12·2469	1·0003

When n is considerable the calculation of the earlier roots becomes troublesome. For very high values of n, $z_n^{(1)} : n$ approximates to a ratio of equality, as may be seen from the consideration that the pitch of the gravest tone of a very acute sector must tend to coincide with that of a long parallel strip, whose width is equal to the greatest width of the sector.

TABLE B.

s	$n = 0$	$n = 1$	$n = 2$	$n = 3$	$n = 4$	$n = 5$
1	2·404	3·832	5·135	6·379	7·586	8·780
2	5·520	7·016	8·417	9·760	11·064	12·339
3	8·654	10·173	11·620	13·017	14·373	15·700
4	11·792	13·323	14·796	16·224	17·616	18·982
5	14·931	16·470	17·960	19·410	20·827	22·220
6	18·071	19·616	21·117	22·583	24·018	25·431
7	21·212	22·760	24·270	25·749	27·200	28·628
8	24·353	25·903	27·421	28·909	30·371	31·813
9	27·494	29·047	30·571	32·050	33·512	34·983

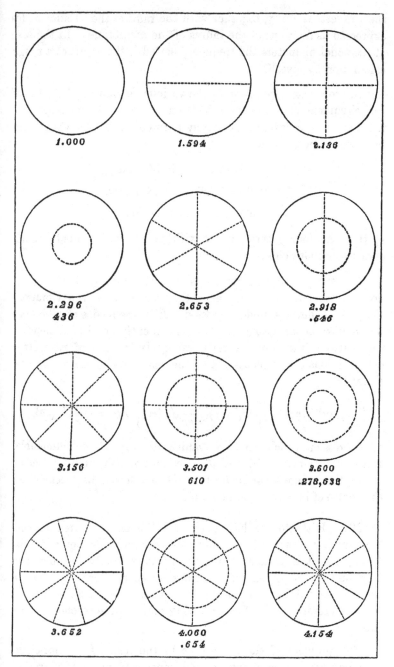

The figures represent the more important normal modes of vibration, and the numbers affixed give the frequency referred to

18—2

the gravest as unity, together with the radii of the circular nodes expressed as fractions of the radius of the membrane. In the case of six nodal diameters the frequency stated is the result of a rough calculation by myself.

The tones corresponding to the various fundamental modes of the circular membrane do not belong to a harmonic scale, but there are one or two approximately harmonic relations which may be worth notice. Thus

$$\tfrac{4}{3} \times 1\cdot594 = 2\cdot125 = 2\cdot136 \text{ nearly,}$$
$$\tfrac{5}{3} \times 1\cdot594 = 2\cdot657 = 2\cdot653 \text{ nearly,}$$
$$2 \times 1\cdot594 = 3\cdot188 = 3\cdot156 \text{ nearly;}$$

so that the four gravest modes with nodal diameters only would give a consonant chord.

The area of the membrane is divided into segments by the nodal system in such a manner that the sign of the vibration changes whenever a node is crossed. In those modes of vibration which have nodal diameters there is evidently no displacement of the centre of inertia of the membrane. In the case of symmetrical vibrations the displacement of the centre of inertia is proportional to

$$\int_0^a J_0\left(\kappa r\right) r\, dr = -\int_0^a \left\{ J_0''\left(\kappa r\right) + \frac{1}{\kappa r}\, J_0'\left(\kappa r\right) \right\} r\, dr = -\frac{a}{\kappa}\, J_0'\left(\kappa a\right),$$

an expression which does not vanish for any of the admissible values of κ, since $J_0'\left(z\right)$ and $J_0\left(z\right)$ cannot vanish simultaneously. In all the symmetrical modes there is therefore a displacement of the centre of inertia of the membrane.

207. Hitherto we have supposed the circular area of the membrane to be complete, and the circumference only to be fixed; but it is evident that our theory virtually includes the solution of other problems, for example—some cases of a membrane bounded by two concentric circles. The *complete* theory for a membrane in the form of a ring requires the second Bessel's function.

The problem of the membrane in the form of a semi-circle may be regarded as already solved, since any mode of vibration of which the semi-circle is capable must be applicable to the complete circle also. In order to see this, it is only necessary to attribute to any point in the complementary semi-circle the opposite motion to that which obtains at its optical image in the bounding diameter. This line will then require no constraint to keep it nodal. Similar considerations apply to any sector whose angle is an aliquot part of two right angles.

When the opening of the sector is arbitrary, the problem may be solved in terms of Bessel's functions of fractional order. If the fixed radii are $\theta = 0$, $\theta = \beta$, the particular solution is

$$w = P J_{\frac{\nu\pi}{\beta}} (\kappa r) \ \sin\frac{\nu\pi\theta}{\beta} \cos (pt - \epsilon) \ \ldots\ldots\ldots (1),$$

where ν is an integer. We see that if β be an aliquot part of π, $\nu\pi \div \beta$ is integral, and the solution is included among those already used for the complete circle.

An interesting case is when $\beta = 2\pi$, which corresponds to the problem of a complete circle, of which the radius $\theta = 0$ is constrained to be nodal.

Fig. 38.

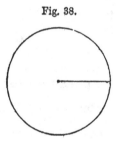

We have

$$w = P J_{\frac{1}{2}\nu} (\kappa r) \ \sin \tfrac{1}{2}\nu\theta \ \cos (pt - \epsilon).$$

When ν is even, this gives, as might be expected, modes of vibration possible without the constraint; but, when ν is odd, new modes make their appearance. In fact, in the latter case the descending series for J terminates, so that the solution is expressible in finite terms. Thus, when $\nu = 1$,

$$w = P \frac{\sin \kappa r}{\sqrt{\kappa r}} \sin \tfrac{1}{2}\theta \ \cos (pt - \epsilon) \ \ldots\ldots\ldots\ldots(2).$$

The values of κ are given by

$$\sin \kappa a = 0, \quad \text{or} \quad \kappa a = m\pi.$$

Thus the circular nodes divide the fixed radius into equal parts, and the series of tones form a harmonic scale. In the case of the gravest mode, the whole of the membrane is at any moment deflected on the same side of its equilibrium position. It is remarkable that the application of the constraint to the radius $\theta = 0$ makes the problem easier than before.

If we take $\nu = 3$, the solution is

$$w = P\frac{1}{\sqrt{\kappa r}}\left(\frac{\sin \kappa r}{\kappa r} - \cos \kappa r\right)\sin \tfrac{3}{2}\theta \; \cos (pt - \epsilon)\ldots\ldots(3).$$

Fig. 39.

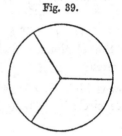

In this case the nodal radii are

$$\theta = \frac{2\pi}{3}, \quad \theta = \frac{4\pi}{3};$$

and the possible tones are given by the equation

$$\tan \kappa a = \kappa a \ldots\ldots\ldots\ldots\ldots\ldots(4).$$

To calculate the roots of $\tan x = x$ we may assume

$$x = (m + \tfrac{1}{2})\pi - y = X - y,$$

where y is a positive quantity, which is small when x is large.

Substituting this, we find $\cot y = X - y$,
whence

$$y = \frac{1}{X}\left(1 + \frac{y}{X} + \frac{y^2}{X^2} + \ldots\right) - \frac{y^3}{3} - \frac{2y^5}{15} - \frac{17y^7}{315} - \ldots$$

This equation is to be solved by successive approximation. It will readily be found that

$$y = X^{-1} + \frac{2}{3}X^{-3} + \frac{13}{15}X^{-5} + \frac{146}{105}X^{-7} + \ldots$$

so that the roots of $\tan x = x$ are given by

$$x = X - X^{-1} - \frac{2}{3} X^{-3} - \frac{13}{15} X^{-5} - \frac{146}{105} X^{-7} - \ldots\ldots\ldots(5),$$

where $\qquad\qquad X = (m + \tfrac{1}{2})\,\pi.$

In the first quadrant there is no root after zero since $\tan x > x$, and in the second quadrant there is none because the signs of x and $\tan x$ are opposite. The first root after zero is thus in the third quadrant, corresponding to $m = 1$. Even in this case the series converges sufficiently to give the value of the root with considerable accuracy, while for higher values of m it is all that could be desired. The actual values of $x : \pi$ are 1·4303, 2·4590, 3·4709, 4·4747, 5·4818, 6·4844, &c.

208. The effect on the periods of a slight inequality in the density of the circular membrane may be investigated by the general method § 90, of which several examples have already been given. It will be sufficient here to consider the case of a small load M attached to the membrane at a point whose radius vector is r'.

We will take first the symmetrical types $(n = 0)$, which may still be supposed to apply notwithstanding the presence of M. The kinetic energy T is (6) § 204 altered from

$$\tfrac{1}{2}\rho\pi a^2 J_0'^2(\kappa_{m0}a)\,\dot{\phi}_{m0}{}^2 \ \text{to}\ \tfrac{1}{2}\rho\pi a^2 J_0'^2(\kappa_{m0}a)\,\dot{\phi}_{m0}{}^2 + \tfrac{1}{2}M\dot{\phi}_{m0}{}^2 J_0{}^2(\kappa_{m0}r'),$$

and therefore

$$p_{m0}{}^2 : P_{m0}{}^2 = 1 - \frac{M}{\rho\pi a^2}\frac{J_0{}^2(\kappa_{m0}r')}{J_0'^2(\kappa_{m0}a)} \ldots\ldots\ldots\ldots\ldots(1),$$

where $P_{m0}{}^2$ denotes the value of $p_{m0}{}^2$, when there is no load.

The unsymmetrical normal types are not fully determinate for the unloaded membrane; but for the present purpose they must be taken so as to make the resulting periods a maximum or minimum, that is to say, so that the effect of the load is the greatest and least possible. Now, since a load can never raise the pitch, it is clear that the influence of the load is the least possible, viz. zero, when the type is such that a nodal diameter (it is indifferent which) passes through the point at which the load is attached. The unloaded membrane must be supposed to have two coincident periods, of which *one* is unaltered by the addition of the load. The other type is to be chosen, so that the alteration of period is as great as possible, which will evidently be the case when the radius vector r' bisects the angle between two adjacent nodal diameters. Thus, if r' correspond to $\theta = 0$, we are to take

$$w = \phi_{mn} J_n (\kappa_{mn} r) \cos n\theta ;$$

so that (2) § 204

$$T = \tfrac{1}{4} \rho \pi a^2 \, \dot{\phi}_{mn}^{\ 2} J_n^{\prime 2} (\kappa_{mn} a) + \tfrac{1}{2} M \dot{\phi}_{mn}^{\ 2} J_n^{\ 2} (\kappa_{mn} r').$$

The *altered* $p_{mn}^{\ 2}$ is therefore given by

$$p_{mn}^{\ 2} : P_{mn}^{\ 2} = 1 - \frac{2M}{\rho \pi a^2} \frac{J_n^{\ 2} (\kappa_{mn} r')}{J_n^{\prime 2} (\kappa_{mn} a)} \quad \dots \dots \dots \dots (2).$$

Of course, if r' be such that the load lies on one of the nodal circles, neither period is affected.

For example, let M be at the centre of the membrane. $J_n (0)$ vanishes, except when $n = 0$; and $J_0 (0) = 1$. It is only the symmetrical vibrations whose pitch is influenced by a central load, and for them by (1)

$$p_{m0}^{\ 2} : P_{m0}^{\ 2} = 1 - \frac{M}{J_0^{\prime 2} (\kappa_{m0} a) \, \rho \pi a^2} \quad \dots \dots \dots \dots (3).$$

By (6) § 200　　　　　$J_0' (z) = - J_1 (z),$

so that the application of the formula requires only a knowledge of the values of $J_1 (z)$, when $J_0 (z)$ vanishes, § 200. For the gravest mode the value of $J_0' (\kappa_{m0} a)$ is $\cdot 51903$[1]. When $\kappa_{m0} a$ is considerable,

$$J_1^{\ 2} (\kappa_{m0} a) = 2 \div \pi \kappa_{m0} a$$

approximately; so that for the higher components the influence of the load in altering the pitch increases.

The influence of a small irregularity in disturbing the nodal system may be calculated from the formulæ of § 90. The most obvious effect is the breaking up of nodal diameters into curves of hyperbolic form due to the introduction of subsidiary symmetrical vibrations. In many cases the disturbance is favoured by close agreement between some of the natural periods.

209. We will next investigate how the natural vibrations of a uniform membrane are affected by a slight departure from the exact circular form.

---

[1] The succeeding values are approximately $\cdot 341$, $\cdot 271$, $\cdot 232$, $\cdot 206$, $\cdot 187$, &c.

Whatever may be the nature of the boundary, w satisfies the equation

$$\frac{d^2w}{dr^2} + \frac{1}{r}\frac{dw}{dr} + \frac{1}{r^2}\frac{d^2w}{d\theta^2} + \kappa^2 w = 0 \ldots\ldots\ldots\ldots(1),$$

where κ is a constant to be determined. By Fourier's theorem w may be expanded in the series

$$w = w_0 + w_1 \cos(\theta + \alpha_1) + w_2 \cos 2(\theta + \alpha_2) + \ldots\ldots$$
$$+ w_n \cos n(\theta + \alpha_n) + \ldots\ldots,$$

where w_0, w_1, &c. are functions of r only. Substituting in (1), we see that w_n must satisfy

$$\frac{d^2w_n}{dr^2} + \frac{1}{r}\frac{dw_n}{dr} + \left(\kappa^2 - \frac{n^2}{r^2}\right)w_n = 0,$$

of which the solution is

$$w_n \propto J_n(\kappa r);$$

for, as in § 200, the other function of r cannot appear.

The general expression for w may thus be written

$$w = A_0 J_0(\kappa r) + J_1(\kappa r)(A_1 \cos\theta + B_1 \sin\theta)$$
$$+ \ldots + J_n(\kappa r)(A_n \cos n\theta + B_n \sin n\theta) + \ldots\ldots\ldots(2).$$

For all points on the boundary w is to vanish.

In the case of a nearly circular membrane the radius vector is nearly constant. We may take $r = a + \delta r$, δr being a small function of θ. Hence the boundary condition is

$$0 = A_0 [J_0(\kappa a) + \kappa\delta r J_0'(\kappa a)] + \ldots\ldots$$
$$+ [J_n(\kappa a) + \kappa\delta r J_n'(\kappa a)][A_n \cos n\theta + B_n \sin n\theta]$$
$$+ \ldots\ldots\ldots\ldots\ldots\ldots\ldots\ldots\ldots\ldots\ldots\ldots\ldots\ldots\ldots\ldots(3),$$

which is to hold good for all values of θ.

Let us consider first those modes of vibration which are nearly symmetrical, for which therefore approximately

$$w = A_0 J_0(\kappa r).$$

All the remaining coefficients are small relatively to A_0, since the type of vibration can only differ a little from what it would be, were the boundary an exact circle. Hence if the squares of the small quantities be omitted, (3) becomes

$$A_0 \left[J_0 (\kappa a) + \kappa \delta r J_0' (\kappa a) \right] + J_1 (\kappa a) \left[A_1 \cos \theta + B_1 \sin \theta \right]$$
$$+ \ldots + J_n (\kappa a) \left[A_n \cos n\theta + B_n \sin n\theta \right] + \ldots = 0 \ldots\ldots(4).$$

If we integrate this equation with respect to θ between the limits 0 and 2π, we obtain

$$2\pi J_0 (\kappa a) + J_0' (\kappa a) \int_0^{2\pi} \kappa \delta r \, d\theta = 0,$$

or
$$J_0 \left\{ \kappa a + \kappa \int_0^{2\pi} \delta r \frac{d\theta}{2\pi} \right\} = 0 \ldots\ldots\ldots\ldots(5),$$

which shews that the pitch of the vibration is approximately the same as if the radius vector had uniformly its *mean value*.

This result allows us to form a rough estimate of the pitch of any membrane whose boundary is not extravagantly elongated. If σ denote the area, so that $\rho\sigma$ is the mass of the whole membrane, the frequency of the gravest tone is approximately

$$2\pi \times 2\text{·}404 \times \sqrt{\frac{\pi T_1}{\sigma \rho}} \ldots\ldots\ldots\ldots(6).$$

In order to investigate the altered type of vibration, we may multiply (4) by $\cos n\theta$, or $\sin n\theta$, and then integrate as before. Thus

$$\left. \begin{array}{l} A_0 J_0' (\kappa a) \displaystyle\int_0^{2\pi} \kappa \delta r \cos n\theta \, d\theta + \pi A_n J_n(\kappa a) = 0 \\[2ex] A_0 J_0' (\kappa a) \displaystyle\int_0^{2\pi} \kappa \delta r \sin n\theta \, d\theta + \pi B_n J_n(\kappa a) = 0 \end{array} \right\} \ldots\ldots(7),$$

which determine the ratios $A_n : A_0$ and $B_n : A_0$.

If
$$\delta r = \delta r_0 + \delta r_1 + \ldots + \delta r_n + \ldots$$

be Fourier's expansion, the final expression for w may be written,

$$w : A_0 = J_0 (\kappa r)$$
$$- \kappa J_0' (\kappa a) \left\{ \frac{J_1 (\kappa r) \, dr_1}{J_1 (\kappa a)} + \ldots + \frac{J_n (\kappa r) \, \delta r_n}{J_n (\kappa a)} + \ldots \right\} \ldots\ldots(8).$$

When the vibration is not approximately symmetrical, the question becomes more complicated. The normal modes for the truly circular membrane are to some extent indeterminate, but the irregularity in the boundary will, in general, remove the indeterminateness. The position of the nodal diameters must be taken, so that the resulting periods may have maximum or minimum values. Let us, however, suppose that the approximate type is

$$w = A_\nu J_\nu (\kappa r) \cos \nu \theta \dots\dots\dots\dots\dots (9),$$

and afterwards investigate how the initial line must be taken in order that this form may hold good.

All the remaining coefficients being treated as small in comparison with A_ν, we get from (4)

$$A_0 J_0 (\kappa a) + \dots + A_\nu [J_\nu (\kappa a) + \kappa \delta r J_\nu' (\kappa a)] \cos \nu \theta$$
$$+ B_\nu J_\nu (\kappa a) \sin \nu \theta + \dots\dots$$
$$+ J_n (\kappa a) [A_n \cos n\theta + B_n \cos n\theta] + \dots = 0 \dots\dots (10).$$

Multiplying by $\cos \nu \theta$ and integrating,

$$\pi J_\nu (\kappa a) + \kappa J_\nu' (\kappa a) \int_0^{2\pi} \delta r \cos^2 \nu \theta \, d\theta = 0,$$

or

$$J_\nu \left[\kappa a + \kappa \int_0^{2\pi} \delta r \cos^2 \nu \theta \, \frac{d\theta}{\pi} \right] = 0,$$

which shews that the effective radius of the membrane is

$$a + \int_0^{2\pi} \delta r \cos^2 \nu \theta \, \frac{d\theta}{\pi} \dots\dots\dots\dots (11).$$

The ratios of A_n and B_n to A_ν may be found as before by integrating equation (10) after multiplication by $\cos n\theta$, $\sin n\theta$.

But the point of greatest interest is the pitch. The initial line is to be so taken as to make the expression (11) a maximum or minimum. If we refer to a line fixed in space by putting $\theta - \alpha$ instead of θ, we have to consider the dependence on α of the quantity

$$\int_{0.}^{2\pi} \delta r \cos^2 \nu (\theta - \alpha) \, d\theta,$$

which may also be written

$$\cos^2 \nu \alpha \int_0^{2\pi} \delta r \cos^2 \nu \theta \, d\theta + 2 \cos \nu \alpha \sin \nu \alpha \int_0^{2\pi} \delta r \cos \nu \theta \sin \nu \theta \, d\theta$$
$$+ \sin^2 \nu \alpha \int_0^{2\pi} \delta r \sin^2 \nu \theta \, d\theta \dots\dots\dots\dots (12),$$

and is of the form

$$A \cos^2 \nu\alpha + 2B \cos \nu\alpha \sin \nu\alpha + C \sin^2 \nu\alpha,$$

A, B, C being independent of α. There are accordingly two admissible positions for the nodal diameters, one of which makes the period a maximum, and the other a minimum. The diameters of one set bisect the angles between the diameters of the other set.

There are, however, cases where the normal modes remain indeterminate, which happens when the expression (12) is independent of α. This is the case when δr is constant, or when δr is proportional to $\cos \nu\theta$. For example, if δr were proportional to $\cos 2\theta$, or in other words the boundary were slightly elliptical, the nodal system corresponding to $n = 2$ (that consisting of a pair of perpendicular diameters) would be arbitrary in position, at least to this order of approximation. But the single diameter, corresponding to $n = 1$, must coincide with one of the principal axes of the ellipse, and the periods will be different for the two axes.

210. We have seen that the gravest tone of a membrane, whose boundary is approximately circular, is nearly the same as that of a mechanically similar membrane in the form of a circle of the same mean radius or area. If the area of a membrane be given, there must evidently be some form of boundary for which the pitch (of the principal tone) is the gravest possible, and this form can be no other than the circle. In the case of approximate circularity an analytical demonstration may be given, of which the following is an outline.

The general value of w being

$$w = A_0 J_0 (\kappa r) + \ldots + J_n (\kappa r) (A_n \cos n\theta + B_n \sin n\theta) + \ldots\ldots (1),$$

in which for the present purpose the coefficients A_1, B_1,... are small relatively to A_0, we find from the condition that w vanishes when $r = a + \delta r$,

$$A_0 J_0 (\kappa a) + \kappa A_0 J_0' (\kappa a) \delta r + \tfrac{1}{2} \kappa^2 A_0 J_0'' (\kappa a) . (\delta r)^2 + \ldots\ldots$$

$$+ \Sigma \left[\{J_n(\kappa a) + \kappa J_n' (\kappa a) \delta r + \ldots\}\{A_n \cos n\theta + B_n \sin n\theta\} \right] = 0 \ldots (2).$$

Hence, if

$$\delta r = \alpha_1 \cos \theta + \beta_1 \sin \theta + \ldots + \alpha_n \cos n\theta + \beta_n \sin n\theta + \ldots\ldots (3),$$

we obtain on integration with respect to θ from 0 to 2π,

$$2A_0 J_0 + \tfrac{1}{2}\kappa^2 A_0 J_0'' \overset{n=\infty}{\underset{n=1}{\Sigma}} (\alpha_n^2 + \beta_n^2)$$

$$+ \kappa \overset{n=\infty}{\underset{n=1}{\Sigma}} [(\alpha_n A_n + \beta_n B_n) J_n'] = 0 \quad\dots\dots\dots \text{(4)},$$

from which we see, as before, that if the squares of the small quantities be neglected, $J_0(\kappa a) = 0$, or that to this order of approximation the mean radius is also the effective radius. In order to obtain a closer approximation we first determine $A_n : A_0$ and $B_n : A_0$ by multiplying (2) by $\cos n\theta$, $\sin n\theta$, and then integrating between the limits 0 and 2π. Thus

$$A_n J_n = -\kappa\alpha_n A_0 J_0', \quad B_n J_n = -\kappa\beta_n A_0 J_0' \quad\dots\dots\dots \text{(5)}.$$

Substituting these values in (4), we get

$$J_0(\kappa a) = \tfrac{1}{2}\kappa^2 \overset{n=\infty}{\underset{n=1}{\Sigma}} \left[(\alpha_n^2 + \beta_n^2) \left\{ \frac{J_n' J_0'}{J_n} - \tfrac{1}{2} J_0'' \right\} \right] \quad\dots\dots \text{(6)}.$$

Since J_0 satisfies the fundamental equation

$$J_0'' + \frac{1}{\kappa a} J_0' + J_0 = 0 \quad\dots\dots\dots\dots\dots\dots \text{(7)},$$

and in the present case $J_0 = 0$ approximately, we may replace J_0'' by $-\dfrac{1}{\kappa a} J_0'$. Equation (6) then becomes

$$J_0(\kappa a) = \tfrac{1}{2}\kappa^2 J_0' \overset{n=\infty}{\underset{n=1}{\Sigma}} \left[(\alpha_n^2 + \beta_n^2) \left\{ \frac{J_n'}{J_n} + \frac{1}{2\kappa a} \right\} \right] \quad\dots\dots \text{(8)}.$$

Let us now suppose that $a + da$ is the equivalent radius of the membrane, so that

$$J_0[\kappa(a + da)] = J_0(\kappa a) + J_0'(\kappa a)\,\kappa\,da = 0.$$

Then by (8) we find

$$da = -\tfrac{1}{2}\kappa \overset{n=\infty}{\underset{n=1}{\Sigma}} \left[(\alpha_n^2 + \beta_n^2) \left\{ \frac{J_n'}{J_n} + \frac{1}{2\kappa a} \right\} \right] \quad\dots\dots \text{(9)}.$$

Again, if $a + da'$ be the radius of the truly circular membrane of equal area,

$$da' = \frac{1}{4a} \overset{n=\infty}{\underset{n=1}{\Sigma}} (\alpha_n^2 + \beta_n^2) \quad\dots\dots\dots \text{(10)};$$

so that

$$da' - da = \frac{1}{2a} \sum_{n=1}^{n=\infty} \left[(\alpha_n{}^2 + \beta_n{}^2) \left\{ 1 + \kappa a \frac{J_n{}'(\kappa a)}{J_n(\kappa a)} \right\} \right] \quad \ldots \ldots (11).$$

The question is now as to the sign of the right-hand member. If $n = 1$, and z be written for κa,

$$1 + z \frac{J_n{}'(z)}{J_n(z)}$$

vanishes approximately by (7), since in general $J_1 = - J_0{}'$, and in the present case $J_0(z) = 0$ nearly. Thus $da' - da = 0$, as should evidently be the case, since the term in question represents merely a displacement of the circle without an alteration in the form of the boundary. When $n = 2$, (8) § 200,

$$J_2 = \frac{2}{z} J_1 - J_0,$$

from which and (7) we find that, when $J_0 = 0$,

$$\frac{J_2{}'}{J_2} = \frac{z^2 - 4}{2z} \quad \ldots \ldots \ldots \ldots \ldots \ldots \ldots (12),$$

whence

$$da' - da = \frac{1}{2a} (\alpha_2{}^2 + \beta_2{}^2) \left(\frac{z^2}{2} - 1 \right) \quad \ldots \ldots \ldots \ldots (13),$$

which is positive, since $z = 2\cdot404$.

We have still to prove that

$$1 + z \frac{J_n{}'(z)}{J_n(z)}$$

is positive for integral values of n greater than 2, when $z = 2\cdot404$. For this purpose we may avail ourselves of a theorem given in Riemann's *Partielle Differentialgleichungen*, to the effect that neither J_n nor $J_n{}'$ has a root (other than zero) less than n. The differential equation for J_n may be put into the form

$$\frac{d^2 J_n(z)}{d (\log z)^2} + (z^2 - n^2) J_n(z) = 0;$$

while initially J_n and $J_n{}'$ $\left(\text{as well as } \dfrac{dJ_n}{d \log z} \right)$ are positive. Accordingly $\dfrac{dJ_n}{d \log z}$ begins by increasing and does not cease to do so before $z = n$, from which it is clear that within the range $z = 0$ to

$z = n$, neither J_n nor J_n' can vanish. And since J_n and J_n' are both positive until $z = n$, it follows that, when n is an integer greater than 2·404, $da' - da$ is positive. We conclude that, unless α_2, β_2, α_3, ... all vanish, da' is greater than da, which shews that in the case of any membrane of approximately circular outline, the circle of equal area exceeds the circle of equal pitch.

We have seen that a good estimate of the pitch of an approximately circular membrane may be obtained from its area alone, but by means of equation (9) a still closer approximation may be effected. We will apply this method to the case of an ellipse, whose semi-axis major is R and eccentricity e.

The polar equation of the boundary is

$$r = R \left\{ 1 - \tfrac{1}{4} e^2 - \tfrac{7}{64} e^4 + \ldots\ldots + \tfrac{1}{4} e^2 \cos 2\theta + \ldots\ldots \right\} \ldots\ldots (14);$$

so that in the notation of this section

$$a = R \left(1 - \tfrac{1}{4} e^2 - \tfrac{7}{64} e^4 \right), \quad \alpha_2 = \tfrac{1}{4} e^2 R.$$

Accordingly by (9)

$$da = - \frac{e^4 R}{32} \cdot \kappa R \cdot \left\{ \frac{J_2'(z)}{J_2(z)} + \frac{1}{2z} \right\},$$

or by (12), since $\kappa R = z = 2\cdot404$,

$$da = - \frac{2\cdot779}{64} e^4 R.$$

Thus the radius of the circle of equal pitch is

$$a + da = R \left\{ 1 - \frac{1}{4} e^2 - \frac{9\cdot779 \, e^4}{64} \right\} \ldots\ldots\ldots\ldots (15),$$

in which the term containing e^4 should be correct.

The result may also be expressed in terms of e and the area σ. We have

$$R = \sqrt{\frac{\sigma}{\pi}} \left(1 + \frac{1}{4} e^2 + \frac{5}{32} e^4 \right),$$

and thus

$$a + da = \sqrt{\frac{\sigma}{\pi}} \left(1 - \frac{3\cdot779}{64} e^4 \right) \ldots\ldots\ldots\ldots (16),$$

from which we see how small is the influence of a moderate eccentricity, when the area is given.

211.   When the fixed boundary of a membrane is neither straight nor circular, the problem of determining its vibrations presents difficulties which in general could not be overcome without the introduction of functions not hitherto discussed or tabulated. A partial exception must be made in favour of an elliptic boundary ; but for the purposes of this treatise the importance of the problem is scarcely sufficient to warrant the introduction of complicated analysis. The reader is therefore referred to the original investigation of M. Mathieu[1]. It will be sufficient to mention here that the nodal system is composed of the confocal ellipses and hyperbolas.

Soluble cases may be invented by means of the general solution

$$w = A_0 J_0(\kappa r) + \ldots + (A_n \cos n\theta + B_n \sin n\theta) J_n(\kappa r) + \ldots\ldots$$

For example we might take

$$w = J_0(\kappa r) - \lambda J_1(\kappa r) \cos \theta,$$

and attaching different values to λ, trace the various forms of boundary to which the solution will then apply.

Useful information may sometimes be obtained from the theorem of § 88, which allows us to prove that any contraction of the fixed boundary of a vibrating membrane must cause an elevation of pitch, because the new state of things may be conceived to differ from the old merely by the introduction of an additional constraint. Springs, without inertia, are supposed to urge the line of the proposed boundary towards its equilibrium position, and gradually to become stiffer. At each step the vibrations become more rapid, until they approach a limit, corresponding to infinite stiffness of the springs and absolute fixity of their points of application. It is not necessary that the part cut off should have the same density as the rest, or even any density at all.

For instance, the pitch of a regular polygon is intermediate between those of the inscribed and circumscribed circles. Closer limits would however be obtained by substituting for the circumscribed circle that of equal area according to the result of § 210. In the case of the hexagon, the ratio of the radius of the circle of equal area to that of the circle inscribed is 1·050, so that the mean

[1] Liouville, 1868.

of the two limits cannot differ from the truth by so much as $2\frac{1}{2}$ per cent. In the same way we might conclude that the sector of a circle of 60° is a graver form than the equilateral triangle obtained by substituting the chord for the arc of the circle.

The following table giving the relative frequency in certain calculable cases for the gravest tone of membranes under similar mechanical conditions and of *equal area* (σ), shews the effect of a greater or less departure from the circular form.

Circle..	$2\cdot404 \cdot \sqrt{\pi} = 4\cdot261.$
Square..	$\sqrt{2} \cdot \pi = 4\cdot443.$
Quadrant of a circle........................	$\dfrac{5\cdot135}{2} \cdot \sqrt{\pi} = 4\cdot551.$
Sector of a circle 60°.......................	$6\cdot379 \sqrt{\dfrac{\pi}{6}} = 4\cdot616.$
Rectangle 3 × 2..............................	$\sqrt{\dfrac{13}{6}} \cdot \pi = 4\cdot624.$
Equilateral triangle.........................	$2\pi \cdot \sqrt{\tan 30°} = 4\cdot774.$
Semicircle......................................	$3\cdot832 \sqrt{\dfrac{\pi}{\cdot2}} = 4\cdot803.$
Rectangle 2 × 1.............................. Right-angled isosceles triangle........... }	$\pi \sqrt{\dfrac{5}{2}} = 4\cdot967.$
Rectangle 3 × 1..............................	$\pi \sqrt{\dfrac{10}{3}} = 5\cdot736.$

For instance, if a square and a circle have the same area, the former is the more acute in the ratio $4\cdot443 : 4\cdot261$.

For the circle the absolute frequency is

$$2\pi \times 2\cdot404\, c \sqrt{\frac{\pi}{\sigma}}, \quad \text{where} \quad c = \sqrt{T_1} \div \sqrt{\rho}.$$

In the case of similar forms the frequency is inversely as the linear dimension.

212. The theory of the free vibrations of a membrane was first successfully considered by Poisson[1]. His theory in the case of the rectangle left little to be desired, but his treatment

---

[1] *Mém. de l'Académie*, t. VIII. 1829.

of the circular membrane was restricted to the symmetrical vibrations. Kirchhoff's solution of the similar, but much more difficult, problem of the circular plate was published in 1850, and Clebsch's *Theory of Elasticity* (1862) gives the general theory of the circular membrane including the effects of stiffness and of rotatory inertia. It will therefore be seen that there was not much left to be done in 1866; nevertheless the memoir of Bourget already referred to contains a useful discussion of the problem accompanied by very complete numerical results, the whole of which however were not new.

213. In his experimental investigations M. Bourget made use of various materials, of which paper proved to be as good as any. The paper is immersed in water, and after removal of the superfluous moisture by blotting paper is placed upon a frame of wood whose edges have been previously coated with glue. The contraction of the paper in drying produces the necessary tension, but many failures may be met with before a satisfactory result is obtained. Even a well stretched membrane requires considerable precautions in use, being liable to great variations in pitch in consequence of the varying moisture of the atmosphere. The vibrations are excited by organ-pipes, of which it is necessary to have a series proceeding by small intervals of pitch, and they are made evident to the eye by means of a little sand scattered on the membrane. If the vibration be sufficiently vigorous, the sand accumulates on the nodal lines, whose form is thus defined with more or less precision. Any inequality in the tension shews itself by the circles becoming elliptic.

The principal results of experiment are the following:—

A circular membrane cannot vibrate in unison with every sound. It can only place itself in unison with sounds more acute than that heard when the membrane is gently tapped.

As theory indicates, these possible sounds are separated by less and less intervals, the higher they become.

The nodal lines are only formed distinctly in response to certain definite sounds. A little above or below confusion ensues, and when the pitch of the pipe is decidedly altered, the membrane remains unmoved. There is not, as Savart supposed, a continuous transition from one system of nodal lines to another.

The nodal lines are circles or diameters or combinations of circles and diameters, as theory requires. However, when the number of diameters exceeds two, the sand tends to heap itself confusedly towards the middle of the membrane, and the nodes are not well defined.

The same general laws were verified by MM. Bernard and Bourget in the case of square membranes[1]; and these authors consider that the results of theory are decisively established in opposition to the views of Savart, who held that a membrane was capable of responding to any sound, no matter what its pitch might be. But I must here remark that the distinction between forced and free vibrations does not seem to have been sufficiently borne in mind. When a membrane is set in motion by aerial waves having their origin in an organ-pipe, the vibration is properly speaking *forced*. Theory asserts, not that the membrane is only capable of vibrating with certain defined frequencies, but that it is only capable of so vibrating *freely*. When however the period of the force is not approximately equal to one of the natural periods, the resulting vibration may be insensible.

In Savart's experiments the sound of the pipe was two or three octaves higher than the gravest tone of the membrane, and was accordingly never far from unison with one of the series of over tones. MM. Bourget and Bernard made the experiment under more favourable conditions. When they sounded a pipe somewhat lower in pitch than the gravest tone of the membrane, the sand remained at rest, but was thrown into vehement vibration as unison was approached. So soon as the pipe was decidedly higher than the membrane, the sand returned again to rest. A modification of the experiment was made by first tuning a pipe about a third higher than the membrane when in its natural condition. The membrane was then heated until its tension had increased sufficiently to bring the pitch above that of the pipe. During the process of cooling the pitch gradually fell, and the point of coincidence manifested itself by the violent motion of the sand, which at the beginning and end of the experiment was sensibly at rest.

M. Bourget found a good agreement between theory and observation with respect to the radii of the circular nodes, though the test was not very precise, in consequence of the sensible width of

---

[1] *Ann. de Chim.* LX. 449—479. 1860.

the bands of sand; but the relative pitch of the various simple tones deviated considerably from the theoretical estimates. The committee of the French Academy appointed to report on M. Bourget's memoir suggest as the explanation the want of perfect fixity of the boundary. It should also be remembered that the theory proceeds on the supposition of perfect flexibility—a condition of things not at all closely approached by an ordinary membrane stretched with a comparatively small force. But perhaps the most important disturbing cause is the resistance of the air, which acts with much greater force on a membrane than on a string or bar in consequence of the large surface exposed. The gravest mode of vibration, during which the displacement is at all points in the same direction, might be affected very differently from the higher modes, which would not require so great a transference of air from one side to the other.

CHAPTER X.

VIBRATIONS OF PLATES.

214. IN order to form according to Green's method the equations of equilibrium and motion for a thin solid plate of uniform isotropic material and constant thickness, we require the expression for the potential energy of bending. It is easy to see that for each unit of area the potential energy V is a positive homogeneous symmetrical quadratic function of the two principal curvatures. Thus, if ρ_1, ρ_2 be the principal radii of curvature, the expression for V will be

$$A \left(\frac{1}{\rho_1^2} + \frac{1}{\rho_2^2} + \frac{2\mu}{\rho_1 \rho_2} \right) \dots\dots\dots\dots\dots (1),$$

where A and μ are constants, of which A must be positive, and μ must be numerically less than unity. Moreover if the material be of such a character that it undergoes no lateral contraction when a bar is pulled out, the constant μ must vanish. This amount of information is almost all that is required for our purpose, and we may therefore content ourselves with a mere statement of the relations of the constants in (1) with those by means of which the elastic properties of bodies are usually defined.

From Thomson and Tait's *Natural Philosophy*, §§ 639, 642, 720, it appears that, if b be the thickness, q Young's modulus, and μ the ratio of lateral contraction to longitudinal elongation when a bar is pulled out, the expression for V is

$$V = \frac{qb^3}{24\,(1-\mu^2)} \left\{ \frac{1}{\rho_1^2} + \frac{1}{\rho_2^2} + \frac{2\mu}{\rho_1\rho_2} \right\}$$

$$= \frac{qb^3}{24\,(1-\mu^2)} \left\{ \left(\frac{1}{\rho_1} + \frac{1}{\rho_2} \right)^2 - \frac{2\,(1-\mu)}{\rho_1\rho_2} \right\} \ldots\ldots (2)\,^1.$$

If w be the small displacement perpendicular to the plane of the plate at the point whose rectangular coordinates in the plane of the plate are x, y,

$$\frac{1}{\rho_1} + \frac{1}{\rho_2} = \nabla^2 w, \qquad \frac{1}{\rho_1\rho_2} = \frac{d^2w}{dx^2}\frac{d^2w}{dy^2} - \left(\frac{d^2w}{dxdy} \right)^2,$$

and thus for a unit of area, we have

$$V = \frac{qb^3}{24\,(1-\mu^2)} \left[(\nabla^2 w)^2 - 2\,(1-\mu) \left\{ \frac{d^2w}{dx^2}\frac{d^2w}{dy^2} - \left(\frac{d^2w}{dxdy} \right)^2 \right\} \right]\ (3),$$

which quantity has to be integrated over the surface (S) of the plate.

215. We proceed to find the variation of V, but it should be previously noticed that the second term in V, namely $\iint \frac{dS}{\rho_1\rho_2}$, represents the *total curvature* of the plate, and is therefore dependent only on the state of things at the edge.

$$\delta V = \frac{qb^3}{12\,(1-\mu^2)} \iint \left\{ \nabla^2 w \cdot \nabla^2 \delta w - (1-\mu)\,\delta\frac{1}{\rho_1\rho_2} \right\} dS \ldots\ldots(1);$$

so that we have to consider the two variations

$$\iint \nabla^2 w \cdot \nabla^2 \delta w \cdot dS \quad \text{and} \quad \iint \delta\frac{1}{\rho_1\rho_2}\,dS.$$

---

[1] The following comparison of the notations used by the principal writers may save trouble to those who wish to consult the original memoirs.

Young's modulus $= E$ (Clebsch) $= M$ (Thomson) $= \frac{9n\kappa}{3\kappa+n}$ (Thomson)

$= \frac{n\,(3m-n)}{m}$ (Thomson) $= q$ (Kirchhoff and Donkin) $= 2K\frac{1+3\theta}{1+\theta}$ (Kirchhoff).

Ratio of lateral contraction to longitudinal elongation

$= \mu$ (Clebsch and Donkin) $= \sigma$ (Thomson) $= \frac{m-n}{2m}$ (Thomson) $= \frac{\theta}{1+2\theta}$ (Kirchhoff).

Poisson assumed this ratio to be $\frac{1}{4}$, and Wertheim $\frac{1}{3}$.

Now by Green's theorem

$$\iint \nabla^2 w \cdot \nabla^2 \delta w \cdot dS = \iint \nabla^4 w \cdot \delta w \cdot dS$$

$$- \int \frac{d\nabla^2 w}{dn} \cdot \delta w \cdot ds + \int \nabla^2 w \frac{d\delta w}{dn} ds \ \dots\dots\dots (2),$$

in which ds denotes an element of the boundary, and $\frac{d}{dn}$ denotes differentiation with respect to the normal of the boundary drawn outwards.

The transformation of the second part is more difficult. We have

$$\delta \iint \frac{dS}{\rho_1 \rho_2} = \iint \left\{ \frac{d^2 w}{dx^2} \frac{d^2 \delta w}{dy^2} + \frac{d^2 w}{dy^2} \frac{d^2 \delta w}{dx^2} - 2 \frac{d^2 w}{dxdy} \frac{d^2 \delta w}{dxdy} \right\} dS.$$

The quantity under the sign of integration may be put into the form

$$\frac{d}{dy} \left(\frac{d\delta w}{dy} \frac{d^2 w}{dx^2} - \frac{d\delta w}{dx} \frac{d^2 w}{dxdy} \right) + \frac{d}{dx} \left(\frac{d\delta w}{dx} \frac{d^2 w}{dy^2} - \frac{d\delta w}{dy} \frac{d^2 w}{dxdy} \right)$$

Now, if F be any function of x and y,

$$\left. \begin{aligned} \iint \frac{dF}{dy} dx dy &= \int F \sin\theta \, ds \\ \iint \frac{dF}{dx} dx dy &= \int F \cos\theta \, ds \end{aligned} \right\} \ \dots\dots\dots (3),$$

where θ is the angle between x and the normal drawn outwards, and the integration on the right-hand side extends round the boundary. Using these, we find

$$\delta \iint \frac{dS}{\rho_1 \rho_2} = \int ds \sin\theta \left\{ \frac{d\delta w}{dy} \frac{d^2 w}{dx^2} - \frac{d\delta w}{dx} \frac{d^2 w}{dxdy} \right\}$$

$$+ \int ds \cos\theta \left\{ \frac{d\delta w}{dx} \frac{d^2 w}{dy^2} - \frac{d\delta w}{dy} \frac{d^2 w}{dxdy} \right\}.$$

If we substitute for $\frac{d\delta w}{dx}$, $\frac{d\delta w}{dy}$ their values in terms $\frac{d\delta w}{dn}$, $\frac{d\delta w}{ds}$, from the equations (see Fig. 40)

$$\left. \begin{aligned} \frac{d\delta w}{dx} &= \frac{d\delta w}{dn} \cos\theta - \frac{d\delta w}{ds} \sin\theta \\ \frac{d\delta w}{dy} &= \frac{d\delta w}{dn} \sin\theta + \frac{d\delta w}{ds} \cos\theta \end{aligned} \right\} \ \dots\dots\dots (4),$$

Fig. 40.

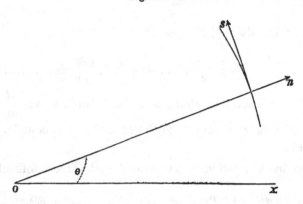

we obtain

$$\delta \iint \frac{dS}{\rho_1 \rho_2} = \int ds \frac{d\delta w}{dn} \left\{ \sin^2 \theta \frac{d^2 w}{dx^2} + \cos^2 \theta \frac{d^2 w}{dy^2} - 2 \sin \theta \cos \theta \frac{d^2 w}{dxdy} \right\}$$
$$+ \int ds \frac{d\delta w}{ds} \left\{ \cos \theta \sin \theta \left(\frac{d^2 w}{dx^2} - \frac{d^2 w}{dy^2} \right) + (\sin^2 \theta - \cos^2 \theta) \frac{d^2 w}{dxdy} \right\} \dots (5).$$

The second integral by a partial integration with respect to s may be put into the form

$$\int \delta w \frac{d}{ds} \left\{ \cos \theta \sin \theta \left(\frac{d^2 w}{dy^2} - \frac{d^2 w}{dx^2} \right) + (\cos^2 \theta - \sin^2 \theta) \frac{d^2 w}{dxdy} \right\} ds.$$

Collecting and rearranging our results, we find

$$\delta V = \frac{q b^3}{12 (1 - \mu^2)} \left[\iint \nabla^4 w \, \delta w \, dS \right.$$
$$- \int \delta w \, ds \left\{ \frac{d\nabla^2 w}{dn} + (1 - \mu) \frac{d}{ds} \left(\cos \theta \sin \theta \left(\frac{d^2 w}{dy^2} - \frac{d^2 w}{dx^2} \right) \right. \right.$$
$$\left. + (\cos^2 \theta - \sin^2 \theta) \frac{d^2 w}{dxdy} \right) \right\}$$
$$+ \int \frac{d\delta w}{dn} \, ds \left\{ \mu \nabla^2 w + (1 - \mu) \left(\cos^2 \theta \frac{d^2 w}{dx^2} + \sin^2 \theta \frac{d^2 w}{dy^2} \right. \right.$$
$$\left. \left. \left. + 2 \cos \theta \sin \theta \frac{d^2 w}{dxdy} \right) \right\} \right] \dots (6).$$

There will now be no difficulty in forming the equations of motion. If ρ be the volume density, and $Z \rho b \, dS$ the transverse force acting on the element dS,

$$\delta V - \iint Z\rho b\, \delta w\, dS + \iint \rho b\, \ddot{w}\, \delta w\, dS = 0 \quad \ldots\ldots\ldots\ldots (7)\,[1]$$

is the general variational equation, which must be true whatever function (consistent with the constitution of the system) δw may be supposed to be. Hence by the principles of the Calculus of Variations

$$\frac{qb^2}{12\rho(1-\mu^2)}\, \nabla^4 w - Z + \ddot{w} = 0 \quad \ldots\ldots\ldots\ldots (8),$$

at every point of the plate.

If the edges of the plate be free, there is no restriction on the hypothetical boundary values of δw and $\dfrac{d\delta w}{dn}$, and therefore the coefficients of these quantities in the expression for δV must vanish. The conditions to be satisfied at a free edge are thus

$$\left. \begin{aligned} &\frac{d\nabla^2 w}{dn} + (1-\mu)\frac{d}{ds}\left\{\cos\theta\sin\theta\left(\frac{d^2 w}{dy^2} - \frac{d^2 w}{dx^2}\right)\right. \\ &\qquad\qquad\left. + (\cos^2\theta - \sin^2\theta)\frac{d^2 w}{dx\,dy}\right\} = 0 \\ &\mu\nabla^2 w + (1-\mu)\left\{\cos^2\theta\frac{d^2 w}{dx^2} + \sin^2\theta\frac{d^2 w}{dy^2}\right. \\ &\qquad\qquad\left. + 2\cos\theta\sin\theta\frac{d^2 w}{dx\,dy}\right\} = 0 \end{aligned} \right\} \ \ldots\ldots (9).$$

If the whole circumference of the plate be clamped, $\delta w = 0$, $\dfrac{d\delta w}{dn} = 0$, and the satisfaction of the boundary conditions is already secured. If the edge be 'supported'[2], $\delta w = 0$, but $\dfrac{d\delta w}{dn}$ is arbitrary. The second of the equations (9) must in this case be satisfied by w.

216. The boundary equations may be simplified by getting rid of the extrinsic element involved in the use of Cartesian co-ordinates. Taking the axis of x parallel to the normal of the bounding curve, we see that we may write

$$\cos^2\theta\frac{d^2 w}{dx^2} + \sin^2\theta\frac{d^2 w}{dy^2} + 2\cos\theta\sin\theta\frac{d^2 w}{dx\,dy} = \frac{d^2 w}{dn^2}$$

Also
$$\nabla^2 w = \frac{d^2 w}{dn^2} + \frac{d^2 w}{d\sigma^2} \quad \ldots\ldots \ldots\ldots\ldots\ldots (1),$$

---

[1] The rotatory inertia is here neglected.   [2] Compare § 162.

where σ is a fixed axis coinciding with the tangent at the point under consideration. In general $\frac{d^2w}{d\sigma^2}$ differs from $\frac{d^2w}{ds^2}$. To obtain the relation between them, we may proceed thus. Expand w by Maclaurin's theorem in ascending powers of the small quantities n and σ, and substitute for n and σ their values in terms of s, the arc of the curve.

Thus in general

$$w = w_0 + \frac{dw}{dn_0} n + \frac{dw}{d\sigma_0} \sigma + \tfrac{1}{2} \frac{d^2w}{dn_0^2} n^2 + \frac{d^2w}{dn_0 d\sigma_0} n\sigma + \tfrac{1}{2} \frac{d^2w}{d\sigma_0^2} \sigma^2 + \ldots,$$

while on the curve $\sigma = s + \text{cubes}$, $n = -\tfrac{1}{2} \frac{s^2}{\rho} + \ldots$, where ρ is the radius of curvature. Accordingly for points on the curve,

$$w = w_0 - \tfrac{1}{2} \frac{dw}{dn_0} \frac{s^2}{\rho} + \frac{dw}{d\sigma_0} s + \tfrac{1}{2} \frac{d^2w}{d\sigma_0^2} s^2 + \text{cubes of } s,$$

and therefore

$$\frac{d^2w}{ds^2} = \frac{d^2w}{d\sigma^2} - \frac{1}{\rho} \frac{dw}{dn} \quad \ldots\ldots\ldots\ldots\ldots\ldots (2);$$

whence from (1)

$$\nabla^2 w = \frac{d^2w}{dn^2} + \frac{1}{\rho} \frac{dw}{dn} + \frac{d^2w}{ds^2} \quad \ldots\ldots\ldots\ldots\ldots (3).$$

We conclude that the second boundary condition in (9) § 215 may be put into the form

$$\frac{d^2w}{dn^2} + \mu \left(\frac{1}{\rho} \frac{dw}{dn} + \frac{d^2w}{ds^2} \right) = 0 \quad \ldots\ldots\ldots\ldots\ldots (4).$$

In the same way by putting $\theta = 0$, we see that

$$\cos\theta \sin\theta \left(\frac{d^2w}{dy^2} - \frac{d^2w}{dx^2} \right) + (\cos^2\theta - \sin^2\theta) \frac{d^2w}{dx\,dy}$$

is equivalent to $\frac{d^2w}{dn\,d\sigma}$, where it is to be understood that the axes of n and σ are fixed. The first boundary condition now becomes

$$\frac{d}{dn} \nabla^2 w + (1 - \mu) \frac{d}{ds} \left(\frac{d^2w}{dn\,d\sigma} \right) = 0 \quad \ldots\ldots\ldots\ldots (5).$$

If we apply these equations to the rectangle whose sides are parallel to the coordinate axes, we obtain as the conditions to be satisfied along the edges parallel to y,

$$\left.\begin{array}{c}\dfrac{d}{dx}\left\{\dfrac{d^2w}{dx^2}+(2-\mu)\dfrac{d^2w}{dy^2}\right\}=0\\[2mm]\dfrac{d^2w}{dx^2}+\mu\dfrac{d^2w}{dy^2}=0\end{array}\right\}\ \ldots\ldots\ldots\ldots(6).$$

In this case the distinction between σ and s disappears, and ρ, the radius of curvature, is infinitely great. The conditions for the other pair of edges are found by interchanging x and y. These results may be obtained equally well from (9) § 215 directly, without the preliminary transformation.

217. If we suppose $Z = 0$, and write

$$\frac{qb^2}{12\rho(1-\mu^2)}=c^4\ \ldots\ldots\ldots\ldots\ldots(1),$$

the general equation becomes

$$\ddot{w}+c^4\nabla^4 w=0\ \ldots\ldots\ldots\ldots\ldots(2),$$

or, if $w \propto \cos(pt-\epsilon)$,

$$\nabla^4 w=\kappa^4 w\ \ldots\ldots\ldots\ldots\ldots(3),$$

where

$$\kappa^4=p^2\div c^4\ \ldots\ldots\ldots\ldots\ldots(4).$$

Any two values of w, u and v, corresponding to the same boundary conditions, are *conjugate*, that is to say

$$\iint uv\, dS=0\ \ldots\ldots\ldots\ldots\ldots(5),$$

provided that the periods be different. In order to prove this from the ordinary differential equation (3), we should have to retrace the steps by which (3) was obtained. This is the method adopted by Kirchhoff for the circular disc, but it is much simpler and more direct to use the variational equation

$$\delta V+\rho b\iint \ddot{w}\,\delta w\, dS=0\ \ldots\ldots\ldots\ldots(6)$$

in which w refers to the actual motion, and δw to an arbitrary displacement consistent with the nature of the system. δV is a symmetrical function of w and δw, as may be seen from § 215, or from the general character of V (§ 94.)

If we now suppose in the first place that $w = u$, $\delta w = v$, we have

$$\delta V = \rho b p^2 \iint uv\,dS;$$

and in like manner if we put $w = v$, $\delta w = u$, which we are equally entitled to do,

$$\delta V = \rho b p'^2 \iint uv\,dS,$$

whence

$$(p^2 - p'^2) \iint uv\,dS = 0 \dots\dots\dots\dots\dots(7).$$

This demonstration is valid whatever may be the form of the boundary, and whether the edge be clamped, supported, or free, in whole or in part.

As for the case of membranes in the last Chapter, equation (7) may be employed to prove that the admissible values of p^2 are real; but this is evident from physical considerations.

218.   For the application to a circular disc, it is necessary to express the equations by means of polar coordinates. Taking the centre of the disc as pole, we have for the general equation to be satisfied at all points of the area

$$(\nabla^4 - \kappa^4)\,w = 0\dots\dots\dots\dots\dots\dots(1),$$

where (§ 200) $$\nabla^2 = \frac{d^2}{dr^2} + \frac{1}{r}\frac{d}{dr} + \frac{1}{r^2}\frac{d^2}{d\theta^2}.$$

To express the boundary condition (§ 216) for a free edge $(r = a)$, we have

$$\frac{d}{dn}\nabla^2 w = \frac{d}{dr}\nabla^2 w, \quad \frac{d}{ds}\left(\frac{d^2 w}{dnd\sigma}\right) = \frac{d}{ad\theta}\frac{d}{dr}\left(\frac{dw}{rd\theta}\right), \quad \frac{d^2 w}{ds^2} = \frac{d^2 w}{a^2 d\theta^2},$$

ρ = radius of curvature = a; and thus

$$\left.\begin{array}{c} \dfrac{d}{dr}\left(\dfrac{d^2 w}{dr^2} + \dfrac{1}{r}\dfrac{dw}{dr}\right) + \dfrac{d^2}{d\theta^2}\left(\dfrac{2-\mu}{a^2}\dfrac{dw}{dr} - \dfrac{3-\mu}{a^3}\,w\right) = 0 \\[2mm] \dfrac{d^2 w}{dr^2} + \mu\left(\dfrac{1}{a}\dfrac{dw}{dr} + \dfrac{1}{a^2}\dfrac{d^2 w}{d\theta^2}\right) = 0 \end{array}\right\}\dots\dots(2).$$

After the differentiations are performed, r is to be made equal to a.

If w be expanded in Fourier's series

$$w = w_0 + w_1 + \dots + w_n + \dots,$$

each term separately must satisfy (2), and thus, since

$$w_n \propto \cos(n\theta - \alpha),$$

$$\left. \begin{aligned} \frac{d}{dr}\left(\frac{d^2 w_w}{dr^2} + \frac{1}{r}\frac{dw_w}{dr}\right) - n^2\left(\frac{2-\mu}{a^2}\frac{dw_n}{dr} - \frac{3-\mu}{a^3}w_n\right) = 0 \\ \frac{d^2 w_n}{dr^2} + \mu\left(\frac{1}{a}\frac{dw_n}{dr} - \frac{n^2}{a^2}w_n\right) = 0 \end{aligned} \right\} \dots\dots(3).$$

The superficial differential equation may be written

$$(\nabla^2 + \kappa^2)(\nabla^2 - \kappa^2)w = 0,$$

which becomes for the general term of the Fourier expansion

$$\left(\frac{d^2}{dr^2} + \frac{1}{r}\frac{d}{dr} - \frac{n^2}{r^2} + \kappa^2\right)\left(\frac{d^2}{dr^2} + \frac{1}{r}\frac{d}{dr} - \frac{n^2}{r^2} - \kappa^2\right)w_n = 0,$$

shewing that the complete value of w_n will be obtained by adding together, with arbitrary constants prefixed, the general solutions of

$$\left(\frac{d^2}{dr^2} + \frac{1}{r}\frac{d}{dr} - \frac{n^2}{r^2} \pm \kappa^2\right)w_n = 0\dots\dots\dots(4).$$

The equation with the upper sign is the same as that which obtains in the case of the vibrations of circular membranes, and as in the last Chapter we conclude that the solution applicable to the problem in hand is $w_n \propto J_n(\kappa r)$, the second function of r being here inadmissible.

In the same way the solution of the equation with the lower sign is $w_n \propto J_n(i\kappa r)$, where $i = \sqrt{-1}$ as usual.

The simple vibration is thus

$$w_n = \cos n\theta \{\alpha J_n(\kappa r) + \beta J_n(i\kappa r)\} + \sin n\theta \{\gamma J_n(\kappa r) + \delta J_n(i\kappa r)\}.$$

The two boundary equations will determine the admissible values of κ and the values which must be given to the ratios $\alpha : \beta$ and $\gamma : \delta$. From the form of these equations it is evident that we must have $\qquad \alpha : \beta = \gamma : \delta,$

and thus w_n may be expressed in the form

$$w_n = P\cos(n\theta - \alpha)\{J_n(\kappa r) + \lambda J_n(i\kappa r)\}\cos(pt - \epsilon)\dots\dots\dots(5).$$

As in the case of a membrane the nodal system is composed of the n diameters symmetrically distributed round the centre, but otherwise arbitrary, denoted by

$$\cos(n\theta - \alpha) = 0 \quad \dots \dots \dots \dots \dots \dots (6),$$

together with the concentric circles, whose equation is

$$J_n(\kappa r) + \lambda J_n(i\kappa r) = 0 \dots \dots \dots \dots \dots \dots (7).$$

219. In order to determine λ and κ we must introduce the boundary conditions. When the edge is free, we obtain from (3) § 218.

$$\left.\begin{aligned}
-\lambda &= \frac{n^2(\mu-1)\{\kappa a\, J_n{}'(\kappa a) - J_n(\kappa a)\} - \kappa^3 a^3 J_n{}'(\kappa a)}{n^2(\mu-1)\{i\kappa a J_n{}'(i\kappa a) - J_n(i\kappa a)\} + i\kappa^3 a^3 J_n{}'(i\kappa a)} \\[2mm]
-\lambda &= \frac{(\mu-1)\{\kappa a J_n{}'(\kappa a) - n^2 J_n(\kappa a)\} - \kappa^2 a^2 J_n(\kappa a)}{(\mu-1)\{i\kappa a J_n{}'(i\kappa a) - n^2 J_n(i\kappa a)\} + \kappa^2 a^2 J_n(i\kappa a)}
\end{aligned}\right\} \quad \dots(1),$$

in which use has been made of the differential equations satisfied by $J_n(\kappa r)$, $J_n(i\kappa r)$. In each of the fractions on the right the denominator may be derived from the numerator by writing $i\kappa$ in place of κ. By elimination of λ the equation is obtained whose roots give the admissible values of κ.

When $n = 0$, the result assumes a simple form, viz.

$$2(1-\mu) + i\kappa a \frac{J_0(i\kappa a)}{J_0{}'(i\kappa a)} + \kappa a \frac{J_0(\kappa a)}{J_0{}'(\kappa a)} = 0 \dots \dots \dots \dots (2).$$

This, of course, could have been more easily obtained by neglecting n from the beginning.

The calculation of the lowest root for each value of n is troublesome, and in the absence of appropriate tables must be effected by means of the ascending series for the functions $J_n(\kappa r)$, $J_n(i\kappa r)$. In the case of the higher roots recourse may be had to the semiconvergent descending series for the same functions. Kirchhoff finds

$$\tan(\kappa a - \tfrac{1}{2}n\pi) = \frac{\dfrac{B}{8\kappa a} + \dfrac{C}{(8\kappa a)^2} - \dfrac{D}{(8\kappa a)^3} + \dots}{A + \dfrac{B}{8\kappa a} + \dfrac{D}{(8\kappa a)^3} + \dots} \quad \dots \dots \dots (3),$$

where

$A = \gamma = (1-\mu)^{-1}$,

$B = \gamma(1 - 4n^2) - 8$,

$C = \gamma(1 - 4n^2)(9 - 4n^2) + 48(1 + 4n^2)$,

$D = -\gamma\tfrac{1}{3}\{(1 - 4n^2)(9 - 4n^2)(13 - 4n^2)\} + 8(9 + 136n^2 + 80n^4)$.

When κa is great,

$$\tan (\kappa a - \tfrac{1}{2} n\pi) = 0 \quad \text{approx.};$$

whence

$$\kappa a = \tfrac{1}{2}\pi\, (n + 2h) \dots\dots\dots\dots\dots (4),$$

where h is an integer.

It appears by a numerical comparison that h is identical with the number of circular nodes, and (4) expresses a law discovered by Chladni, that the frequencies corresponding to figures with a given number of nodal diameters are, with the exception of the lowest, approximately proportional to the squares of consecutive even or uneven numbers, according as the number of the diameters is itself even or odd. Within the limits of application of (4), we see also that the pitch is approximately unaltered, when any number is subtracted from h, provided twice that number be added to n. This law, of which traces appear in the following table, may be expressed by saying that towards raising the pitch nodal circles have twice the effect of nodal diameters. It is probable, however, that, strictly speaking, no two normal components have exactly the same pitch.

h	$n = 0$			$n = 1$		
	Ch.	P.	W.	Ch.	P.	W.
0	...	...	...	...	...	...
1	Gis	Gis +	A +	b	h −	c −
2	gis′ +	b′ −	b′ +	e″ +	f″ +	fis″ +

h	$n = 2$			$n = 3$		
	Ch.	P.	W.	Ch.	P.	W.
0	C	C	C	d	dis −	dis −
1	g′	gis′ +	a′ −	d″.dis″	dis″ +	e″ −

The table, extracted from Kirchhoff's memoir, gives the pitch of the more important overtones of a free circular plate, the gravest being assumed to be C[1]. The three columns under the heads *Ch, P, W* refer respectively to the results as observed by Chladni and as calculated from theory with Poisson's and Wertheim's values of μ. A *plus* sign denotes that the actual pitch is a little higher, and a *minus* sign that it is a little lower, than that written.

---

[1] Gis corresponds to G♯ of the English notation, and h to b natural.

The discrepancies between theory and observation are considerable, but perhaps not greater than may be attributed to irregularity in the plate.

220. The radii of the nodal circles in the symmetrical case ($n = 0$) were calculated by Poisson, and compared by him with results obtained experimentally by Savart. The following numbers are taken from a paper by Strehlke[1], who made some careful measurements. The radius of the disc is taken as unity.

	Observation.	Calculation.
One circle ...	0·67815	0·68062.
Two circles...	$\begin{cases} 0\text{·}39133 \\ 0\text{·}84149 \end{cases}$	$\begin{matrix} 0\text{·}39151. \\ 0\text{·}84200. \end{matrix}$
Three circles	$\begin{cases} 0\text{·}25631 \\ 0\text{·}59107 \\ 0\text{·}89360 \end{cases}$	$\begin{matrix} 0\text{·}25679. \\ 0\text{·}59147. \\ 0\text{·}89381. \end{matrix}$

The calculated results appear to refer to Poisson's value of μ, but would vary very little if Wertheim's value were substituted.

The following table gives a comparison of Kirchhoff's theory (n not zero) with measurements by Strehlke made on less accurate discs.

Radii of Circular Nodes.

	Observation.				Calculation.	
					$\mu = \frac{1}{4}$ (P.).	$\mu = \frac{1}{3}$ (W.).
$n=1,\ h=1$	0·781	0·783	0·781	0·783	0·78136	0·78088
$n=2,\ h=1$	0·79	0·81	0·82		0·82194	0·82274
$n=3,\ h=1$	0·838	0·842			0·84523	0·84681
$n=1,\ h=2$	0·488	0·492			0·49774	0·49715
	0·869	0·869			0·87057	0·87015

221. When the plate is truly symmetrical, whether uniform or not, theory indicates, and experiment verifies, that the position of the nodal diameters is arbitrary, or rather dependent only on the manner in which the plate is supported. By varying the place of support, any desired diameter may be made nodal. It is generally otherwise when there is any sensible departure from exact symmetry. The two modes of vibration, which originally,

[1] Pogg. *Ann.* xcv. p. 577. 1855.

in consequence of the equality of periods could be combined in any proportion without ceasing to be simple harmonic, are now separated and affected with different periods. At the same time the position of the nodal diameters becomes determinate, or rather limited to two alternatives. The one set is derived from the other by rotation through half the angle included between two adjacent diameters of the same set. This supposes that the deviation from uniformity is small; otherwise the nodal system will no longer be composed of approximate circles and diameters at all. The cause of the deviation may be an irregularity either in the material or in the thickness or in the form of the boundary. The effect of a small load at any point may be investigated as in the parallel problem of the membrane § 208. If the place at which the load is attached does not lie on a nodal circle, the normal types are made determinate. The diametral system corresponding to one of the types passes through the place in question, and for this type the period is unaltered. The period of the other type is increased.

The most general motion of the uniform circular plate is expressed by the superposition, with arbitrary amplitudes and phases, of the normal components already investigated. The determination of the amplitude and phase to correspond to arbitrary initial displacements and velocities is effected precisely as in the corresponding problem for the membrane by the aid of the characteristic property of the normal functions proved in § 217.

The two other cases of a circular plate in which the edge is either clamped or *supported* would be easier than the preceding in their theoretical treatment, but are of less practical interest on account of the difficulty of experimentally realising the conditions assumed. The general result that the nodal system is composed of concentric circles, and diameters symmetrically distributed, is applicable to all the three cases.

222. We have seen that in general Chladni's figures as traced by sand agree very closely with the circles and diameters of theory; but in certain cases deviations occur, which are usually attributed to irregularities in the plate. It must however be remembered that the vibrations excited by a bow are not strictly speaking free, and that their periods are therefore liable to a certain modification. It may be that under the action of the bow two or more normal component vibrations coexist. The whole

motion may be simple harmonic in virtue of the external force, although the natural periods would be a little different. Such an explanation is suggested by the regular character of the figures obtained in certain cases.

Another cause of deviation may perhaps be found in the manner in which the plates are supported. The requirements of theory are often difficult to meet in actual experiment. When this is so, we may have to be content with an imperfect comparison; but we must remember that a discrepancy may be the fault of the experiment as well as of the theory.

223. The first attempt to solve the problem with which we have just been occupied is due to Sophie Germain, who succeeded in obtaining the correct differential equation, but was led to erroneous boundary conditions. For a free plate the latter part of the problem is indeed of considerable difficulty. In Poisson's memoir 'Sur l'équilibre et le mouvement des corps élastiques[1],' that eminent mathematician gave *three* equations as necessary to be satisfied at all points of a free edge, but Kirchhoff has proved that in general it would be impossible to satisfy them all. It happens, however, that an exception occurs in the case of the symmetrical vibrations of a circular plate, when one of the equations is true identically. Owing to this peculiarity, Poisson's theory of the symmetrical vibrations is correct, notwithstanding the error in his view as to the boundary conditions. In 1850 the subject was resumed by Kirchhoff[2], who first gave the *two* equations appropriate to a free edge, and completed the theory of the vibrations of a circular disc.

224. The correctness of Kirchhoff's boundary equations has been disputed by Mathieu[3], who, without explaining where he considers Kirchhoff's error to lie, has substituted a different set of equations. He proves that if u and u' be two normal functions, so that $w = u \cos pt$, $w = u' \cos p't$ are possible vibrations, then

$$(p^2 - p'^2) \iint uu' dx dy$$

$$= c^4 \int ds \left\{ u' \frac{d\nabla^2 u}{dn} - \nabla^2 u \frac{du'}{dn} - \nabla^2 u' \frac{du}{dn} + u \frac{d\nabla^2 u'}{dn} \right\} \ldots\ldots(1).$$

[1] *Mém. de l'Acad. d. Sc. à Par.* 1829.

[2] *Crelle*, t. XL. p. 51. Ueber das Gleichgewicht und die Bewegung einer elastichen Scheibe.

[3] *Liouville*, t. XIV. 1869.

This follows, if it be admitted that u, u' satisfy respectively the equations

$$c^4 \nabla^4 u = p^2 u, \qquad c^4 \nabla^4 u' = p'^2 u'.$$

Since the left-hand member is zero, the same must be true of the right-hand member; and this, according to Mathieu, cannot be the case, unless at all points of the boundary both u and u' satisfy one of the four following pairs of equations:

$$\left. \begin{matrix} u = 0 \\ \dfrac{du}{dn} = 0 \end{matrix} \right\}, \qquad \left. \begin{matrix} \nabla^2 u = 0 \\ \dfrac{d\nabla^2 u}{dn} = 0 \end{matrix} \right\}, \qquad \left. \begin{matrix} u = 0 \\ \nabla^2 u = 0 \end{matrix} \right\}, \qquad \left. \begin{matrix} \dfrac{du}{dn} = 0 \\ \dfrac{d\nabla^2 u}{dn} = 0 \end{matrix} \right\}$$

The second pair would seem the most likely for a free edge, but it is found to lead to an impossibility. Since the first and third pairs are obviously inadmissible, Mathieu concludes that the fourth pair of equations must be those which really express the condition of a free edge. In his belief in this result he is not shaken by the fact that the corresponding conditions for the free end of a bar would be

$$\frac{du}{dx} = 0, \qquad \frac{d^3 u}{dx^3} = 0,$$

the first of which is contradicted by the roughest observation of the vibration of a large tuning fork.

The fact is that although any of the four pairs of equations would secure the evanescence of the boundary integral in (1), it does not follow conversely that the integral can be made to vanish in no other way; and such a conclusion is negatived by Kirchhoff's investigation. There are besides innumerable other cases in which the integral in question would vanish, all that is really necessary being that the boundary appliances should be either at rest, or devoid of inertia.

225. The vibrations of a rectangular plate, whose edge is *supported*, may be easily investigated theoretically, the normal functions being identical with those applicable to a membrane of the same shape, whose boundary is fixed. If we assume

$$w = \sin \frac{m\pi x}{a} \; \sin \frac{n\pi y}{b} \; \cos pt \ldots\ldots\ldots\ldots(1),$$

we see that at all points of the boundary,

$$w = 0, \qquad \frac{d^2w}{dx^2} = 0, \qquad \frac{d^2w}{dy^2} = 0,$$

which secure the fulfilment of the necessary conditions (§ 215). The value of p, found by substitution in $c^4 \nabla^4 w = p^2 w$,

is
$$p = c^2 \pi^2 \left(\frac{m^2}{a^2} + \frac{n^2}{b^2} \right) \dots\dots\dots\dots\dots\dots(2),$$

shewing that the analogy to the membrane does not extend to the sequence of tones.

It is not necessary to repeat here the discussion of the primary and derived nodal systems given in Chapter IX. It is enough to observe that if two of the fundamental modes (1) have the same period in the case of the membrane, they must also have the same period in the case of the plate. The derived nodal systems are accordingly identical in the two cases.

The generality of the value of w obtained by compounding with arbitrary amplitudes and phases all possible particular solutions of the form (1) requires no fresh discussion.

Unless the contrary assertion had been made, it would have seemed unnecessary to say that the nodes of a *supported* plate have nothing to do with the ordinary Chladni's figures, which belong to a plate whose edges are free.

The realization of the conditions for a supported edge is scarcely attainable in practice. Appliances are required capable of holding the boundary of the plate at rest, and of such a nature that they give rise to no couples about tangential axes. We may conceive the plate to be held in its place by friction against the walls of a cylinder circumscribed closely round it.

226. The problem of a rectangular plate, whose edges are free, is one of great difficulty, and has for the most part resisted attack. If we suppose that the displacement w is independent of y, the general differential equation becomes identical with that with which we were concerned in Chapter VIII. If we take the solution corresponding to the case of a bar whose ends are free, and therefore satisfying

$$\frac{d^2w}{dx^2} = 0, \qquad \frac{d^3w}{dx^3} = 0,$$

when $x = 0$ and when $x = a$, we obtain a value of w which satisfies the general differential equation, as well as the pair of boundary equations

$$\left. \begin{aligned} \frac{d}{dx}\left\{ \frac{d^2w}{dx^2} + (2-\mu)\frac{d^2w}{dy^2} \right\} &= 0 \\ \frac{d^2w}{dx^2} + \mu\frac{d^2w}{dy^2} &= 0 \end{aligned} \right\} \quad \ldots\ldots\ldots\ldots(1),$$

which are applicable to the edges parallel to y; but the second boundary condition for the other pair of edges, namely

$$\frac{d^2w}{dy^2} + \mu\frac{d^2w}{dx^2} = 0 \quad \ldots\ldots\ldots\ldots\ldots\ldots(2),$$

will be violated, unless $\mu = 0$. This shews that, except in the case reserved, it is not possible for a free rectangular plate to vibrate after the manner of a bar; unless indeed as an approximation, when the length parallel to one pair of edges is so great that the conditions to be satisfied at the second pair of edges may be left out of account.

Although the constant μ (which expresses the ratio of lateral contraction to longitudinal extension when a bar is drawn out) is positive for every known substance, in the case of a few substances—cork, for example—it is comparatively very small. There is, so far as we know, nothing absurd in the idea of a substance for which μ vanishes. The investigation of the problem under this condition is therefore not devoid of interest, though the results will not be strictly applicable to ordinary glass or metal plates, for which the value of μ is about $\frac{1}{3}$.[1]

If u_1, u_2, &c. denote the normal functions for a free bar investigated in Chapter VIII., corresponding to 2, 3, ... nodes, the vibrations of a rectangular plate will be expressed by

$$w = u_1\left(\frac{x}{a}\right), \quad w = u_2\left(\frac{x}{a}\right), \quad \&c.,$$

or

$$w = u_1\left(\frac{y}{b}\right), \quad w = u_2\left(\frac{y}{b}\right), \quad \&c.$$

[1] In order to make a plate of material, for which μ is not zero, vibrate in the manner of a bar, it would be necessary to apply constraining couples to the edges parallel to the plane of bending to prevent the assumption of a contrary curvature. The effect of these couples would be to raise the pitch, and therefore the calculation founded on the type proper to $\mu = 0$ would give a result somewhat higher in pitch than the truth.

In each of these primitive modes the nodal system is composed of straight lines parallel to one or other of the edges of the rectangle. When $b = a$, the rectangle becomes a square, and the vibrations

$$w = u_n \left(\frac{x}{a}\right), \quad w = u_n \left(\frac{y}{a}\right),$$

having necessarily the same period, may be combined in any proportion, while the whole motion still remains simple harmonic. Whatever the proportion may be, the resulting nodal curve will of necessity pass through the points determined by

$$u_n \left(\frac{x}{a}\right) = 0, \quad u_n \left(\frac{y}{a}\right) = 0.$$

Now let us consider more particularly the case of $n = 1$. The nodal system of the primitive mode, $w = u_1 \left(\frac{x}{a}\right)$, consists of a pair of straight lines parallel to y, whose distance from the nearest edge is $\cdot 2242\, a$. The points in which these lines are met by the corresponding pair for $w = u_1 \left(\frac{y}{a}\right)$, are those through which the nodal curve of the compound vibration must in all cases pass. It is evident that they are symmetrically disposed on the diagonals of the square. If the two primitive vibrations be taken equal, but in opposite phases (or, algebraically, with equal and opposite amplitudes), we have

$$w = u_1 \left(\frac{x}{a}\right) - u_1 \left(\frac{y}{a}\right) \dots \dots \dots \dots \dots (3),$$

from which it is evident that w vanishes when $x = y$, that is along the diagonal which passes through the origin. That w will also vanish along the other diagonal follows from the symmetry of the functions, and we conclude that the nodal system of (3) com-

Fig. 41

prises both the diagonals (Fig. 41). This is a well-known mode of vibration of a square plate.

A second notable case is when the amplitudes are equal and their phases the same, so that

$$w = u_1\left(\frac{x}{a}\right) + u_1\left(\frac{y}{a}\right)\dots\dots\dots\dots\dots\dots(4).$$

The most convenient method of constructing graphically the curves, for which $w = $ const., is that employed by Maxwell in similar cases. The two systems of curves (in this instance straight lines) represented by $u_1\left(\frac{x}{a}\right) = $ const., $u_1\left(\frac{y}{a}\right) = $ const., are first laid down, the values of the constants forming an arithmetical progression with the same common difference in the two cases. In this way a network is obtained which the required curves cross diagonally. The execution of the proposed plan requires an inversion of the table given in Chapter VIII., § 178, expressing the march of the function u_1, of which the result is as follows :—

u_1	$x : a$	u_1	$x : a$
+ 1·00	·5000	− ·25	·1871
·75	·3680	·50	·1518
·50	·3106	·75	·1179
·25	·2647	1·00	·0846
·00	·2242	1·25	·0517
		− 1·50	·0190

The system of lines represented by the above values of x (completed symmetrically on the further side of the central line) and the corresponding system for y are laid down in the figure (42). From these the curves of equal displacement are deduced. At the centre of the square we have w a maximum and equal to 2 on the scale adopted. The first curve proceeding outwards is the locus of points at which $w = 1$. The next is the nodal line, separating the regions of opposite displacement. The remaining curves taken in order give the displacements $-1, -2, -3$. The numerically greatest negative displacement occurs at the corners of the square, where it amounts to $2 \times 1·645 = 3·290$.[1]

---

[1] On the nodal lines of a square plate. *Phil. Mag.* August, 1873.

The nodal curve thus constructed agrees pretty closely with the observations of Strehlke[1]. His results, which refer to three carefully worked plates of glass, are embodied in the following polar equations:

$$r = \begin{Bmatrix} \cdot40143 & \cdot0171 \\ \cdot40143 + \cdot0172 \\ \cdot4019 & \cdot0168 \end{Bmatrix} \cos 4t + \begin{Bmatrix} \cdot00127 \\ \cdot00127 \\ \cdot0013 \end{Bmatrix} \cos 8t,$$

Fig. 42.

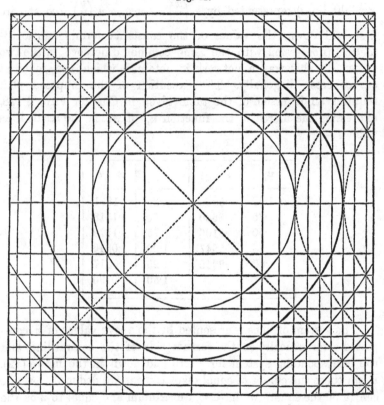

the centre of the square being pole. From these we obtain for the radius vector parallel to the sides of the square ($t = 0$) ·41980, ·41981, ·4200, while the calculated result is ·4154. The radius vector measured along a diagonal is ·3856, ·3855, ·3864, and by calculation ·3900.

[1] Pogg. *Ann.* Vol. CXLVI. p. 319.

By crossing the network in the other direction we obtain the locus of points for which

$$u_1\left(\frac{x}{a}\right) - u_1\left(\frac{y}{a}\right)$$

is constant, which are the curves of constant displacement for that mode in which the diagonals are nodal. The *pitch* of the vibration is (according to theory) the same in both cases.

Fig. 43.

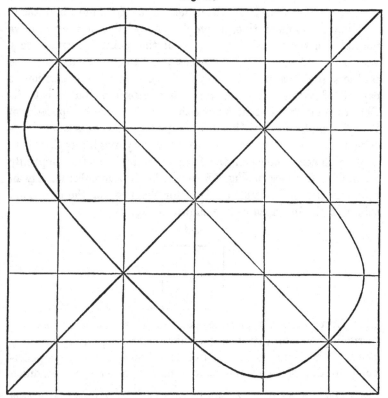

The primitive modes represented by $w = u_2\left(\frac{x}{a}\right)$ or $w = u_2\left(\frac{y}{a}\right)$ may be combined in like manner. Fig. 43 shews the nodal curve for the vibration

$$w = u_2\left(\frac{x}{a}\right) \pm u_2\left(\frac{y}{a}\right) \quad\text{......................}(5).$$

The form of the curve is the same relatively to the other diagonal, if the sign of the ambiguity be altered.

227. The method of superposition does not depend for its application on any particular form of normal function. Whatever the form may be, the mode of vibration, which when $\mu = 0$ passes into that just discussed, must have the same period, whether the approximately straight nodal lines are parallel to x or to y. If the two synchronous vibrations be superposed, the resultant has still the same period, and the general course of its nodal system may be traced by means of the consideration that no point of the plate can be nodal at which the primitive vibrations have the same sign. To determine exactly the line of compensation, a complete knowledge of the primitive normal functions, and not merely of the points at which they vanish, would in general be necessary. Doctor Young and the brothers Weber appear to have had the idea of superposition as capable of giving rise to new varieties of vibration, but it is to Sir Charles Wheatstone[1] that we owe the first systematic application of it to the explanation of Chladni's figures. The results actually obtained by Wheatstone are however only very roughly applicable to a plate, in consequence of the form of normal function implicitly assumed. In place of Fig. 42 (itself, be it remembered, only an approximation) Wheatstone finds for the node of the compound vibration the inscribed square shewn in Fig. 44.

Fig. 44.

This form is really applicable, not to a plate vibrating in virtue of rigidity, but to a stretched membrane, so supported that every point of the circumference is free to move along lines perpendicular to the plane of the membrane. The boundary condition applicable under these circumstances is $\dfrac{dw}{dn} = 0$; and it is easy to shew that the normal functions which involve only one co-ordinate are $w = \cos\left(m\,\dfrac{\pi x}{a}\right)$, or $w = \cos\left(m\,\dfrac{\pi y}{a}\right)$, the origin being at a corner of the square. Thus the vibration

$$w = \cos\frac{2\pi x}{a} + \cos\frac{2\pi y}{a} \quad\ldots\ldots\ldots\ldots\ldots\ldots(1)$$

[1] Phil. Trans. 1833.

has its nodes determined by

$$\cos \frac{\pi (x + y)}{a} \cos \frac{\pi (x - y)}{a} = 0,$$

whence $x + y = \frac{1}{2}a$ or $\frac{3}{2}a$, or $x - y = \pm \frac{1}{2}a$, equations which represent the inscribed square.

If
$$w = \cos \frac{2\pi x}{a} - \cos \frac{2\pi y}{a} \quad\dots\dots\dots\dots\dots(2),$$

the nodal system is composed of the two diagonals. This result, which depends only on the symmetry of the normal functions, is strictly applicable to a square plate.

When $m = 3$,
$$w = \cos \frac{3\pi x}{a} + \cos \frac{3\pi y}{a} \quad\dots\dots\dots\dots\dots(3),$$

and the equations of the nodal lines are

$$x + y = \frac{a}{3},\ a,\ \frac{5a}{3}; \qquad x - y = \pm \frac{a}{3},$$

Fig. 45.

shewn in Fig. 45. If the other sign be taken, we obtain a similar figure with reference to the other diagonal.

When $m = 4$,
$$w = \cos \frac{4\pi x}{a} + \cos \frac{4\pi y}{a} \quad\dots\dots\dots\dots\dots(4),$$

Fig. 46.

giving the nodal lines

$$x + y = \frac{a}{4},\ \frac{3a}{4},\ \frac{5a}{4},\ \frac{7a}{4},\quad x - y = \pm \frac{a}{4},\ \pm \frac{3a}{4} \quad (\text{Fig } 46).$$

With the other sign

$$w = \cos\frac{4\pi x}{a} - \cos\frac{4\pi y}{a} \quad\dots\dots\dots\dots\dots(5),$$

we obtain

$$x + y = \frac{a}{2},\ a,\ \frac{3a}{2},\quad x - y = 0,\ \pm\frac{a}{2}\ (\text{Fig. 47}),$$

Fig. 47.

representing a system composed of the diagonals, together with the inscribed square.

These forms, which are strictly applicable to the membrane, resemble the figures obtained by means of sand on a square plate more closely than might have been expected. The sequence of tones is however quite different. From § 176 we see that, if μ were zero, the interval between the form (43) derived from three primitive nodes, and (41) or (42) derived from two, would be 1·4629 octave; and the interval between (41) or (42) and (46) or (47) would be 2·4358 octaves. Whatever may be the value of μ the forms (41) and (42) should have exactly the same pitch, and the same should be true of (46) and (47). With respect to the first-mentioned pair this result is not in agreement with Chladni's observations, who found a difference of more than a whole tone, (42) giving the higher pitch. If however (42) be left out of account, the comparison is more satisfactory. According to theory ($\mu = 0$), if (41) gave d, (43) should give $g' -$, and (46), (47) should give $g'' +$. Chladni found for (43) $g'{\sharp}+$, and for (46), (47) $g''{\sharp}$ and $g''{\sharp} +$ respectively.

228. The gravest mode of a square plate has yet to be considered. The nodes in this case are the two lines drawn through the middle points of opposite sides. That there must be such a mode will be shewn presently from considerations of symmetry, but neither the form of the normal function, nor the pitch, has yet been determined, even for the particular case of $\mu = 0$. A rough calculation however may be founded on an assumed type of vibration.

If we take the nodal lines for axes, the form $w = xy$ satisfies $\nabla^4 w = 0$, as well as the boundary conditions proper for a free edge at all points of the perimeter except the actual corners. This is in fact the form which the plate would assume if held at rest by four forces numerically equal, acting at the corners perpendicularly to the plane of the plate, those at the ends of one diagonal being in one direction, and those at the ends of the other diagonal in the opposite direction. From this it follows that $w = xy \cos pt$ would be a possible mode of vibration, if the mass of the plate were concentrated equally in the four corners. By (3) § 214, we see that

$$V = \frac{q\, b^3 a^2}{12\,(1+\mu)} \cos^2 pt \quad\quad\quad\ldots\ldots\ldots\ldots\ldots\ldots(1),$$

inasmuch as

$$\frac{d^2 w}{dx^2} = \frac{d^2 w}{dy^2} = 0, \quad\quad \frac{d^2 w}{dx\,dy} = \cos pt.$$

For the kinetic energy, if ρ be the volume density, and M the additional mass at each corner,

$$T = \tfrac{1}{2} p^2 \sin^2 pt \left\{ \int_{-\frac{1}{2}a}^{+\frac{1}{2}a} \int_{-\frac{1}{2}a}^{+\frac{1}{2}a} \rho b x^2 y^2 \, dx\, dy + \tfrac{1}{4} M a^4 \right\}$$

$$= \tfrac{1}{2} p^2 \sin^2 pt \left\{ \frac{\rho b a^6}{16 \times 9} + \frac{a^4}{4} M \right\} \quad\ldots\ldots\ldots\ldots(2).$$

Hence

$$\frac{1}{p^2} = \frac{\rho\,(1+\mu)\,a^4}{24\, q b^2} \left(1 + 36\, \frac{M}{M'} \right) \ldots\ldots\ldots\ldots(3),$$

where M' denotes the mass of the plate without the loads. This result tends to become accurate when M is relatively great; otherwise by § 89 it is sensibly less than the truth. But even when $M = 0$, the error is probably not very great. In this case we should have

$$p^2 = \frac{24\, q b^2}{\rho\,(1+\mu)\,a^4} \quad\quad\ldots\ldots\ldots\ldots\ldots\ldots(4),$$

giving a pitch which is somewhat too high. The gravest mode next after this is when the diagonals are nodes, of which the pitch, if $\mu = 0$, would be given by

$$p'^2 = \frac{q b^2}{\rho a^4} \frac{(4 \cdot 7300)^4}{12} \quad\quad\ldots\ldots\ldots\ldots\ldots\ldots(5),$$

(see § 174).

We may conclude that if the material of the plate were such that $\mu = 0$, the interval between the two gravest tones would be somewhat greater than that expressed by the ratio 1·318. Chladni makes the interval a fifth.

229. That there must exist modes of vibration in which the two shortest diameters are nodes may be inferred from such considerations as the following. In Fig. (48) suppose that GH

Fig. 48.

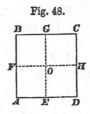

is a plate of which the edges HO, GO are *supported*, and the edges GC, CH free. This plate, since it tends to a definite position of equilibrium, must be capable of vibrating in certain fundamental modes. Fixing our attention on one of these, let us conceive a distribution of w over the three remaining quadrants, such that in any two that adjoin, the values of w are equal and opposite at points which are the images of each other in the line of separation. If the whole plate vibrate according to the law thus determined, no constraint will be required in order to keep the lines GE, FH fixed, and therefore the whole plate may be regarded as free. The same argument may be used to prove that modes exist in which the diagonals are nodes, or in which both the diagonals and the diameters just considered are together nodal.

The principle of symmetry may also be applied to other forms of plate. We might thus infer the possibility of nodal diameters in a circle, or of nodal principal axes in an ellipse. When the

Fig. 49.   Fig. 50.   Fig. 51.

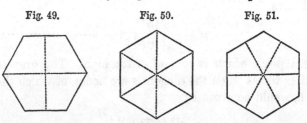

boundary is a regular hexagon, it is easy to see that Figs. (49), (50), (51) represent possible forms.

It is interesting to trace the continuity of Chladni's figures, as the form of the plate is gradually altered. In the circle, for example, when there are two perpendicular nodal diameters, it is a matter of indifference as respects the pitch and the type of vibration, in what position they be taken. As the circle develops into a square by throwing out corners, the position of these diameters becomes definite. In the two alternatives the pitch of the vibration is different, for the projecting corners have not the same efficiency in the two cases. The vibration of a square plate shewn in Fig. (42) corresponds to that of a circle when there is one circular node. The correspondence of the graver modes of a hexagon or an ellipse with those of a circle may be traced in like manner.

230. For plates of uniform material and thickness and of invariable shape, the period of the vibration in any fundamental mode varies as the square of the linear dimension, provided of course that the boundary conditions are the same in all the cases compared. When the edges are clamped, we may go further and assert that the removal of *any* external portion is attended by a rise of pitch, whether the material and the thickness be uniform, or not.

Let AB be a part of a clamped edge (it is of no consequence whether the remainder of the boundary be clamped, or not), and

Fig. 52.

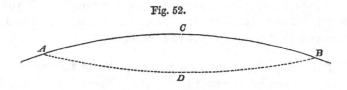

let the piece $ACBD$ be removed, the new edge ADB being also clamped. The pitch of any fundamental vibration is sharper than before the change. This is evident, since the altered vibrations might be obtained from the original system by the introduction of a constraint clamping the edge ADB. The effect of the constraint is to raise the pitch of every component, and the portion $ACBD$ being plane and at rest throughout the motion, may be removed. In order to follow the sequence of changes with greater security from error. it is best to suppose the line of clamping to advance by stages between the two positions ACB, ADB. For example, the pitch of a uniform clamped plate in the form of a regular hexagon is lower than for the inscribed circle and higher than for the circumscribed circle.

When a plate is free, it is not true that an addition to the edge always increases the period. In proof of this it may be sufficient to notice a particular case.

AB is a narrow thin plate, itself without inertia but carrying loads at A, B, C. It is clear that the addition to the breadth

Fig. 53.

indicated by the dotted line would augment the stiffness of the bar, and therefore *lessen* the period of vibration. The same consideration shews that for a uniform free plate of given area there is no lower limit of pitch; for by a sufficient elongation the period of the gravest component may be made to exceed any assignable quantity. When the edges are clamped, the form of gravest pitch is doubtless the circle.

If all the dimensions of a plate, including the thickness, be altered in the same proportion, the period is proportional to the linear dimension, as in every case of a solid body vibrating in virtue of its own elasticity.

The period also varies inversely as the square root of Young's modulus, if μ be constant, and directly as the square root of the mass of unit of volume of the substance.

231. Experimenting with square plates of thin wood whose grain ran parallel to one pair of sides, Wheatstone[1] found that the pitch of the vibrations was different according as the approximately straight nodal lines were parallel or perpendicular to the fibre of the wood. This effect depends on a variation in the flexural rigidity in the two directions. The two sets of vibrations having different periods cannot be combined in the usual manner, and consequently it is not possible to make such a plate of wood vibrate with nodal diagonals. The inequality of periods may however be obviated by altering the ratio of the sides, and then the ordinary mode of superposition giving nodal diagonals is again possible. This was verified by Wheatstone.

[1] *Phil. Trans.* 1833.

A further application of the principle of superposition is due to König[1]. In order that two modes of vibration may combine, it is only necessary that the periods agree. Now it is evident that the sides of a rectangular plate may be taken in such a ratio, that (for instance) the vibration with two nodes parallel to one pair of sides may agree in pitch with the vibration having three nodes parallel to the other pair of sides. In such a case new nodal figures arise by composition of the two primary modes of vibration.

232. When the plate whose vibrations are to be considered is naturally curved, the difficulties of the question are generally much increased. But there is one case in which the complication due to curvature is more than compensated by the absence of a free edge; and this case happens to be of considerable interest, as being the best representative of a bell which at present admits of analytical treatment.

A long cylindrical shell of circular section and uniform thickness is evidently capable of vibrations of a flexural character in which the axis remains at rest and the surface cylindrical, while the motion of every part is perpendicular to the generating lines. The problem may thus be treated as one of two dimensions only, and depends upon the consideration of the potential and kinetic energies of the various deformations of which the section is capable. The same analysis also applies to the corresponding vibrations of a ring, formed by the revolution of a small closed area about an external axis.

The cylinder, or ring, is susceptible of two classes of vibrations depending respectively on extensibility and flexural rigidity, and analogous to the longitudinal and lateral vibrations of straight bars. When, however, the cylinder is thin, the forces resisting bending become small in comparison with those by which extension is opposed; and, as in the case of straight bars, the vibrations depending on bending are graver and more important than those which have their origin in longitudinal rigidity. In the limiting case of an infinitely thin shell (or ring), the flexural vibrations become independent of any extension of the circumference as a whole, and may be calculated on the supposition that each part of the circumference retains its natural length throughout the motion.

[1] Pogg. *Ann.* 1864, cxxii. p. 238.

But although the vibrations about to be considered are analogous to the transverse vibrations of straight bars in respect of depending on the resistance to flexure, we must not fall into the common mistake of supposing that they are exclusively normal. It is indeed easy to see that a motion of a cylinder or ring in which each particle is displaced in the direction of the radius would be incompatible with the condition of no extension. In order to satisfy this condition it is necessary to ascribe to each part of the circumference a tangential as well as a normal motion, whose relative magnitudes must satisfy a certain differential equation. Our first step will be the investigation of this equation.

233. The original radius of the circle being a, let the equilibrium position of any element of the circumference be defined by the vectorial angle θ. During the motion let the polar co-ordinates of the element become

$$r = a + \delta r, \quad \phi = \theta + \delta\theta.$$

If ds represent the arc of the deformed curve corresponding to $ad\theta$, we have

$$(ds)^2 = (a\,d\theta)^2 = (d\delta r)^2 + r^2\,(d\theta + d\delta\theta)^2;$$

whence we find, by neglecting the squares of the small quantities δr, $\delta\theta$,

$$\frac{\delta r}{a} + \frac{d\delta\theta}{d\theta} = 0 \quad \dots\dots\dots\dots\dots\dots\dots(1),$$

as the required relation.

In whatever manner the original circle may be deformed at time t, δr may be expanded by Fourier's theorem in the series

$$\delta r = a \{A_1 \cos\theta + B_1 \sin\theta + A_2 \cos 2\theta + B_2 \sin 2\theta + \dots$$
$$+ A_n \cos n\theta + B_n \sin n\theta + \dots\}\dots\dots\dots\dots(2),$$

and the corresponding tangential displacement required by the condition of no extension will be

$$\delta\theta = -A_1 \sin\theta + B_1 \cos\theta + \dots - \frac{A_n}{n} \sin n\theta + \frac{B_n}{n} \cos n\theta - \dots \dots(3),$$

the constant that might be added to $\delta\theta$ being omitted.

If $\sigma a\, d\theta$ denote the mass of the element $a\, d\theta$, the kinetic energy T of the whole motion will be

$$T = \tfrac{1}{2}\sigma a \int_0^{2\pi} \left\{ \left(\frac{d\delta r}{dt}\right)^2 + a^2 \left(\frac{d\delta\theta}{dt}\right)^2 \right\} d\theta$$

$$= \tfrac{1}{2}\sigma\pi a^3 \left\{ 2(\dot{A}_1^2 + \dot{B}_1^2) + \frac{5}{4}(\dot{A}_2^2 + \dot{B}_2^2) + \ldots \right.$$

$$\left. + \left(1 + \frac{1}{n^2}\right)(\dot{A}_n^2 + \dot{B}_n^2) + \ldots \right\} \ldots\ldots\ldots\ldots(4),$$

the products of the co-ordinates A_n, B_n disappearing in the integration.

We have now to calculate the form of the potential energy V. Let ρ be the radius of curvature of any element ds, then for the corresponding element of V we may take $\tfrac{1}{2}B\, ds\left(\delta\dfrac{1}{\rho}\right)^2$, where B is a constant depending on the material and on the thickness. Thus

$$V = \tfrac{1}{2}Ba \int_0^{2\pi} \left(\delta\frac{1}{\rho}\right)^2 d\theta \ldots\ldots\ldots\ldots\ldots\ldots(5).$$

Now

$$\frac{1}{\rho} = u + \frac{d^2u}{d\phi^2},$$

and

$$u = \frac{1}{r} = \frac{1}{a}\{1 - A_1\cos\phi - B_1\sin\phi - \ldots\},$$

for in the small terms the distinction between ϕ and θ may be neglected.

Hence

$$\delta\frac{1}{\rho} = \frac{1}{a}\Sigma\{(n^2 - 1)(A_n\cos n\phi + B_n\sin n\phi)\},$$

and

$$V = \frac{B}{2a}\int_0^{2\pi}\{\Sigma(n^2 - 1)(A_n\cos n\theta + B_n\sin n\theta)\}^2 d\theta$$

$$= \pi\frac{B}{2a}\Sigma(n^2 - 1)^2(A_n^2 + B_n^2) \ldots\ldots\ldots\ldots\ldots\ldots(6)$$

in which the summation extends to all positive integral values of n.

The term for which $n = 1$ contributes nothing to the potential energy, as it corresponds to a displacement of the circle as a whole, without deformation.

We see that when the configuration of the system is defined as above by the co-ordinates A_1, B_1, &c., the expressions for T and V involve only squares; in other words, these are the *normal* co-ordinates, whose independent harmonic variation expresses the vibration of the system.

If we consider only the terms involving $\cos n\theta$, $\sin n\theta$, we have by taking the origin of θ suitably,

$$\delta r = a A_n \cos n\theta, \quad \delta\theta = -\frac{A_n}{n} \sin n\theta \ldots\ldots\ldots\ldots\ldots(7),$$

while the equation defining the dependence of A_n upon the time is

$$\sigma a^3 \left(1 + \frac{1}{n^2}\right) \ddot{A}_n + \frac{B}{a} (n^2 - 1)^2 A_n = 0 \ldots\ldots\ldots\ldots(8),$$

from which we conclude that, if A_n varies as $\cos(pt - \epsilon)$,

$$p^2 = \frac{B}{\sigma a^3} \cdot \frac{n^2 (n^2 - 1)^2}{n^2 + 1} \ldots\ldots\ldots\ldots\ldots(9).$$

This result was given by Hoppe for a ring in a memoir published in Crelle, Bd. 63, 1871. His method, though more complete than the preceding, is less simple, in consequence of his not recognising explicitly that the motion contemplated corresponds to complete inextensibility of the circumference.

According to Chladni the frequencies of the tones of a ring are as

$$3^2 : 5^2 : 7^2 : 9^2 \ldots\ldots\ldots$$

If we refer each tone to the gravest of the series, we find for the ratios characteristic of the intervals

2·778, 5·445, 9, 13·44, &c.

The corresponding numbers obtained from the above theoretical formulæ, by making n successively equal to 2, 3, 4, &c., are

2·828, 5·423, 8·771, 12·87, &c.,

agreeing pretty nearly with those found experimentally.

234. When $n = 1$, the frequency is zero, as might have been anticipated. The principal mode of vibration corresponds to $n = 2$, and has four nodes, distant from each other by $90°$. These so-called nodes are not, however, places of absolute rest, for the tangential motion is there a maximum. In fact the tangential vibration at these points is half the maximum normal motion. In general for the nth term the maximum tangential motion is $\frac{1}{n}$ of the maximum normal motion, and occurs at the nodes of the latter.

When a bell shaped body is sounded by a blow, the point of application of the blow is a place of maximum normal motion of the resulting vibrations, and the same is true when the vibrations are excited by a violin-bow, as generally in lecture-room experiments. Bells of glass, such as finger-glasses, are however more easily thrown into regular vibration by friction with the wetted finger carried round the circumference. The pitch of the resulting sound is the same as of that elicited by a tap with the soft part of the finger; but inasmuch as the tangential motion of a vibrating bell has been very generally ignored, the production of sound in this manner has been felt as a difficulty. It is now scarcely necessary to point out that the effect of the friction is in the first instance to excite tangential motion, and that the point of application of the friction is the place where the tangential motion is greatest, and therefore where the normal motion vanishes.

235. The existence of tangential vibration in the case of a bell was verified in the following manner. A so-called air-pump receiver was securely fastened to a table, open end uppermost, and set into vibration with the moistened finger. A small chip in the rim, reflecting the light of a candle, gave a bright spot whose motion could be observed with a Coddington lens suitably fixed. As the finger was carried round, the line of vibration was seen to revolve with an angular velocity double that of the finger; and the amount of excursion (indicated by the length of the line of light), though variable, was finite in every position. There was, however, some difficulty in observing the correspondence between the momentary direction of vibration and the situation of the point of excitement. To effect this satisfactorily it was found necessary to apply the friction in the neighbourhood of one point. It then

became evident that the spot moved tangentially when the bell was excited at points distant therefrom 0, 90, 180, or 270 degrees; and normally when the friction was applied at the intermediate points corresponding to 45, 135, 225 and 315 degrees. Care is sometimes required in order to make the bell vibrate in its gravest mode without sensible admixture of overtones.

If there be a small load at any point of the circumference, a slight augmentation of period ensues, which is different according as the loaded point coincides with a node of the normal or of the tangential motion, being greater in the latter case than in the former. The sound produced depends therefore on the place of excitation; in general both tones are heard, and by interference give rise to *beats*, whose frequency is equal to the difference between the frequencies of the two tones. This phenomenon may often be observed in the case of large bells.

END OF VOL. I.

CAMBRIDGE: PRINTED BY C. J. CLAY, M.A., AT THE UNIVERSITY PRESS.

Printed in the United States
By Bookmasters

Printed in the United States
By Bookmasters